Climate Change as a Security Risk

Members of the German Advisory Council on Global Change (WBGU)

(as of 11 May 2007)

Prof Dr Renate Schubert (chair), Economist
Director of the Institute for Environmental Decisions, ETH Zurich (Switzerland)

Prof Dr Hans Joachim Schellnhuber CBE (vice chair), Physicist
Director of the Potsdam Institute for Climate Impact Research and visiting professor at Oxford University, UK

Prof Dr Nina Buchmann, Ecologist
Professor of Grassland Science, Institute of Plant Sciences, ETH Zurich (Switzerland)

Prof Dr Astrid Epiney, Lawyer
Professor of International Law, European Law and Swiss Public Law, Université de Fribourg (Switzerland)

Dr Rainer Grießhammer, Chemist
Director of the Institute for Applied Ecology, Freiburg/Breisgau

Prof Dr Margareta E. Kulessa, Economist
Professor of International Economics, University of Applied Science, Mainz

Prof Dr Dirk Messner, Political Scientist
Director of the German Development Institute, Bonn

Prof Dr Stefan Rahmstorf, Physicist
Professor for Physics of the Oceans at Potsdam University and head of the Climate System Department at the Potsdam Institute for Climate Impact Research

Prof Dr Jürgen Schmid, Aerospace Engineer
Professor at Kassel University, Chairman of the Executive Board of the Institute for Solar Energy Technology

WBGU is an independent, scientific advisory body to the German Federal Government set up in 1992 in the run-up to the Rio Earth Summit. The Council has nine members, appointed for a term of four years by the federal cabinet. The Council is supported by an interministerial committee of the federal government comprising representatives of all ministries and of the federal chancellery. The Council's principal task is to provide scientifically-based policy advice on global change issues to the German Federal Government.

The Council:
- analyses global environment and development problems and reports on these,
- reviews and evaluates national and international research in the field of global change,
- provides early warning of new issue areas,
- identifies gaps in research and initiates new research,
- monitors and assesses national and international policies for sustainable development,
- elaborates recommendations for action, and
- raises public awareness and heightens the media profile of global change issues.

WBGU publishes flagship reports every two years, making its own choice of focal themes. In addition, the German government can commission the Council to prepare special reports and policy papers.
For more information please visit www.wbgu.de

German Advisory Council on Global Change

Climate Change as a Security Risk

London and Sterling, VA

German Advisory Council on Global Change (WBGU)
Secretariat
Reichpietschufer 60-62, 8th Floor
D-10785 Berlin, Germany

http://www.wbgu.de

German edition published in 2007, entitled
Welt im Wandel: Sicherheitsrisiko Klimawandel
Springer-Verlag Berlin, Heidelberg, New York, 2008
ISBN 978-3-540-73247-1

First published by Earthscan in the UK and USA in 2008

Copyright © German Advisory Council on Global Change, 2008

ISBN   978-1-84407-536-2

Printed and bound by Gutenberg Press, Malta
Translation by Christopher Hay, Seeheim-Jugenheim, ecotranslator@t-online.de
Pictures for cover design with kind permission of The Press and Information Office of the Federal Government Germany
except 'Rain meter' (plakboek) and 'Speed Limit 35' (Greg Hounslow).

For a full list of publications please contact:
Earthscan
8-12 Camden High Street
London, NW1 0JH, UK
Ph: +44 (0)20 7387 8558
Fax: +44 (0)20 7387 8998
Email: earthinfo@earthscan.co.uk
Web: www.earthscan.co.uk

22883 Quicksilver Drive, Sterling, VA 20166-2012, USA

Earthscan publishes in association with the International Institute for Environment and Development

A catalogue record for this book is available from the British Library

Library of Congress Cataloging-in-Publication Data
  Wissenschaftlicher Beirat der Bundesregierung Globale Umweltveränderungen (Germany)
  Climate Change as a Security Risk / German Advisory Council on Global Change.
  p. cm.
  Includes bibliographical references (p.).
  ISBN 978-1-84407-536-2
  1. Environmental Conflicts--Environmental Security--Destablilisation of Societies. 2. Climate Change--Natural Disasters--Food Security--Degradation of Freshwater Resources--Migration. 3. Fragile States--Conflict Research--Good Governance--Conflict Prevention. I. Title

QH77.G3 W57 2001
333.95'16'0943--dc21
                                                                                                    2001023313

This book is printed on elemental chlorine-free paper

Council Staff and Acknowledgments

Scientific Staff at the Secretariat

Prof Dr Meinhard Schulz-Baldes
(Secretary-General)

Dr Carsten Loose
(Deputy Secretary-General)

Dr Karin Boschert (since 01.06.2006)

Dr Oliver Deke (since 17.10.2005)

Dipl Umweltwiss Tim Hasler

Dipl Pol Lena Kempmann (until 31.05.2006)

Dr Nina V Michaelis

Dr Benno Pilardeaux
(Media and Public Relations)

Dr Astrid Schulz

Administration, Editorial work and Secretariat

Vesna Karic-Fazlic (Accountant)

Martina Schneider-Kremer, MA (Editorial work)

Margot Weiß (Secretariat)

Scientific Staff to the Council Members

Dipl Phys Jochen Bard (Insitute for Solar Energy Technology, ISET Kassel, since 01.03.2006)

Steffen Bauer, MA (German Development Institute, DIE Bonn, since 01.01.2006)

Dr Gregor Betz (Potsdam Institute for Climate Impact Research, PIK, until 30.09.2005)

Dipl Volksw Julia E Blasch (Institute for Environmental Decisions, ETH Zurich, Switzerland, since 16.10.2006)

Dipl-Phys Gregor Czisch (Insitute for Solar Energy Technology, ISET Kassel, until 28.02.2006)

Dr Georg Feulner (Potsdam Institute for Climate Impact Research, PIK)

Dr Monika Heupel (German Development Institute, DIE Bonn, until 15.10.2005)

Dipl Volksw Kristin Hoffmann (Institute for Environmental Decisions, ETH Zurich, Switzerland, until 16.10.2006)

Dr Susanne Kadner (Potsdam Institute for Climate Impact Research, PIK, 10.01.2006 until 30.04.2006)

Dr Sabina Keller (ETH Zurich, Switzerland)

Dipl Geogr Andreas Manhart (Institute for Applied Ecology, Freiburg)

Dr Franziska Matthies (Potsdam Institute for Climate Impact Research, PIK, until 30.09.2005)

Dipl Volksw Markus Ohndorf (ETH Zurich, Switzerland)

Dr Martin Scheyli (Universität Fribourg, Switzerland)

Dr Ingeborg Schinninger (ETH Zurich, Switzerland, until 31.05.2007)

Dipl-Pol Joachim Schwerd (University of Applied Science, Mainz, until 31.10.2006)

WBGU wishes to thank the authors of commissioned studies for their important contributions and support:
- Dr habil Hans Günter Brauch (AFES-Press, Mosbach): Regional Expertise 'Destabilisierungs- und Konfliktpotenzial prognostizierter Umweltveränderungen in der Region Südeuropa und Nordafrika bis 2020/2050'.
- Dipl Pol Alexander Carius, Dennis Tänzler, Judith Winterstein (Adelphi Consult, Berlin): Weltkarte von Umweltkonflikten: Ansätze zur Typologisierung.
- Dr Martin Cassel-Gintz: Erstellung von GIS-Karten zur Klimaentwicklung – GIS II.
- Prof William A V Clark (University of California, Department of Geography, Los Angeles): Environmentally Induced Migration and Conflict.
- Prof Dr Ernst Giese, Jenniver Sehring, MA (University Gießen, Institute of Geography): Regional Expertise 'Destabilisierungs- und Konfliktpotenzial prognostizierter Umweltveränderungen in der Region Zentralasien bis 2020/2050'.
- Prof Dr Thomas Heberer, Anja-Désirée Senz, MA (University Duisburg, Institute of East Asian Studies): Regional Expertise 'Destabilisierungs- und Konfliktpotenzial prognostizierter Umweltveränderungen in China bis 2020/2050'.
- Larry A Swatuk, PhD, Associate Professor (Harry Oppenheimer Okavango Research Centre, Botswana): Regional Expertise 'Southern Africa, Environmental Change and Regional Security: An Assessment'.
- Martin Wodinski (Climate and Environment Consulting – CEC, Potsdam): Erstellung von GIS-Karten zur Klimaentwicklung – GIS I.
- Prof Aaron T Wolf (Oregon State University, Department of Geosciences, Corvallis): A Long Term View of Water and Security: International Waters, National Issues, and Regional Tensions.

Important written contributions on the political, economic and social impacts of climate change in individual regions of the world were made by Dr Jörg Faust, Jochen Kenneweg and Dr Imme Scholz of the German Development Institute (GDI, Bonn). Dr Imme Scholz and Jochen Kenneweg also advised on other parts of this report.

WBGU gained valuable insights during its intensive workshop held in 2006 in Schmöckwitz from the presentations on 'Introduction to the methodology of futures studies' made by Prof Dr Rolf Kreibich and Dr Robert Gaßner (both Institute for Futures Studies and Technology Assessment – IZT, Berlin).

WBGU further wishes to thank all those who promoted the progress of this report through their comments and advice or by reviewing individual parts:

Dr Ludwig Braun (Bavarian Academy of Sciences and Humanities, Munich), Dr Thomas Fues (German Development Institute – GDI, Bonn), Dr Jörn Grävingholt (German Development Institute – GDI, Bonn), Dipl Geoök Holger Hoff (Stockholm Environment Institute – SEI), Oberst i. G. Roland Kaestner (Führungsakademie der Bundeswehr, Hamburg), Dipl-Pol Stefan Lindemann (German Advisory Council on the Environment – SRU, Berlin), Dr Susanne Neubert (German Development Institute – GDI, Bonn), Dr Manfred Schütze (Institut für Automation und Kommunikation – ifak e.V., Magdeburg), Dr Andreas Stamm (German Development Institute – GDI, Bonn), Dr Denis M. Tull (German Institute for International and Security Affairs – SWP, Berlin), Dr Juan Carlos Villagrán de León (Institute for Environment and Human Security – United Nations University, Bonn), Dr habil Christian Wagner (German Institute for International and Security Affairs – SWP, Berlin), Dipl-Pol Silke Weinlich (Institute for Intercultural and International Studies – InIIS, Bremen), Dipl Ing Elizabeth Zamalloa-Skoddow (German Development Service – DED, Peru).

WBGU thanks Christopher Hay (Übersetzungsbüro für Umweltwissenschaften, Seeheim-Jugenheim, Germany) for his expert translation of this report into English from the German original.

Contents

Council Staff and Acknowledgments V

Contents VII

Boxes XIII

Tables XIV

Figures XV

Acronyms and Abbreviations XVII

Summary for Policy-makers 1

1 Introduction 15

2 Environmental change in security discourse 19

    2.1 Redefining security 19
        2.1.1 Comprehensive security 19
        2.1.2 Human security 20

    2.2 Current security policy strategies 21

    2.3 WBGU's aims and use of terms 22

3 Known conflict impacts of environmental change 25

    3.1 State of conflict research at the interface of environment and security 25
        3.1.1 Environment and conflict research 25
            3.1.1.1 The Toronto group around Homer-Dixon 26
            3.1.1.2 The Zurich group around Bächler and Spillmann 27
            3.1.1.3 The Oslo group around Gleditsch 28
            3.1.1.4 The Irvine group around Matthew 28
            3.1.1.5 The German research scene and WBGU's syndrome approach 29
            3.1.1.6 Fundamental critique of environment and conflict research 29
            3.1.1.7 Key findings from environment and conflict research 30

    3.2 World map of past environmental conflicts 31
        3.2.1 Resource conflicts over land, soil, water and biodiversity 31
        3.2.2 Conflict-related impacts of storm and flood disasters 31

    3.3 War and conflict research 35
        3.3.1 Regime type, political stability and governance structures 35

		3.3.2	Economic factors	36
			3.3.2.1 Economic performance and distributive justice	36
			3.3.2.2 Natural resources	37
		3.3.3	Societal stability and demography	37
			3.3.3.1 Population trends	37
			3.3.3.2 Socio-cultural composition of the population	37
			3.3.3.3 History of conflict	38
		3.3.4	Geographical factors	38
		3.3.5	International distribution of power and interdependencies	39
		3.3.6	Main findings of conflict research	39
	3.4	**Conclusions**		**39**
4	**Rising conflict risks due to state fragility and a changing world order**			**41**
	4.1	**Introduction**		**41**
	4.2	**State fragility and the limits of governance**		**41**
		4.2.1	Characteristics of state fragility	42
		4.2.2	Destabilizing effects of environmental degradation	44
	4.3	**Unstable multipolarity: The political setting of global change**		**45**
		4.3.1	Conflict or cooperation through the transformation of the world order?	46
		4.3.2	Global trends: China, India and the path towards multipolarity	47
		4.3.3	Global governance in the context of Chinese and Indian ascendancy	51
			4.3.3.1 Multipolarity as a threat to multilateralism?	52
			4.3.3.2 General dynamics of global political change	52
			4.3.3.3 China and India as the driving forces of global political change	53
	4.4	**Conclusions**		**54**
5	**Impacts of climate change on the biosphere and human society**			**55**
	5.1	**Changes in climatic parameters**		**55**
		5.1.1	Temperature	55
		5.1.2	Precipitation	57
		5.1.3	Tropical cyclones	59
		5.1.4	Sea-level rise	60
	5.2	**Climate impacts upon human well-being and society**		**63**
		5.2.1	Impacts upon freshwater availability	64
		5.2.2	Climate change impacts upon vegetation and land use	66
		5.2.3	Climate change impacts upon storm and flood events	69
		5.2.4	Indirect economic and social impacts of climate change	70
			5.2.4.1 Climate change impacts in selected economic sectors	70
			5.2.4.2 Climate change impacts on the global economy	71
			5.2.4.3 Climate change impacts on society	72
	5.3	**Non-linear effects and tipping points**		**72**
		5.3.1	Weakening of the North Atlantic Current	73
		5.3.2	Monsoon transformation	74
		5.3.3	Instability of the continental ice sheets	74
		5.3.4	Collapse of the Amazon rainforest	75
		5.3.5	Conclusions	75

6 Conflict constellations ... 77

6.1 Methodology ... 77
6.1.1 Selection and definition ... 77
6.1.2 Using narrative scenarios to identify security risks ... 77
6.1.3 Deriving recommendations for action ... 78

6.2 Conflict constellation 'Climate-induced degradation of freshwater resources' ... 79
6.2.1 Background ... 79
6.2.1.1 Brief description of the conflict constellation ... 79
6.2.1.2 Water crises today and tomorrow ... 79
6.2.2 Causal linkages ... 81
6.2.2.1 From climate change to changes in water availability ... 81
6.2.2.2 From changes in water availability to water crisis ... 82
6.2.2.3 From water crisis to conflict and violence ... 84
6.2.3 Scenarios ... 86
6.2.3.1 Glacier retreat, water crisis and violent conflict in the greater Lima area ... 87
6.2.3.2 Glacier retreat, water crisis and violent confrontation in Central Asia ... 88
6.2.4 Recommendations for action ... 90

6.3 Conflict constellation: 'Climate-induced decline in food production' ... 93
6.3.1 Background ... 93
6.3.1.1 Global food production: Future trends in supply and demand ... 93
6.3.1.2 Changing framework conditions for global food production ... 94
6.3.2 Causal linkages ... 96
6.3.2.1 From environmental change to declining food production ... 96
6.3.2.2 From declining food production to food crisis ... 97
6.3.2.3 From food crisis to destabilization and violence ... 98
6.3.3 Scenario: Agricultural production crisis, food crisis and violence in southern Africa . 100
6.3.4 Recommendations for action ... 102

6.4 Conflict constellation: 'Climate-induced increase in storm and flood disasters' ... 103
6.4.1 Background ... 103
6.4.2 Causal linkages ... 104
6.4.2.1 From environmental change to increase in storm and flood disasters ... 104
6.4.2.2 From more frequent storm and flood disasters to crisis ... 105
6.4.2.3 From crisis to destabilization and violence ... 106
6.4.2.4 The time sequence of disaster-induced conflict mechanisms ... 108
6.4.3 Scenarios ... 110
6.4.3.1 Storm and flood disasters in China ... 110
6.4.3.2 Hurricane risks in the Gulf of Mexico and the Caribbean ... 113
6.4.4 Recommendations for action ... 115

6.5 Conflict constellation: 'Environmentally induced migration' ... 116
6.5.1 Background ... 116
6.5.1.1 Structure of the conflict constellation ... 116
6.5.1.2 Environmentally induced migration as a core element of the conflict constellation ... 117
6.5.2 Causal linkages ... 119
6.5.2.1 From environmental change to migration ... 120
6.5.2.2 From migration to conflict ... 120
6.5.3 Scenarios ... 122
6.5.3.1 Environmentally induced migration and conflict in Bangladesh ... 122
6.5.3.2 Environmentally induced migration and conflicts in North Africa and neighbouring Mediterranean countries ... 124
6.5.4 Recommendations for action ... 127

		6.5.4.1	Avoiding environmentally induced migration	127
		6.5.4.2	Managing environmentally induced migration	127
		6.5.4.3	Supporting developing countries	128
		6.5.4.4	Instruments of international law	129

7 Hotspots of climate change: Selected regions ... 131

7.1 Arctic and Subarctic ... 132
- 7.1.1 Impacts of climate change on the biosphere and human society ... 132
- 7.1.2 Political and economic situation in the region ... 132
- 7.1.3 Conclusions ... 133

7.2 Southern Europe and North Africa ... 133
- 7.2.1 Impacts of climate change on the biosphere and human society ... 133
- 7.2.2 Political and economic situation in the region ... 134
- 7.2.3 Conclusions ... 136

7.3 Sahel zone ... 136
- 7.3.1 Impacts of climate change on the biosphere and human society ... 136
- 7.3.2 Political and economic situation in the region ... 137
- 7.3.3 Conclusions ... 138

7.4 Southern Africa ... 138
- 7.4.1 Impacts of climate change on the biosphere and human society ... 138
- 7.4.2 Political and economic situation in the region ... 139
- 7.4.3 Conclusions ... 140

7.5 Central Asia ... 141
- 7.5.1 Impacts of climate change on the biosphere and human society ... 141
- 7.5.2 Political and economic situation in the region ... 142
- 7.5.3 Conclusions ... 143

7.6 India, Pakistan and Bangladesh ... 143
- 7.6.1 Impacts of climate change on the biosphere and human society ... 143
- 7.6.2 Political and economic situation in the region ... 144
- 7.6.3 Conclusions ... 146

7.7 China ... 146
- 7.7.1 Impacts of climate change on the biosphere and human society ... 146
- 7.7.2 Political and economic situation in the region ... 147
- 7.7.3 Conclusions ... 148

7.8 Caribbean and the Gulf of Mexico ... 149
- 7.8.1 Impacts of climate change on the biosphere and human society ... 149
- 7.8.2 Political and economic situation in the region ... 150
- 7.8.3 Conclusions ... 151

7.9 Andes region ... 151
- 7.9.1 Impacts of climate change on the biosphere and human society ... 151
- 7.9.2 Political and economic situation in the region ... 152
- 7.9.3 Conclusions ... 153

7.10 Amazon region ... 154
- 7.10.1 Impacts of climate change on the biosphere and human society ... 154
- 7.10.2 Political and economic situation in the region ... 155
- 7.10.3 Conclusions ... 156

8 Climate change as a driver of social destabilization and threat to international security 157

8.1 Climate-induced conflict constellations: Analysis and findings 157
- 8.1.1 Key factors determining the emergence and amplification of conflicts 157
- 8.1.2 Reciprocal amplification of conflict constellations 159
- 8.1.3 The new quality of conflicts induced by climate change 162

8.2 International climate policy scenarios and their long-term implications 165
- 8.2.1 'Green Business As Usual' scenario: Too little, too late, too slow 166
- 8.2.2 'International Policy Failure' scenario: Collapse of the multilateral climate regime ... 167
- 8.2.3 'Strong Climate Policy' scenario: Compliance with the 2 °C guard rail 167

8.3 Climate change as a threat to international security 168
- 8.3.1 Possible increase in the number of destabilized states as a result of climate change ... 169
- 8.3.2 Risks for global economic development 170
- 8.3.3 Risks of growing distributional conflicts between the main drivers of climate change and those most affected 171
- 8.3.4 Climate change undermines human rights: Calling emitters to account 173
- 8.3.5 Climate change triggers and intensifies migration 174
- 8.3.6 Climate change overstretches classic security policy 174
- 8.3.7 Summary: Overstretching the capacities of the global governance system 175

9 Research recommendations 177

9.1 Understanding the climate-security nexus – fundamentals 177
- 9.1.1 Climate research 177
- 9.1.2 Environmental and climate impact research 178
- 9.1.3 Early warning systems 179
- 9.1.4 Social destabilization through climate change 179

9.2 Policies to prevent and contain conflict 180
- 9.2.1 Research and policy focused on the long term 180
- 9.2.2 Adaptation strategies in developing countries 182
- 9.2.3 Developing preventive strategies to stabilize fragile states 182
- 9.2.4 International institutions in the context of global change and climate-induced conflicts 183

9.3 Conflict constellations and their prevention 184
- 9.3.1 Degradation of freshwater resources 185
- 9.3.2 Decline in food production 186
- 9.3.3 Increase in storm and flood disasters 186
- 9.3.4 Environmentally induced migration 187

10 Recommendations for action 189

10.1 WBGU's key findings 189

10.2 Scope for action on the part of the German government 191

10.3 The window of opportunity for climate security: 2007–2020 191
- 10.3.1 Fostering a cooperative setting for a multipolar world 192
  - 10.3.1.1 Initiative 1: Shaping global political change 192
  - 10.3.1.2 Initiative 2: Reforming the United Nations 195

	10.3.2	Climate policy as security policy I: Preventing conflict by avoiding dangerous climate change 198
		10.3.2.1 Initiative 3: Ambitiously pursuing international climate policy 198
		10.3.2.2 Initiative 4: Transforming energy systems in the EU 199
		10.3.2.3 Initiative 5: Developing mitigation strategies through partnerships 199
	10.3.3	Climate policy as security policy II: Preventing conflict by implementing adaptation strategies 200
		10.3.3.1 Initiative 6: Supporting adaptation strategies for developing countries 200
		10.3.3.2 Initiative 7: Stabilizing fragile states and weak states that are additionally threatened by climate change 202
		10.3.3.3 Initiative 8: Managing migration through cooperation and further developing international law 204
		10.3.3.4 Initiative 9: Expanding global information and early warning systems 207
	10.3.4	Securing the financing of the initiatives 208
		10.3.4.1 Avoiding dangerous climate change 208
		10.3.4.2 Adapting to unavoidable climate change 209
		10.3.4.3 International conflict prevention 211

10.4 Window missed – mitigation failed: Strategies in the event of destabilization and conflict 213

11 References 215

12 Glossary 235

13 Index 243

Boxes

Box 1	Climate change amplifies mechanisms which lead to insecurity and violence	2
Box 2.2-1	Worldwatch Institute: Changing the oil economy	23
Box 3.1-1	Climate change and environmental change in the past and their impacts on human society	26
Box 4.2-1	Qualitative categorization of state stability	42
Box 4.2-2	State fragility: Destabilizing factors	43
Box 4.2-3	OECD-DAC: Principles for Good International Engagement in Fragile States and Situations	44
Box 4.3-1	Interpretations of the post-1990 world order	51
Box 5.2-1	Water shortages affecting people: Indices	65
Box 6.1-1	Scenarios and forecasts	78
Box 6.2-1	Integrated water resources management	80
Box 6.2-2	Dams and conflict	85
Box 6.3-1	Examples of destabilization and violence resulting from crop failures and food crises	99
Box 6.5-1	Migration – definitions and trends	117
Box 8.3-1	The major newly industrializing countries' possible future share of global greenhouse gas emissions	172
Box 8.3-2	Security threats in the 21st century: A comparison with strategic analyses from classic security policy	175

Tables

Table 2.2-1	Main differences between the security strategies of the United States of America and the European Union	21
Table 8.1-1	Key factors determining the emergence and amplification of conflict constellations	158
Table 8.3-1	Global energy-related CO_2 emissions and selected countries'/groups of countries' shares in these emissions based on the IEA's Alternative Policy Scenario	172
Table 8.3-2	Per capita greenhouse gas emissions for selected countries and groups of countries	173
Table 10.3-1	Overview of the nine initiatives proposed by WBGU for the mitigation of destabilization and conflict risks associated with climate change	193
Table 10.3-2	Overview of the instruments proposed by WBGU to fund the initiatives	209

Figures

Figure 1	Security risks associated with climate change: Selected hotspots	4
Figure 3.2-1	World map of environmental conflicts (1980–2005): Causes and intensity	32
Figure 3.2-2	Environmental conflicts in Central America and the Caribbean (1980–2005)	32
Figure 3.2-3	Environmental conflicts in Africa (1980–2005)	33
Figure 3.2-4	Storm and flood disasters with destabilizing and conflict-inducing consequences	34
Figure 3.2-5	Predicted likelihood of the emergence of new conflicts within five years in relation to per capita income	36
Figure 4.2-1	Weak and fragile states: A global overview	46
Figure 4.3-1	China's, India's, US and EU shares of global primary energy consumption and global energy-related CO_2 emissions	49
Figure 5.1-1	Global-mean temperature development over land and sea up to 2006 based on the NASA data set	56
Figure 5.1-2	Linear temperature trends from measurements on land in the period 1975–2004	57
Figure 5.1-3	Climatic water balance mean for the period 1961–1990	58
Figure 5.1-4	Relative changes in precipitation towards the end of the century as compared to 1990 under SRES scenario A1B	60
Figure 5.1-5	Percentage change in maximum dry periods under scenario A1B in a simulation by the Max Planck Institute (MPI), Hamburg	61
Figure 5.1-6	Future dynamics of drought risk	61
Figure 5.1-7	Percentage changes in annual extreme precipitation under scenario A1B	62
Figure 5.1-8	Risk from tropical storms: storm tracks and intensities over the past 150 years	62
Figure 5.1-9	Global sea-level rise as measured by tide gauge (brown) and satellite (black)	63
Figure 5.2-1	Projections of populations suffering severe water stress under three of the SRES scenarios	65
Figure 5.2-2	Current global distribution of the water scarcity indicator (Box 5.2-1)	66
Figure 5.2-3	Terrestrial ecosystems that will be affected by changes in the event of a global average temperature increase of 3°C (based on HADCM-GCM)	67
Figure 5.2-4	Global soil degradation by severity and rate of progression	68
Figure 5.2-5	Share of agriculture in GDP and per capita income (2004)	70
Figure 5.3-1	Map with some of the tipping elements in the climate system discussed in the text	73
Figure 6.2-1	Water use by sector	79
Figure 6.2-2	Areas of 'physical' and 'economic' water scarcity	81
Figure 6.2-3	Conflict constellation 'Climate-induced degradation of freshwater resources': Key factors and interactions	83
Figure 6.3-1	Global food production 1960–2003	93
Figure 6.3-2	Conflict constellation: 'Climate-induced decline in food production': Key factors and interactions	97
Figure 6.4-1	Tropical cyclone threat to urban agglomerations	105
Figure 6.4-2	Characteristic time sequence of disaster-induced conflict mechanisms	108
Figure 6.4-3	Conflict constellation: 'Climate-induced increase in storm and flood disasters': Key factors and interactions	109
Figure 6.5-1	Conflict constellation: 'Environmentally induced migration': Key factors and interactions	119
Figure 8.1-1	Conflict constellations as drivers of international destabilization	160

Figure 8.1-2	Climate status and Climate future	161
Figure 8.1-3	Security risks associated with climate change: Selected hotspots	163
Figure 8.1-4	Consequences of climate change for ecosystems and economic sectors at different levels of warming	164
Figure 8.2-1	Emissions reductions required in order to avoid global warming of more than 2 °C	165
Figure 8.3-1	Rough breakdown of global greenhouse gas emissions in 2004	172

Acronyms and Abbreviations

AA	Auswärtiges Amt [Federal Foreign Office, Germany]
AIDS	Acquired Immune Deficiency Syndrome
APEC	Asia-Pacific Economic Cooperation
ASEAN	Association of South-East Asian Nations
BMBF	Bundesministerium für Bildung und Forschung [Federal Ministry of Education and Research, Germany]
BMU	Bundesministerium für Umwelt, Naturschutz und Reaktorsicherheit [Federal Ministry for Environment, Nature Conservation and Reactor Safety, Germany]
BMZ	Bundesministerium für wirtschaftliche Zusammenarbeit und Entwicklung [Federal Ministry for Economic Cooperation and Development, Germany]
BNE	Bruttonationaleinkommen
CBD	Convention on Biological Diversity
CDM	Clean Development Mechanism (Kyoto Protocol)
CEC	Central and Eastern European Countries
CFU	Carbon Finance Unit (World Bank)
CGIAR	Consultative Group on International Agricultural Research
CIESIN	Center for International Earth Science Information Network (University Columbia)
CILSS	Comité Inter Etats de Lutte Contre la Sécheresse dans le Sahel
COMCAD	Center on Migration, Citizenship and Development (Universität Bielefeld)
COP	Conference of the Parties
COSIMO	Conflict Simulation Model (HIIK)
CRED	Collaborating Centre for Research on the Epidemiology of Disasters (WHO)
CSCE	Conference on Security and Co-operation in Europe
CSD	Commission on Sustainable Development (UN)
DAC	Development Assistance Committee (OECD)
DIE	Deutsches Institut für Entwicklungspolitik [German Development Institute, Bonn]
DR-CAFTA	Dominican Republic – Central American Free Trade Agreement
EACH-FOR	Environmental Change and Forced Migration Scenarios (EU Research Project)
ECOMAN	Environmental Change Consensus Building and Resource Management in the Horn of Africa (Nachfolgeprojekt ENCOP)
ECONILE	Environment and Cooperation in the Nile Basin (ENCOP Project)
ECOSOC	Economic and Social Council (UN)
EMP	Europäisch-Mediterrane Partnerschaft
ENCOP	Environment and Conflicts Project (ETH Zurich, Swisspeace, Bern)
ENSO	El Niño and Southern Oscillation
ENVSEC	Environment Security Initiative (UNDP, UNEP, OSZE)
ETS	European Emissions Trading System
EU	European Union
EWC	International Conference on Early Warning (UN)
FAO	Food and Agriculture Organization of the United Nations
FEMA	Federal Emergency Management Administration (USA)

FIPs	Five Interested Parties (WTO)
G8	Group of Eight
GAM	Gerakan Aceh Merdeka [Free Aceh Movement]
GASP	Common Foreign Security Policy (EU)
GDP	Gross Domestic Product
GECHS	Global Environmental Change and Human Security (IHDP Project)
GEF	Global Environment Facility (UNDP, UNEP, World Bank)
GIS	Geographical Information System
GLASOD	Global Assessment of Human Induced Soil Degradation (UNEP)
GPS	Global Positioning System
GTZ	Deutsche Gesellschaft für Technische Zusammenarbeit [German Society on Development Cooperation]
HDI	Human Development Index
HIIK	Heidelberger Institut für Internationale Konfliktforschung [Heidelberg Institute for International Conflict Research]
HIV	Human Immunodeficiency Virus
ICLEI	International Council for Local Environmental Initiatives
ICRC	International Committee of the Red Cross
ICSU	International Council for Science
IDM	International Dialogue on Migration (IOM)
IDPs	Internally Displaced Persons (UNHCR)
IDS	Institute for Development Studies (UK)
IEA	International Energy Agency
ICJ	International Court of Justice
IHDP	International Human Dimensions Programme on Global Environmental Change (ISSC, ICSU)
IIASA	International Institute for Applied Systems Analysis (Laxenburg, Austria)
IMF	International Monetary Fund
IMISCOE	International Migration, Integration and Social Cohesion (EU)
IOM	International Organization for Migration
IPCC	Intergovernmental Panel on Climate Change (WMO, UNEP)
IPPC	International Plant Protection Convention (FAO)
ISDR	International Strategy for Disaster Reduction (UN)
ISS	Institute for Security Studies, Paris
ISSC	International Social Science Council (UNESCO)
IWRM	Integrated Water Resources Management
KfW	German Development Bank
LTTE	Liberation Tigers of Tamil Eelam
MA	Millennium Ecosystem Assessment (UN)
MDGs	Millennium Development Goals (UN)
NAFTA	North American Free Trade Agreement
NASA	National Aeronautics and Space Administration, USA
NATO	North Atlantic Treaty Organisation
NCCR	National Centre of Competence in Research North-South, Switzerland
NEPAD	New Partnership for Africa's Development (OAU)
NGO	Non-governmental Organization
NOAA	National Oceanic and Atmospheric Administration, USA
OAU	Organisation of African Unity
OECD	Organisation for Economic Co-operation and Development
OSZE	Organisation für Sicherheit und Zusammenarbeit in Europa
PRIO	International Peace Research Institute Oslo (Norway)
PRSP	Poverty Reduction Strategy Paper (Governments, IMF, World Bank)
RNE	Rat für Nachhaltige Entwicklung [German Council for Sustainable Development]
SEPA	China's State Environmental Protection Administration

SOZ	Shanghai Cooperation Organisation
SRES	Special Report on Emissions Scenarios (IPCC)
SRU	Sachverständigenrat für Umweltfragen [Council of Environmental Experts, Germany]
UN	United Nations
UNCCD	United Nations Convention to Combat Desertification in Countries Experiencing Serious Drought and/or Desertification, Particularly in Africa
UNDP	United Nations Development Programme
UNEP	United Nations Environment Programme
UNESCO	United Nations Educational, Scientific and Cultural Organization
UNFCCC	United Nations Framework Convention on Climate Change
UNHCR	United Nations High Commissioner on Refugees
UNICEF	United Nations Children's Fund
UNU	United Nations University
UNU-EHS	Institute for Environment and Human Security
USGS	United States Geological Survey
WBGU	Wissenschaftlicher Beirat der Bundesregierung Globale Umweltveränderungen [German Advisory Council on Global Change]
WFP	World Food Programme (UN)
WHO	World Health Organization (UN)
WTO	World Trade Organization

Summary for Policy-makers

A new security policy challenge

The core message of WBGU's risk analysis is that without resolute counteraction, climate change will overstretch many societies' adaptive capacities within the coming decades. This could result in destabilization and violence, jeopardizing national and international security to a new degree. However, climate change could also unite the international community, provided that it recognizes climate change as a threat to humankind and soon sets the course for the avoidance of dangerous anthropogenic climate change by adopting a dynamic and globally coordinated climate policy. If it fails to do so, climate change will draw ever-deeper lines of division and conflict in international relations, triggering numerous conflicts between and within countries over the distribution of resources, especially water and land, over the management of migration, or over compensation payments between the countries mainly responsible for climate change and those countries most affected by its destructive effects.

In order to avoid these developments, an ambitious global climate policy must be put into operation over the next 10-15 years. An effective international climate protection regime must ensure that global greenhouse gas emissions are halved by the mid 21st century. This major international policy challenge arises in parallel to a far-reaching shift in the centres of power of the political world order, which will be dominated by the ascendancy of new powers such as China and India and the United States' simultaneous relative loss of power. The lessons of history suggest that this transition will be accompanied by turbulence in the international system which may make it more difficult to achieve the necessary breakthroughs in multilateral climate policy. In order to provide a counterbalance, the European Union must take a leading role in global climate policy and convince both the USA and the newly ascendant Asian powers of the importance of concerted efforts to avoid dangerous climate change.

That is the backdrop against which WBGU, in this flagship report, summarizes the state-of-the-art of science on the subject of "Climate Change as a Security Risk". It is based on the findings of research into environmental conflicts, the causes of war, and of climate impact research. It appraises past experience but also ventures to cast a glance far into the future in order to assess the likely impacts of climate change on societies, nation-states, regions and the international system.

Climate change is only just beginning, but its impacts will steadily intensify in the coming decades. WBGU shows that *firstly*, climate change could exacerbate existing environmental crises such as drought, water scarcity and soil degradation, intensify land-use conflicts and trigger further environmentally induced migration. Rising global temperatures will jeopardize the bases of many people's livelihoods, especially in the developing regions, increase vulnerability to poverty and social deprivation, and thus put human security at risk. Particularly in weak and fragile states with poorly performing institutions and systems of government, climate change is also likely to overwhelm local capacities to adapt to changing environmental conditions and will thus reinforce the trend towards general instability that already exists in many societies and regions (Box 1). In general it can be said that the greater the warming, the greater the security risks to be anticipated.

Secondly, new conflict constellations are likely to occur. Sea-level rise and storm and flood disasters could in future threaten cities and industrial regions along the coasts of China, India and the USA. The melting of the glaciers would jeopardize water supply in the Andean and Himalayan regions.

Thirdly, unabated climate change could cause large-scale changes in the Earth System such as the dieback of the Amazon rainforest or the loss of the Asian monsoon, which could have incalculable consequences for the societies concerned.

Overall, WBGU considers that climate-induced interstate wars are unlikely to occur. However, climate change could well trigger national and international distributional conflicts and intensify problems already hard to manage such as state failure, the erosion of social order, and rising violence. In the worst-

> **Box 1**
>
> **Climate change amplifies mechanisms which lead to insecurity and violence**
>
> POLITICAL INSTABILITY AND CONFLICTS
> Societies in transition from authoritarian to democratic systems are especially vulnerable to crises and conflicts. Climate change will affect many of these countries, putting them under additional pressure to adapt their societies during such phases of transition. This linkage could be significant for many African countries, for example, as well as for China.
>
> WEAK GOVERNANCE STRUCTURES AND CONFLICTS
> Violent conflicts are a very frequent feature of weak and fragile states, of which there are currently about 30, and which are characterized by the permanent weakening or even the dissolution of their state structures. The impacts of climate change will particularly affect those regions of the world in which states with weak steering and problem-solving capacities already predominate. Climate change could thus lead to the further proliferation of weak and fragile statehood and increase the probability of violent conflicts occurring.
>
> ECONOMIC PERFORMANCE AND TENDENCY TO VIOLENCE
> Empirical studies show that poor countries are far more prone to conflict than affluent societies. Climate change will result in tangible economic costs for developing countries in particular: a drop in agricultural yields, extreme weather events and migratory movements can all impede economic development. Climate change can thus reinforce obstacles to development and heighten poverty, thereby increasing the risk of conflicts occurring in these societies.
>
> DEMOGRAPHICS AND CONFLICT
> Wherever high population growth and density, resource scarcity (farmland, water) and a low level of economic development occur in tandem, there is an increased risk of conflict. In many countries and regions which are already affected by high population growth and density as well as poverty, climate change will exacerbate resource scarcity and thus heighten the risk of conflict.
>
> SPILLOVER RISK IN CONFLICT REGIONS
> Conflicts which are initially limited to local or national level often destabilize neighbour countries, e.g. through refugee flows, arms trafficking or combatant withdrawal. Conflicts thus have a spillover effect. The social impacts of climate change can transcend borders, thereby swiftly expanding the geographical extent of crisis and conflict regions.

affected regions, this could lead to the proliferation of destabilization processes with diffuse conflict structures. These dynamics threaten to overstretch the established global governance system, thus jeopardizing international stability and security.

Climate change as a threat to international security

Climate-induced conflict constellations

WBGU identifies four conflict constellations in which critical developments can be anticipated as a result of climate change and which may occur with similar characteristics in different regions of the world. "Conflict constellations" are defined as typical causal linkages at the interface of environment and society, whose dynamic can lead to social destabilization and, in the end, to violence.
- *Conflict constellation "Climate-induced degradation of freshwater resources":* 1.1 thousand million people are currently without access to safe drinking water. The situation could worsen for hundreds of millions of people as climate change alters the variability of precipitation and the quantity of available water. At the same time, demand for water is increasing due to the world's growing population and its mounting aspirations. This dynamic triggers distributional conflicts and poses major challenges to water management systems in the countries concerned. For example, regions which depend on melt water from mountain glaciers – which are at risk from climate change – will require new water management strategies and infrastructures, as well as political efforts to avert national or even transboundary conflicts over the distribution of increasingly scarce water resources. However, the countries which will suffer the greatest water stress are generally those which already lack the political and institutional framework necessary for the adaptation of water and crisis management systems. This could overstretch existing conflict resolution mechanisms, ultimately leading to destabilization and violence.
- *Conflict constellation "Climate-induced decline in food production":* More than 850 million people worldwide are currently undernourished. This situation is likely to worsen in future as a result of climate change, as food insecurity in the lower latitudes, i.e. in many developing countries, will increase with a temperature rise of just 2 °C (relative to the 1990 baseline). With global warming of 2–4 °C, a drop in agricultural productivity is anticipated worldwide. This trend will be substantially reinforced by desertification, soil salinization or water scarcity. In South Asia and North Africa, for example, the areas suitable for agriculture are already largely exploited. This may well trigger regional food crises and further undermine the economic performance of weak and unstable states, thereby encouraging or exacerbating desta-

bilization, the collapse of social systems, and violent conflicts.
- *Conflict constellation "Climate-induced increase in storm and flood disasters":* Climate change is likely to result in further sea-level rise and more intensive storms and heavy precipitation. This will greatly increase the risk of natural disasters occurring in many cities and industrial regions in coastal zones. Those risks will be further amplified by deforestation along the upper reaches of rivers, land subsidence in large urban areas and the ever greater spatial concentration of populations and assets. Storm and flood disasters have already contributed to conflict in the past, especially during phases of domestic political tension, e.g. in Central America, India and China. Conflicts are likely to occur more frequently in future, firstly because regions especially at risk from storm and flood disasters, such as Central America and Southern Africa, generally have weak economic and political capacities, making adaptation and crisis management much more difficult. Secondly, frequent storm and flood disasters along the densely populated east coasts of India and China could cause major damage and trigger and/or intensify migration processes that are difficult to control.
- *Conflict constellation "Environmentally induced migration":* Experience has shown that migration can greatly increase the likelihood of conflict in transit and target regions. It can be assumed that the number of environmental migrants will substantially rise in future due to the impacts of climate change. In developing countries in particular, the increase in drought, soil degradation and growing water scarcity in combination with high population growth, unstable institutions, poverty or a high level of dependency on agriculture means that there is a particularly significant risk of environmental migration occurring and increasing in scale. Most environmental migration is initially likely to occur within national borders. Transboundary environmental migration will mainly take the form of south-south migration, but Europe and North America must also expect substantially increased migratory pressure from regions most at risk from climate change. The question as to which states will have to bear the costs of environmentally induced migration in future also contains conflict potential.

Regional hotspots

The social impacts of climate change will vary in the different regions of the world. A glance at the world map shown in Figure 1, entitled "Security risks associated with climate change", shows selected regional hotspots identified as a result of WBGU's analysis:

North Africa: The potential for political crisis and migratory pressure will intensify as a result of the interaction between increasing drought and water scarcity, high population growth, a drop in agricultural potential and poor political problem-solving capacities. The populous Nile Delta will be at risk from sea-level rise and salinization in agricultural areas.

Sahel zone: Climate change will cause additional environmental stress and social crises (e.g. drought, harvest failure, water scarcity) in a region already characterized by weak states (e.g. Somalia, Chad), civil wars (e.g. Sudan, Niger) and major refugee flows (Sudan: more than 690,000 people; Somalia: more than 390,000 people).

Southern Africa: Climate change could further weaken the economic potential of this region, whose countries already belong to the poorest in the world in most cases. It could also worsen the conditions for human security and overstretch the capacities of states in the region.

Central Asia: Above-average warming and glacial retreat will exacerbate the water, agricultural and distributional problems in a region which is already characterized by political and social tensions, burgeoning Islamism, civil war (Tajikistan) and conflicts over access to water and energy resources.

India, Pakistan, Bangladesh: The impacts of climate change will be especially severe in this region: glacial retreat in the Himalayas will jeopardize the water supply for millions of people, changes to the annual monsoon will affect agriculture, and sea-level rise and cyclones will threaten human settlements around the populous Bay of Bengal. These dynamics will increase the social crisis potential in a region which is already characterized by cross-border conflicts (India/Pakistan), unstable governments (Bangladesh/Pakistan) and Islamism.

China: Climate change will intensify the existing environmental stress (e.g. air and water pollution, soil degradation) due to the increase in heat waves and droughts, which will worsen desertification and water scarcity in some parts of the country. Sea-level rise and tropical cyclones will threaten the economically significant and populous east coast. The government's steering capacities could be overwhelmed by the rapid pace of modernization, environmental and social crises and the impacts of climate change.

Caribbean and the Gulf of Mexico: Increased frequency of more intense hurricanes could overwhelm the economic and political problem-solving capacities in the region (especially in Central America).

Andean region and Amazonia: Faster glacial retreat in the Andes will worsen the region's water problems.

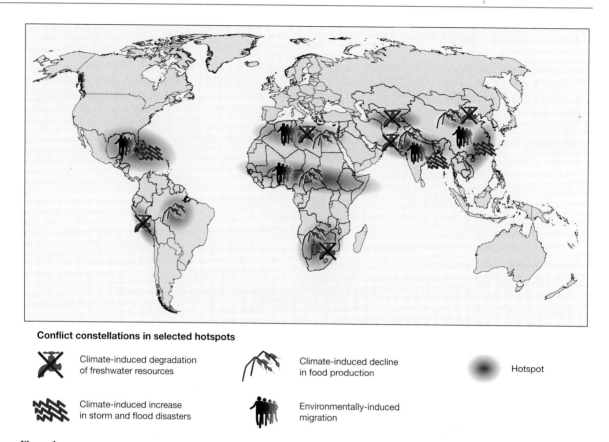

Figure 1
Security risks associated with climate change: Selected hotspots. The map only shows the regions which are dealt with in this report and which could develop into crisis hotspots.
Source: WBGU

The collapse of the Amazon rainforest, which cannot be ruled out, would radically alter South America's natural environment, with incalculable economic and social consequences.

Unstable multipolarity: The international policy setting of climate change

WBGU assumes that China and India in particular, due to their population size and economic dynamics, will gain more global political significance in the near future. The United States of America – currently the world's only superpower – is likely to experience a relative loss of power at the same time. The ascendancy of China and India therefore marks a major shift in the centres of power of the political world order, which will move from a unipolar to a multipolar system. A glance back at history shows that transitions from one type of world order to another rarely take place peacefully. The ensuing political, institutional and socio-economic turbulence and adaptation requirements can trigger major conflicts of interests within the international community and increase countries' vulnerability to armed conflict. That is not to say that the transformation processes which are anticipated in the international arena in the future will necessarily be violent. They will, however, absorb valuable time and resources which would then no longer be available for effective climate policy, for example.

Global politics over the next two decades will therefore have to master two challenges in parallel: the shift in the centres of power of the political world order, and the global turnaround towards effective climate policy. For both challenges, the stabilization and further development of the multilateral system are essential. Ultimately, the future interaction of old and new global political actors will be one of the factors that crucially determine whether and how the global challenges and risks arising in the 21st century can be managed successfully, and which role the "rest of the world" can play in this context. Climate policy is a case in point: without constructive cooperation between the OECD countries and the new drivers of global change, it will not be possible to limit climate change in a way which avoids destabilizing social impacts and threats to international security.

Six threats to international stability and security

In light of current knowledge about the social impacts of climate change, WBGU identifies the following six key threats to international security and stability which will arise if climate change mitigation fails:

1. *Possible increase in the number of weak and fragile states as a result of climate change:* Weak and fragile states have inadequate capacities to guarantee the core functions of the state, notably the state's monopoly on the use of force, and therefore already pose a major challenge for the international community. So far, however, the international community has failed to summon the political will or provide the necessary financial resources to support the long-term stabilization of these countries. Moreover, the impacts of unabated climate change would hit these countries especially hard, further limiting and eventually overstretching their problem-solving capacities. Conflict constellations may also be mutually reinforcing, e.g. if they extend beyond the directly affected region through environmental migration and thus destabilize other neighbouring states. This could ultimately lead to the emergence of "failing subregions" consisting of several simultaneously overstretched states, creating "black holes" in world politics that are characterized by the collapse of law and public order, i.e. the pillars of security and stability. It is uncertain at present whether, against the backdrop of more intensive climate impacts, the international community would be able to curb this erosion process effectively.

2. *Risks for global economic development:* Climate change will alter the conditions for regional production processes and supply infrastructures. Regional water scarcity will impede the development of irrigated agriculture and other water-intensive sectors. Drought and soil degradation will result in a drop in agricultural yields. More frequent extreme events such as storms and flooding put industrial sites and the transport, supply and production infrastructures in coastal regions at risk, forcing companies to relocate or close production sites. Depending on the type and intensity of the climate impacts, this could have a significant and adverse effect on the global economy. Unabated climate change is likely to result in substantially reduced rates of growth. This will increasingly limit the economic scope, at national and international level, to address the urgent challenges associated with the Millennium Development Goals.

3. *Risks of growing international distributional conflicts between the main drivers of climate change and those most affected:* Climate change is mainly caused by the industrialized and newly industrializing countries. The major differences in the per capita emissions of industrialized and developing/newly industrializing countries are increasingly regarded as an "equity gap", especially as the rising costs of climate change are mainly being borne by the developing countries. The greater the damage and the burden of adaptation in the South, the more intensive the distributional conflicts between the main drivers of climate change and those most affected will become. The worst affected countries are likely to invoke the "polluter pays" principle, so international controversy over a global compensation regime for climate change will probably intensify. Beside today's industrialized countries, the major ascendant economies whose emissions are increasing substantially, notably China but also India and Brazil, for example, will also be called to account by the developing countries in future. A key line of conflict in global politics in the 21st century would therefore divide not only the industrialized and the developing countries, but also the rapidly growing newly industrializing countries and the poorer developing countries. The international community is ill-prepared at present for this type of distributional conflict.

4. *The risk to human rights and the industrialized countries' legitimacy as global governance actors:* Unabated climate change could threaten livelihoods, erode human security and thus contribute to the violation of human rights. Against the backdrop of rising temperatures, growing awareness of social climate impacts and inadequate climate change mitigation efforts, the CO_2-emitting industrialized countries and, in future, buoyant economies such as China could increasingly be accused of knowingly causing human rights violations, or at least doing so in *de facto* terms. The international human rights discourse in the United Nations is therefore also likely to focus in future on the threat that climate impacts pose to human rights. Unabated climate change could thus plunge the industrialized countries in particular into crises of legitimacy and limit their international scope for action.

5. *Triggering and intensification of migration:* Migration is already a major and largely unresolved international policy challenge. Climate change and its social impacts will affect growing numbers of people, so the number of migration hotspots around the world will increase. The associated conflict potential is considerable, especially as "environmental migrants" are currently not provided for in international law. Disputes over compensation payments and the financing of systems

to manage refugee crises will increase. In line with the "polluter pays" principle, the industrialized countries will have to face up to their responsibilities. If global temperatures continue to rise unabated, migration could become one of the major fields of conflict in international politics in future.

6. *Overstretching of classic security policy:* The future social impacts of unabated climate change are unlikely to trigger "classic" interstate wars; instead, they will probably lead to an increase in destabilization processes and state failure with diffuse conflict structures and security threats in politically and economically overstretched states and societies. The specific conflict constellations, the failure of disaster management systems after extreme weather events and increasing environmental migration will be almost impossible to manage without support from police and military capacities, and therefore pose a challenge to classic security policy. In this context, a well-functioning cooperation between development and security policy will be crucial, as civilian conflict management and reconstruction assistance are reliant on a minimum level of security. At the same time, the largely unsuccessful operations by highly equipped military contingents which have aimed to stabilize and bring peace to weak and fragile states since the 1990s show that "classic" security policy's capacities to act are limited. A climate-induced increase in the number of weak and fragile states or even the destabilization of entire subregions would therefore overstretch conventional security policy.

Overstretching the capacities of the global governance system

The greater the scale of climate change, the greater the probability that in the coming decades, climate-induced conflict constellations will impact not only on individual countries or subregions but also on the global governance system as a whole. These new global risk potentials can only be countered by policies that aim to manage global change. Every one of the six threats to international stability and security, outlined above, is itself hard to manage. The interaction between these threats intensifies the challenges for international politics. It is almost inconceivable that in the coming years, a global governance system could emerge with the capacity to respond effectively to the conflict constellations identified by WBGU. Against the backdrop of globalization, unabated climate change is likely to overstretch the capacities of a still insufficient global governance system.

As the climate-induced security risks of the 21st century have their own specific characteristics, they will be difficult to mitigate through classic military interventions. Instead, an intelligent and well-crafted global governance strategy to mitigate these new security risks would initially consist of an effective climate policy, which would then evolve into a core element of preventive security policy in the coming decades. The more climate change advances, the more important adaptation strategies in the affected countries will become, and these must be supported by international development policy. At international level, the focus will be on global diplomacy to contain climate-induced conflicts, as well as on the development of compensation mechanisms for those affected by climate change, global migration policy, and measures to stabilize the world economy. The opportunities to establish a well-functioning global governance architecture will narrow as global temperatures rise, revealing a vicious circle: climate change can only be combated effectively through international cooperation, but with advancing climate change, the basis for constructive multilateralism will diminish. Climate change thus poses a challenge to international security, but classic, military-based security policy will be unable to make any major contributions to resolving the impending climate crises.

Recommendations

As yet, there is little sign of climate change manifesting itself in the form of conflict constellations and social crises. Globally averaged surface temperatures have so far increased by 0.8°C relative to the pre-industrial value. Without more intensified mitigation efforts, it must be assumed that by the end of the 21st century, globally averaged surface temperatures will rise by 2–7°C relative to the pre-industrial value, depending on the amount of greenhouse gases emitted and the uncertainties in the climate system. In WBGU's view, climate policy thus becomes preventive security policy, for if climate policy is successful in limiting the rise in globally averaged surface temperatures to no more than 2°C relative to the pre-industrial value, the climate-induced threat to international security would likely be averted. Conversely, WBGU anticipates that in the event of mitigation efforts failing, climate-induced security risks will begin to manifest themselves in various regions of the world from around 2025–2040. The key challenge is to take resolute climate policy action within the next 10–15 years, in order to avert the socio-economic distortions and implications for international security that will otherwise intensify in subsequent decades.

Fostering a cooperative setting for a multipolar world

Initiative 1: Shaping global political change

In order to ensure the acceptance and, above all, the constructive participation of the ascendant new world powers China and India, a multilateral order is needed which is viewed as fair by all countries. Germany can act as a pioneer here by undertaking the important and necessary advocacy work within the EU and working pro-actively at international level for the adoption of confidence-building measures. One option, for example, is to initiate and institutionalize a theme-specific process, modelled on the Conference on Security and Co-operation in Europe (CSCE) and aimed at confidence-building worldwide.

Germany and the EU should invest to a far greater extent than before in a coherent, future-oriented common foreign and security policy and set aside national egotisms. One issue to be explored is whether convening a world conference to consider the implications of the anticipated shift in the centres of power of the political world order could help foster a positive climate of cooperation. The diffuse uncertainty in the face of geopolitical change could perhaps then be channelled constructively. The aim would be to generate a positive mood that is conducive to a fresh start, emphasizing and building on the opportunities afforded by the anticipated changes.

Climate policy and energy policy offer ideal fields of action for Europe to play a pioneering international role. More intensive efforts to achieve resolute, fair and targeted international cooperation in the fields of climate protection and poverty reduction would also consolidate multilateral institutions as a whole and thus contribute to peaceful development in the world.

Initiative 2: Reforming the United Nations

As environmentally induced conflicts and the associated security issues are likely to increase in significance, the question which arises is which role the United Nations and its various institutions should play in managing the ensuing problems. In general, WBGU is in favour of better coordinating the efforts of the relevant organizations and programmes under the auspices of the UN and significantly enhancing their role in the interests of prevention.

Reflecting on the role and tasks of the UN Security Council
In WBGU's view, the impacts of unabated climate change, severe environmental degradation and environmentally induced conflicts can be regarded as a threat to international security and world peace. Presumably, therefore, the Security Council is authorized to take action in cases of widespread destruction of environmental goods and grave violations of international environmental law, and can apply appropriate sanctions against the states responsible. The Security Council now having debated in depth the security policy implications of climate change for the first time in April 2007, the question which arises is whether and how the Security Council's mandate can be appropriately adapted to meet these challenges. One option is to invoke the principle of the "responsibility to protect" by means of which the United Nations claims high moral authority. The Security Council could perhaps charge the UN Peacebuilding Commission, newly established in 2005, with addressing the specific tasks arising from this principle.

Upgrading the United Nations Environment Programme
WBGU reaffirms its recommendation that the United Nations Environment Programme (UNEP) be strengthened and upgraded by granting it the status of a UN specialized agency. Until that happens, UNEP and the Environmental Management Group should be actively supported by the member states in order to improve coordination of the numerous institutions engaged in international environmental policy and link environmental themes more closely with the United Nations' work in the economic and social fields. To this end, adequate medium- and long-term financing for UNEP should also be guaranteed.

Strengthening the United Nations' development capacities
WBGU reiterates its call for the establishment, in the long term, of a high-level Council on Global Development and Environment within the UN system, which ideally would replace the largely ineffective Economic and Social Council (ECOSOC). In the short term, WBGU recommends that policy be guided by the pragmatic proposals made by the High-Level Panel on System-wide Coherence and that a UN Sustainable Development Board be established, reporting to ECOSOC. The Board should be granted substantial political authority at the level of the heads of state and government and exercise joint supervision of relevant UN programmes, thus curbing the fragmentation of the UN development system.

Climate policy as security policy I: Preventing conflict by avoiding dangerous climate change

WBGU has made recommendations in various previous reports and policy papers on the specific form that an effective climate protection policy should take. For that reason, the following initiatives merely briefly outline, in key words, the topical and important fields of action for climate change mitigation.

Initiative 3: Ambitiously pursuing international climate policy

MAKING THE 2 °C GUARD RAIL AN INTERNATIONAL STANDARD
Specific international targets with a long-term focus increase the prospects of implementing a successful climate policy which initiates the global technological revolution and the shift in attitudes that are necessary to stabilize the concentration of greenhouse gases in the atmosphere at a level that would prevent dangerous anthropogenic interference with the climate system. At international level, a consensus must therefore be reached on quantifying the ultimate objective of the United Nations Framework Convention on Climate Change (UNFCCC) as set out in its Article 2. To this end, WBGU recommends the adoption, as an international standard, of a global temperature guard rail limiting the rise in near-surface air temperature to a maximum of 2 °C relative to the pre-industrial value. This will require a 50 % reduction in global greenhouse gas emissions by 2050 compared with a 1990 baseline.

GEARING THE KYOTO PROTOCOL TOWARDS THE LONG TERM
The mechanism established under Article 9 UNFCCC to review the Kyoto Protocol should be utilized for the ambitious further development of this Protocol and its compliance mechanisms. In WBGU's view, equal per capita allocation of emission entitlements on a global basis is the allocation formula which should be aimed for in the long term. All countries must ultimately play a part in achieving this goal. For the second commitment period of the Kyoto Protocol, the industrialized countries should adopt ambitious goals in the order of a 30 % effective reduction in greenhouse gas emissions by 2020 against the 1990 baseline. In order to integrate newly industrializing and developing countries into mitigation efforts to a greater extent, WBGU recommends the adoption of a more flexible approach to the setting of reduction commitments and clear differentiation within this country group.

CONSERVING NATURAL CARBON STOCKS
Preserving the natural carbon stocks of terrestrial ecosystems should be a key goal of future climate protection policy alongside the reduction of greenhouse gas emissions from the use of fossil fuels. Tropical forest conservation should be a particular priority in this context.

Initiative 4: Transforming energy systems in the EU

STRENGTHENING THE EU'S LEADING ROLE
In order to be a credible negotiating partner within the climate process, the European Union should achieve its Kyoto commitments and set more far-reaching and ambitious reduction targets for the future. In WBGU's view, a 30 % reduction target for greenhouse gas emissions by 2020 compared with the 1990 baseline and an 80 % reduction target by 2050 are appropriate.

IMPROVING AND IMPLEMENTING THE ENERGY POLICY FOR EUROPE
In WBGU's view, the proposals for an Energy Policy for Europe, presented by the European Commission in January 2007, point in the right direction and their basic elements should be adopted and rigorously implemented by the Member States. Binding targets, threshold values and timetables are essential to make the Energy Policy for Europe more specific. However, WBGU also sees a need for improvement in relation to certain expansion targets and individual technological options. Overall, the proposals should be geared more strongly towards sustainability criteria such as those proposed by WBGU in its report on sustainable energy systems.

TRIGGERING AN EFFICIENCY REVOLUTION
The proposals set out in the Energy Efficiency Action Plan, as well as existing directives and regulations, provide a sound basis for the necessary improvements in energy efficiency. The potential energy savings of 20 % by 2020, cited in the Action Plan and endorsed by the European Council, should be increased substantially through binding European rules, ambitious national targets and the rigorous enforcement of existing legislation. This applies especially to buildings, cars and product standards. Here, dynamic standards should be set which progressively lead to a reduction in energy input and emissions and thus establish long-term objectives for technological development as well.

EXPANDING RENEWABLES
WBGU proposes that in addition to the targets put forward in the Energy Policy for Europe and reaf-

firmed by the European Council, a binding target of 40 % of renewables in electricity generation by 2020 be adopted, along with a figure of 25 % of renewables in primary energy production. However, renewables expansion should not take place at the expense of other dimensions of sustainability; this applies especially to bioenergy or hydropower. Key prerequisites for the efficient integration of renewables are unimpeded access to the (national) grids and their fusion into a high-capacity trans-European grid.

Initiative 5: Developing mitigation strategies through partnerships

ESTABLISHING CLIMATE PROTECTION AS A CROSS-CUTTING THEME IN DEVELOPMENT COOPERATION
In development cooperation, path dependencies of emissions-intensive technologies should be avoided, and high priority should be granted to the promotion of sustainable energy systems in order to overcome energy poverty. To this end, climate protection must be integrated as a cross-cutting theme into poverty reduction strategies from the outset. A further key field of action for climate protection in developing countries is the avoidance of emissions from land use changes, especially deforestation. Within the German Ministry for Economic Cooperation and Development (BMZ) and the institutions of German Technical Cooperation (GTZ) and Financial Cooperation (KfW), and also within the framework of donor coordination in the European Union, poverty reduction and climate protection strategies should be "joined up" more systematically and far more rigorously than before.

AGREEING DECARBONIZATION PARTNERSHIPS WITH NEWLY INDUSTRIALIZING COUNTRIES
Germany and the EU should enter into strategic decarbonization partnerships with those newly industrializing countries that are likely to play an important role in the future world's energy sector. The aim should be to move energy systems and energy efficiency towards sustainability, thus providing innovative impetus and acting as a role model on a worldwide basis. Especially China and India could be partners in this area.

AGREEING AN INNOVATION PACT WITHIN THE FRAMEWORK OF G8+5
The G8+5 forum should be utilized for the development of joint targets for the promotion of climate-compatible technologies and products. This group, comprising the world's leading industrial nations and newly industrializing countries, represents the heavyweights in the global political arena and accounts for around two-thirds of global greenhouse gas emissions. On the basis of national Road Maps charting the transformation of national energy systems in the interests of climate protection, a joint Road Atlas for the Decarbonization of Energy Systems could then be produced. By adopting joint parameters for efficiency and CO_2 emissions standards, and promoting comprehensive technological cooperation, the G8+5 countries have the potential to become the driving force in the transformation of the world's energy systems.

Climate policy as security policy II: Preventing conflict by implementing adaptation strategies

Initiative 6: Supporting adaptation strategies for developing countries

Climate change will hit developing countries especially hard. Timely adaptation measures should therefore be an integral element of their national policies. However, most developing countries lack the skills and capacities to implement effective adaptation measures. Moreover, the impacts of climate change will increase the vulnerability of weak and fragile states and further reduce their adaptive capacities. This has yet to be fully recognized by many German and international development institutions.

ADAPTING WATER RESOURCES MANAGEMENT TO CLIMATE CHANGE AND AVOIDING WATER CRISES
- *Promoting international cooperation on the provision of information:* In order to adapt water resources management to the impacts of climate change, it is essential to draw on the findings of regional models which take account of climate change. International cooperation is vital to facilitate developing countries' access to current scientific data on the regional impacts of climate change on water availability. One issue which should be explored is whether a universally accessible database could be established and maintained by the international community for this purpose. In order to avoid water conflicts, cooperation on transboundary water management should be encouraged for regions sharing waters.
- *Reorienting water management towards action under increased uncertainty:* For effective action to be taken, there is often no need to await the development of appropriate forecasting models. Measures which improve adaptation to existing climate variability can often be applied to adaptation to future climate impacts as well. This is especially true of measures to improve the efficiency of

water management, local water storage capacity, systems for the distribution of stored water, and demand management. Integrated water resources management offers a suitable framework here.

GEARING AGRICULTURE TO CLIMATE CHANGE
- *Strengthening and reorienting rural development:* Greater account must be taken of climate change in the FAO scenarios. At the same time, in view of the anticipated drop in agricultural yields, development cooperation should focus to a greater extent on the development of rural regions. However, it is not enough simply to invest more resources in strengthening the agricultural sector. Instead, a new qualitative focus is required in agricultural development strategies in light of climate change.
- *Reforming world agricultural markets:* The reform of world agricultural markets should be pursued vigorously in order to generate opportunities for market access and production incentives in the developing countries. However, liberalization leads to price increases which can have an extremely adverse effect on Low-Income Food-Deficit Countries. For that reason, it is particularly important to establish compensation mechanisms for these countries, akin to those already in place in the WTO or the Bretton Woods institutions. The German Government should endeavour to ensure that such compensation mechanisms are adequately resourced.
- *Taking account of many developing countries' growing dependency on food imports:* The liberalization of the agricultural markets and short-term compensation payments will not solve the long-term supply and demand problems faced by many developing countries. A number of developing countries will experience major drops in agricultural yields and growing dependency on farm imports, not least as a result of climate change. For that reason, international climate policy should focus to a greater extent on this issue as well. One option which could be considered is whether those countries which are the main drivers of climate change should pay compensation to other adversely affected states for world market price increases and climate-related drops in agricultural yields.

STRENGTHENING DISASTER PREVENTION
- *Developing cross-sectoral approaches in development cooperation:* Development cooperation should develop and implement cross-sectoral strategies for the prevention of disaster risks to a greater extent, focussing especially on emergency planning, adaptation of land-use planning, establishment of clear decision-making structures at an early stage, and the inclusion of disaster prevention in education programmes. Early warning systems should also be embedded in development programmes.
- *Integrating disaster risks into development strategies to a greater extent:* Disaster prevention should be taken into account from the outset in the preparation of Poverty Reduction Strategy Papers and in the major poverty reduction programmes.
- *Reviewing disaster prevention in industrialized countries:* Disaster prevention should not be limited to the developing countries. Industrialized countries are also vulnerable to disasters. WBGU recommends a review of disaster prevention systems in the industrialized countries, especially in light of the challenges posed by ongoing climate change.

Initiative 7: Stabilizing fragile states and weak states that are additionally threatened by climate change

It is likely that the additional problems caused by climate change will impede the stabilization of weak and fragile states, and may even trigger further destabilization. Crisis prevention costs far less than crisis management at a later stage. The implications of climate change for the scale, longevity and financing of possible German contributions to the stabilization of fragile states should be taken into account to a greater extent in the Action Plan "Civilian Crisis Prevention, Conflict Resolution and Post-Conflict Peace-Building". The debate should be conducted first and foremost within the European Union framework. In this context, WBGU recommends, in particular, the operationalization of the Solana Strategy in line with the Barcelona Report, which prioritizes crisis prevention with the aim of avoiding military intervention as far as possible.

The German Government should therefore continue to play an active role in the Fragile States Group set up by the OECD's Development Assistance Committee and drive forward the implementation and further development of its Principles for Good International Engagement in Fragile States and Situations. In particular, WBGU recommends that the German Government endeavours to ensure that appropriate account is taken, in this context, of the environmental impacts and risks arising from climate change. Specifically, fragile states' capacities to manage environmental risks must be maintained and reinforced, and if necessary re-established, even under difficult political and economic conditions.

Initiative 8: Managing migration through cooperation and further developing international law

Developing comprehensive international strategies for migration

In order to manage environmentally induced migration, a comprehensive migration policy strategy is required which takes account of the interests of all stakeholders. Its long-term objectives must be geared towards the interests of the destination, transit and home countries alike. In WBGU's view, an approach which focuses primarily on the industrialized countries' internal security – current EU policy being a case in point – is too one-sided, reactive and, at best, only effective in the short term. Prevention strategies do not feature in the numerous bilateral readmission agreements between the industrialized nations and countries of origin. WBGU recommends that at future international migration forums, environmentally induced migration feature on the agenda and that appropriate plans be developed to deal with this issue. Focussing solely on economically motivated migration is not enough. Germany and the EU must step up their engagement in this area.

Integrating migration policy into development cooperation

In the Least Developed Countries, unabated climate change would increase the risk of people being forced to abandon their home regions due to the collapse of their natural life-support systems. Development cooperation can help to strengthen the adaptive capacities of people living in absolute poverty and thus make it easier for them to remain in their homes. However, development strategies must take greater account of foreseeable climate impacts at local level. It can be assumed that climate-induced migration within and between affected states will increase in future, opening up a new field of action in development cooperation. The importance of a comprehensive, pro-active and development-oriented migration policy is increasingly being recognized at political level as well.

Enshrining the protection of environmental migrants in international law

Environmental migrants currently do not fit into the agreed categories of international refugee and migration law, even though a strong increase in environmentally induced migratory movements is anticipated. Under current international refugee law, states have no specific obligations in relation to the treatment of environmental migrants, nor are any other legal mechanisms in place for the protection of the affected individuals. In the interests of improving the legal status and protection of environmental migrants, it is important to consider ways of closing this gap in international law. WBGU recommends that rather than adopting an additional protocol to the existing United Nations Convention Relating to the Status of Refugees, vigorous efforts be made at this stage to establish a cross-sectoral multilateral Convention aiming at the issue of environmental migrants. UNHCR should be involved as fully as possible in negotiations on the adoption of the requisite international agreement. This agreement should institutionalize the cooperation between UNHCR and the bodies established within the framework of the participating conventions. Furthermore, the United Nations' efforts to protect internally displaced persons, which have already begun, should be intensified.

Initiative 9: Expanding global information and early warning systems

Both the gradual changes caused by climate change and the natural disasters which are expected to occur with increasing frequency could destabilize the affected regions and, in extreme cases, constitute a major risk factor for national and international security. Global information and early warning systems can therefore do much to mitigate these adverse effects and make a major contribution to conflict and crisis prevention.

On the one hand, these systems should provide timely information and warning in advance of extreme events and crises. The German Government, which has been active in this area for many years, should continue to participate in the development of a global early warning system. The system should not be confined to individual risks but should address threats to human security on a comprehensive basis. This early warning system should provide information about all types of natural hazard, epidemics and technological risks, and also take account of slowly advancing environmental changes.

On the other hand, the system must provide processed data on expected regional climate impacts, especially for developing countries which lack adequate capacities of their own to model and evaluate these data. This type of database should collate regional forecasts, with all their uncertainties, and make them accessible in an easy-to-understand format for users.

In order to establish this type of global information and early warning system, the activities of existing UN institutions (e.g. WMO, FAO, UNDP, UNEP, UNFCCC) and other forums such as ISDR or IPCC must be properly coordinated.

Financing the initiatives

The prevention of environmentally induced security risks not only requires resolute political action by the relevant national and international actors, but also adequate financial resources to implement the measures.

Avoiding dangerous climate change

Climate protection is worthwhile: The global costs of effective climate protection are far lower than the costs of inaction. What is required now is international coordination in order to ensure that the financial resources are channelled into efficient mitigation measures.

TRANSFORMING ENERGY SYSTEMS WORLDWIDE
In order to initiate the necessary transformation of energy systems in the developing countries, the existing multilateral funds (e.g. Global Environment Facility, Carbon Finance Unit) should be boosted by better and more reliable financing. Additional sources of funding can be harnessed through new financing instruments such as the introduction of emissions-dependent user charges for aviation and shipping, unless these emissions are already covered by other regulatory schemes. In the longer term, a system of internationally tradable quotas for renewable energies can also generate revenue. Financial resources can also be mobilized by restructuring existing budgets: subsidies for fossil fuels can be progressively reduced, freeing up funds which can then be channelled into the promotion and global deployment of renewable energies.

CONSERVING TERRESTRIAL CARBON STOCKS
The protection of terrestrial carbon stocks, especially the tropical forests, should be a further funding priority. A large proportion of this forest stock is located in developing countries, but is under threat from overexploitation and deforestation. The industrialized countries should actively promote the conservation of these forests. The UNFCCC process to reduce deforestation in developing countries offers a good starting point and should be pursued as a matter of urgency. In particular, the Annex I countries under the UNFCCC regime should provide incentives, in the form of financial compensation for loss of income from alternative land use, to encourage these countries to refrain from deforestation.

Adaptation to unavoidable climate change

Developing countries generally contribute very little to anthropogenic climate change, but they still have to adopt comprehensive adaptation measures which they often cannot afford due to a lack of capital. For that reason, adaptation measures in these countries should be co-financed by the international community.

BOOSTING OFFICIAL DEVELOPMENT ASSISTANCE
The funding of Official Development Assistance (ODA) is still failing to reach the target of 0.7 % of gross national income agreed by the United Nations. In May 2005, the European Union's development ministers set a new intermediate target for development aid of 0.56 % of donor countries' gross national income by 2010, which would put Europe on course to reach the UN's 0.7 % target by 2015. This timetable must be rigorously adhered to.

DEVELOPING A UNFCCC ADAPTATION STRATEGY
WBGU recommends that a comprehensive strategy be developed to promote adaptation in the developing and newly industrializing countries. The Funds so far established under the UNFCCC and the Kyoto Protocol are inadequate to meet the challenges described above, both in terms of their volume and their institutional structures. The financial contributions made by individual states to this strategy should be based on their contribution to global warming and their economic capacities. In the short term, more resources should be made available to the Least Developed Countries Fund and the adaptation "window" of the Special Climate Change Fund.

STRENGTHENING MICROFINANCE
Microfinancing institutions and instruments (e.g. microcredits or microinsurance) should be expanded with resources from international development cooperation. Despite great hopes that microinsurance, for example, could be a suitable instrument to guard against climate-induced natural disasters, microfinancing cannot replace – but at best can only supplement – international financial assistance.

ESTABLISHING AN ENVIRONMENTAL MIGRATION FUND
A new international environmental migration fund should provide the financial basis for measures to deal with environmental migrants. The International Dialogue on Migration launched by the International Organization for Migration in 2001 offers an appropriate platform for this purpose. Fair and efficient burden-sharing between those countries which are affected by environmental migration and those

which are not should satisfy the "polluter pays" principle, described above, and the "ability-to-pay" principle by linking contributions to the Fund to the level of country-specific greenhouse gas emissions and other indicators such as gross domestic product.

Financing international conflict prevention

ADOPTING AN INTEGRATED APPROACH TO THE FINANCING OF CRISIS PREVENTION, DEVELOPMENT COOPERATION AND MILITARY SPENDING

Due to the clear overlaps between civilian crisis prevention and development cooperation, WBGU takes the view that there is no need for an additional funding target for crisis prevention. Instead, the political focus should be geared entirely towards compliance with the existing timetable for increasing ODA. WBGU proposes that security spending be critically reviewed, especially as regards its effectiveness for international peacebuilding, and adjusted accordingly. The German Government should drive forward the international debate and negotiating processes within the EU, NATO and beyond. Military budgets should be restructured in favour of preventive measures in the field of development cooperation. As military spending is realigned towards preventive security policy, the need for funding in the "classic" areas of military spending will be reduced.

STRENGTHENING THE FINANCIAL INSTITUTIONS IN THE UN SYSTEM

The mechanisms to finance international crisis prevention and peacebuilding regimes at UN level are inadequately resourced, in WBGU's view. The German Government should support the Central Emergency Response Fund with appropriate contributions and lobby for a binding schedule for the financing of this Fund. It should also continue to take an active role in financing the UN Trust Fund for the Consolidation of Peace and lobby for the adoption of rules to ensure regular contributions to the Fund in future.

If climate protection fails: Strategies in the event of destabilization and conflict

If climate protection fails and the 2°C guard rail is not adhered to, the international community must prepare itself to deal with climate-induced conflicts such those described as "conflict constellations". In any event, a pro-active climate protection policy must remain in place to mitigate greenhouse gas emissions, with the aim of keeping global warming as close to the 2°C guard rail as possible. Due to the anticipated high costs of mitigation and adaptation, economic policy should also develop strategies to avert the possible destabilization of the global economy as a result of climate change. In the field of development policy, the need to manage water and food crises and storm and flood disasters would also substantially increase. In view of the growing number of weak and fragile states and an increasingly degraded natural environment, development cooperation would be called upon more and more frequently to prevent human development from dropping back, rather than advancing development as is currently the case.

The increase in migration worldwide – both within developing regions and between North and South – would absorb considerable political and economic capacities. Overall, major disruptions in international relations could be anticipated, not least in the North-South context. In order to avert destabilization and the escalation of conflicts, the crisis management potential of the world's leading powers should be pooled, the multilateral institutional architecture strengthened, and substantial additional resources mobilized. If climate protection policy fails and these efforts are not made, it is likely that from the mid 21st century local and regional conflicts will proliferate and the international system will be destabilized, threatening global economic development and completely overstretching global governance structures. In order to avoid these dangerous developments, the appropriate climate policy course must be set now.

Introduction

> *It is not predicting the future that matters, but being prepared for it*
> PERICLES, GREEK STATESMAN, 493–429 BC

Climate change is advancing rapidly. Without resolute counteraction, a global increase in temperature of 2–7 °C relative to pre-industrial levels can be expected to occur by 2100. This will cause more frequent and more severe extreme weather events such as heavy rains, drought, heatwaves and storms. There is also a danger of tropical cyclones not only becoming stronger but also occurring with greater frequency in extratropical regions. At the same time, sea levels are continuing to rise. These direct impacts of climate change will have far-reaching effects upon societies and the lives of people around the world. If climate change continues unabated, agricultural yields will decline significantly in many regions of the world, especially Africa and South Asia, and poverty will grow accordingly; drought will make it difficult for many millions of people worldwide to gain access to clean drinking water; extreme weather events will continue to gain destructive force and may confront governments and societies with major issues of adaptation, for instance in Central America, but also on the east coasts of China and India. Many states that are already weak and fragile will be faced with additional 'environmental stress'. Comprehensive changes in biogeophysical conditions will jeopardize the livelihood bases of people in the particularly affected regions of the world and will trigger migration. The present report examines whether the emerging trends may contribute in the future to a destabilization of societies, regions or even the whole international system.

In the policy arena, environmental degradation and climate change are increasingly perceived as challenges to international policy and security:

- The issue is high on the political agenda in Germany. Federal Chancellor Merkel stressed in a security policy address: 'We have devastating regional conflicts, dire poverty problems, especially in our neighbouring continent of Africa, and massive migration. We know that conflicts over the distribution of increasingly scarce resources can cause ever greater unrest and violence, as can environmental problems. These are matters of oil and gas, of climatic changes, of potable water. All these aspects are the source of conflicts with a very high potential for violence' (address by German Federal Chancellor Angela Merkel to Impulse 21 – Berlin Forum on Security Policy, 10 November 2006).
- Former UN Secretary-General Kofi Annan has referred to climate change as a 'threat to peace and security'. Annan has stressed that the international community must devote just as much attention to climate change as it does to preventing war and the proliferation of weapons of mass destruction (opening address on 15 November 2006 to the 12th session of the Conference of the Parties to the UNFCCC in Nairobi).
- In April 2007, upon the initiative of the United Kingdom, the United Nations Security Council debated climate change in depth for the first time. In her speech, UK Foreign Secretary Beckett compared emerging climate change to the looming threat of war in the period before 1939. 'An unstable climate risks some of the drivers of conflict – such as migratory pressures and competition for resources – getting worse' noted Beckett. The Foreign Secretary went on to explain that it is essential for the UN body to take account of climate change in order to maintain international peace and security (UK Foreign & Commonwealth Office, 2007).
- In April 2007, high-ranking retired US generals published a report terming climate change a serious threat to the security of the USA that will promote extremism and terrorism, especially in unstable regions (CNA Corporation, 2007).

WHICH SECURITY RISKS DOES CLIMATE CHANGE ENTAIL?

The question of whether environmental changes actually threaten international stability and security is thus already on the international policy agenda. The present report from WBGU draws together the present state of knowledge and science by collating the findings of research on the causes of conflict and war, evaluating past experience of conflicts triggered by environmental degradation, and exploring the impacts of climate change upon people and societies. WBGU pursues the question of whether the emerging forms of environmental change may lead to a destabilization of societies and overstretching of political systems and hence ultimately to violence, and might thus confront the international system with new challenges.

Research conducted in recent decades shows that land degradation, water shortage and resource competition, when combined with other conflict-amplifying factors, have indeed caused violence and conflict in the past. The review of 73 empirically well recorded 'environmental conflicts' which occurred between 1980 and 2005, however, also showed that these were limited to a regional scope and did not present any serious threat to international security. The decisive question is therefore whether global climate change is fundamentally altering this scene. WBGU comes to the conclusion in its report that this will indeed be the case if global warming continues unchecked. Climate change can then become an international security risk.

This fresh perspective on the environment-security nexus requires a broadening of the timeframe under analysis. Climate change is only just beginning, but its impacts will mount steadily in coming decades if global warming is not slowed by effective climate policy. It is thus apparent that large-scale disruptions relevant to security are only to be anticipated in future decades. WBGU therefore had to review past experience of conflicts induced by environmental degradation, while at the same time looking far into the future in order to anticipate the gradually unfolding impacts of climate change upon national societies, world regions and the international system. In order to identify the key linkages between climate change and its potential destabilizing and conflict-generating effects, as well as to pinpoint regional hotspots and to develop strategic approaches aimed at putting policy processes on track, WBGU has based this report upon the following guiding questions:
- In which circumstances might climate change trigger or amplify security problems such as unrest, civil war, a collapse of societies and states or cross-border conflict?
- Can typical conflict constellations and mechanisms be identified that might be triggered by climate change?
- Which regions are particularly susceptible to climate-induced conflict?
- In what manner will climate change threaten international stability and security?
- How can the risk of such conflicts be prevented through strategic shifts in development trajectories, and how can cooperation be fostered at both national and international level?
- What can and must Germany and Europe do to cope with the new challenges?

As yet, environmental changes have triggered conflict and violence only in isolated cases. There is empirical evidence, for example, of outbreaks of violence and anarchy in the wake of storm and flood disasters. However, the manner and rate of climate change today are without precedent in the history of humankind. Fundamental changes in the biosphere are confronting humanity with entirely new challenges. Today's civilization – with a population numbering some 6.5 thousand million, a finely woven global infrastructure, global flows of trade, information and transport, differentiation among industrialized, newly industrializing and developing countries, and disparate capacities for resolving problems and conflicts – may be threatened by climate impacts for whose management no historical models exist.

IDENTIFYING FUTURE CONFLICT POTENTIAL
WBGU's report shows how unabated climate change could increasingly undermine human security in many regions of the world, and explores the conditions under which human insecurity may heighten susceptibility to social destabilization and violence. The aim cannot be to deliver a precise prediction of future conflicts or disasters. The purpose is rather to identify the mechanisms which encourage conflict under conditions of rapid and intense climate change and to do so early enough for remedial measures to be taken proactively. WBGU identifies four conflict constellations that may be driven by climate change: 'Climate-induced degradation of freshwater resources', 'Climate-induced decline in food production', 'Climate-induced increase in storm and flood disasters' – and 'Environmentally induced migration', which is triggered by the former three constellations.

In order to depict and extrapolate these conceivable conflict constellations, the report develops a number of fictitious yet plausible narrative scenarios. Positive and negative scenarios illustrate potential development pathways which are free of violence or, conversely, involve a substantial probability of violence. The positive scenarios are important because they show how, under conditions of rapid climate change, ways can be found to defuse emerging social crisis, violent conflict and destabilization of the international system in good time. Thus, in concert with the analysis of conflict constellations, the scenarios provide a basis for recommendations on action to prevent future climate-induced security risks.

Not all countries or regions of the world are exposed equally to the risks of destabilization and violent conflict arising from climate change. The thin line between stability and instability, and between security and insecurity, will be determined by the extent of global climate change, its impacts upon societies and regions, and their specific capacities to adapt to and tackle the problems. Agriculture-based economies are more susceptible to climate change impacts than service-based economies; rich societies are better able to handle the costs of climate change than developing countries; highly capable governments can manage the consequences of environmental degradation and climate change better than weak states; well-organized civil societies are better able to take precautionary action than fragmented societies which may already be characterized by violence. Climate change thus widens social, political and economic disparities in the world society. Yet even countries with high problem-solving capacity are not entirely secure, for massive immigration from regions experiencing severe problems related to environmental upheaval and conflict would present considerable additional security challenges.

INTERNATIONAL ACTION IN A WORLD IN TRANSITION

It is characteristic of the challenges presented by climate change that a delicate equation links the present need to act and the future occurrence of harmful impacts that are anticipated and need to be prevented. Only if action is taken today will it be possible to contain violent conflicts and social crises induced by climate change. Prevention is necessary and possible: climate-induced security risks can still be avoided if resolute action is taken in the coming two decades, keeping global warming within the 2 °C guard rail. The challenge is to muster the political will to make far-sighted strategic decisions. Political systems are better able to find answers to short-term problems than to act prospectively. Moreover, climate change will play out against the backdrop of a fundamental shift in the centres of power of the political world order. The dominant position of the USA will most likely give way in the coming decades to a system that is more multipolar in nature. China and India currently constitute the newly emerging centres of the global economy and of global politics alongside the USA, Europe and Russia. The question thus arises whether the new and old powers in international politics, who are at the same time those principally responsible for climate change, will be preoccupied in future with power disputes and struggles to assert dominance in world politics, or whether they will succeed in conceiving of climate change as a common global challenge and hence work together to craft an effective global climate policy.

In this report, WBGU develops a package of recommendations for action and research to mitigate the risks of conflict set out above. The core message is that it is essential to make strategic decisions in the next two decades in order to change course away from trajectories that are highly risky and would entail scarcely controllable destabilization and crisis in societies, world regions and the international system. WBGU warns anew that this change of course must be made as a matter of urgency, even if, from the present perspective, the costs appear more prominent than the benefits – future benefits will outweigh present costs many times over. Europe now has a great opportunity. It can cast itself as a pioneer of sustainable climate policy and can thus gain leverage in one of the key arenas of global policy. A global transformation of energy systems is the lynchpin of efforts to mitigate climate change successfully. Tapping efficiency potential and boosting renewable energies are key elements of this transformation. If Europe were to succeed in integrating its policies on climate, technology and innovation in a way that could provide a beacon for the future, this would in the long term enhance its position in the global economy.

Environmental change in security discourse

Broadening the concept of security has been a theme of international policy discourse ever since the end of the Cold War. It is in this context that potential threats to security from environmental change have entered the security policy debate. WBGU has prepared the present report in response to an upsurge of interest in Germany, and indeed across Europe, in the possible links between environment and security. Public debate on the subject is highly diffuse in character and is being conducted on the basis of a sketchy set of facts. WBGU's intention in presenting this report is to strengthen the empirical basis of the debate and, in addition, to develop a long-term perspective that goes beyond *ad hoc* interpretations of current and past 'environmental conflicts'.

This chapter explores the relevance of security with regard to global structural policy, with special attention being paid to the environmental dimension of security. A first step is to define a concept of security that is as precise as possible.

2.1
Redefining security

The concept of security and the question of how it is best conceptualized are among the most controversial issues in international politics. Due to the inherent flexibility of the concept, security can be interpreted in different ways, depending on the interests of security experts, intellectuals and politicians (Dalby, 1992). The approach adopted by Brock (2004) can be taken to be largely undisputed, where security means the absence of war and yet cannot be equated with peace.

In classical terms security means the integrity of territorially organized sovereign nation states within the system of international law as represented by the United Nations since the end of the Second World War. Thus framed, security is the preservation of nation state integrity in the face of external threats in an anarchic world of states; the task of guaranteeing security is seen as being ultimately a military one. With the end of the Cold War and increasing globalization it came to be universally recognized that insecurity, instability and violence are brought about not by military aggression alone, but may have complex political, economic, socio-cultural and ecological origins (Biermann et al., 1998). This led to calls for a re-assessment of security and corresponding policy adjustments, e.g. the much quoted essay by Tuchman Mathews entitled 'Redefining Security' (1989). In academia, too, intense debate about the concept of security developed after the end of the Cold War (Dalby, 1992; Lipschutz, 1995). In a certain sense, the 'peace discourse' of the 1970s and 1980s, which had proceeded in a setting defined by the Cold War, was gradually transformed into a 'security discourse' (Brock, 2004).

2.1.1
Comprehensive security

The policy debates have since given rise to new, broader concepts of security, which have been made the basis of both NATO's and the German government's security strategies. According to these new concepts, security policy is not only a question of military capacity, but is based on the ability to defuse political and socio-economic crises that threaten to cross the threshold to violence, and to do so as early as possible using non-military and, if necessary, military means. In addition to classical foreign and economic policy, the non-military methods increasingly include development and environmental policy measures. The concept explicitly includes interests in securing strategic resources, which serve to safeguard affluence in the industrialized countries. Thus, mineral resources, such as oil, gas and ores, as well as other strategic resources such as freshwater access and safe shipping routes, have become assets of significance in terms of security policy.

When the concept of security is extended in this way, non-military areas of policy come to be identified as potentially security relevant. Public concerns, such as environmental degradation and poverty, which previously had been regarded as 'soft' policy fields, have been accorded greater importance as a result. At the same time, areas of non-military policy have come to be 'securitized' (Waever, 1995; Brock, 1997). In this regard, critics point to the vagueness of the concept and its lack of a clear referent: if the concept of security is used in many different contexts, it becomes harder to identify security risks, responsibilities and appropriate responses (Deudney, 1990, 1991). This lack of clarity also underlies the political ambiguity of a wider security concept: the concept can be used 'to underscore demands for non-military conflict resolution just as much as to justify the expansion of military security policy abroad and the curtailment of civil rights at home' (Brock, 2004). This makes it possible for new military interventions, including so-called pre-emptive wars, to be justified under the guise of comprehensive security. Such a trend may even lead ultimately to greater insecurity. The original strength of the new security debate could therefore turn into precisely the opposite.

2.1.2
Human security

In 1994 the United Nations Development Programme (UNDP) introduced the concept of 'human security', which places the focus on individuals' security needs and has since played a major role in shaping the international security discourse. In a wide-ranging conceptualization, security is no longer seen merely as 'freedom from fear', but also as 'freedom from want'. According to this definition, economic crises and pandemics, such AIDS, malaria or tuberculosis, are to be considered security risks just as much as violent conflicts and wars. This approach was taken up and given concrete form in a commission headed by former UN High Commissioner for Refugees Ogata and winner of the Nobel Prize for economics Sen (UNDP, 1994; Commission on Human Security, 2003). It is in the nature of the organization that for the United Nations, the state remains the key guarantor of security. However, the commission's report explicitly notes the fact that states often fail to meet their security obligations and sometimes even pose a threat themselves to their own populations: 'The focus must broaden from the state to the security of people – to human security' (Commission on Human Security, 2003).

In the context of debates around environmental challenges with relevance for security policy, proponents of the concept of human security stress that a concept of security which refers to collective actors, such as states and societies, is not capable of encompassing all relevant threats, challenges, vulnerabilities and risks that exist (Altvater, 2003; Brauch, 2005).

Environment-related conflicts do not usually involve longstanding clashes of interest between organized groups, but are rather discrete disputes that occur between people whose intention is to safeguard their means of survival. From the point of view of human security, environmental security would have to be geared not towards the nation state but towards the ecosystem in which humans themselves represent the real threat to security in their role as users and polluters of the environment (Brauch, 2005, 2006).

However, a range of objections have been raised against the concept of human security. One is that the UNDP approach plays down security risks, such as wars, and the important role of nation states in their significance, and that it blurs the boundary between human security and human development. Clarity can be gained only by restricting the concept of security to 'freedom from direct physical violence'. Even then, there is a danger that the concept of human security, with its positive connotations, could be instrumentalized to construct new sources of legitimacy for military interventions aimed at safeguarding security (from environmental risks, AIDS, migration, etc.). Last but not least, when the concept of security is watered down in this way, it leads to a loss of analytical clarity from a research perspective as well (Section 9.2).

Overall, then, the debates about comprehensive security and its policy implications remain controversial due to the inevitably Janus-faced character of the issues. On the one hand, an expanded concept of security could counteract the further militarization of international politics, if the drive to achieve security were increasingly understood to encompass economic, social and ecological concerns in the context of sustainable development. On the other hand, however, it allows international structural policy to become militarized, as environmental policy and other non-military matters are turned into security concerns. This may lead, among other things, to a situation in which non-civilian actors, such as secret services or police and border protection authorities, are increasingly entrusted with tasks that lie outside their original area of responsibility (see the Patriot Act in the USA) or in which resources intended for development cooperation are used to co-finance military operations in the context of humanitarian interventions. Similarly, it is to be feared that the 'colonization' of international environmental policy by the dominant industrialized powers will further reinforce the North-South conflict.

Both tendencies – the endeavour to develop non-military supporting measures for security policy and military back-up for global structural policy – are reflected in the more recent debate about development strategies for weak and fragile states (Faust and

Messner, 2004; Klingebiel and Roehder, 2004, 2005; Schneckener, 2004).

2.2
Current security policy strategies

Since the events of 11 September 2001 security discourse in international politics has been dominated by the fight against global terrorism. This is reflected clearly in the new security strategies that have arisen on both sides of the Atlantic (Daase, 2002; Hippler, 2003; Berenskoetter, 2005), with the USA and the European Union setting differing priorities (Table 2.2-1). The National Security Strategy of the USA clearly emphasizes military aspects, even including pre-emptive military action, whereas the European security strategy argues more along political lines and is oriented towards the objective of avoiding conflict. While the strategy of the USA also contains provisions for increasing foreign and development assistance, its overall content and style is oriented towards military offensiveness: 'our best defense is a good offense' (White House, 2002).

In contrast to this, the European security strategy, formulated in December 2003 under the supervision of Solana (Secretary-General and High Representative for the Common Foreign and Security Policy of the EU), emphasizes the possibilities of multilateral diplomacy. The strategy calls, above all, for a strengthening of the multilateral world order and of international law; it also calls for conflict prevention measures to be designed for the long term and to be based on both non-military cooperation and political pressure, which in the last resort – as in the case of humanitarian disasters, genocide or acute state failure – is to be complemented by robust military intervention within the framework of the international legal order (ESS, 2003; Faust and Messner, 2004).

The general principles formulated in the European security strategy have been lent greater precision, however, in a European defence paper commissioned by the EU member states and prepared by the Paris Institute for Security Studies (ISS). This paper notes that there is an increasing need for expeditionary forces that are flexible and capable of being deployed at short notice, as well as for occupation forces that can be deployed over the long term (ISS, 2004). The ISS paper bears the unmistakeable trademark of a military policy-based approach, suggesting that the EU's security policy think tank has yet not accommodated the broader, more non-military security approach. Thus the transatlantic differences in approach are no longer quite so pronounced.

In addition to the European defence paper, though, the report commissioned by Solana entitled 'A Human Security Doctrine for Europe' and prepared by the Study Group on Europe's Security Capabilities (2004) has also become a central component of the debate over European security policy. This report – also known as the 'Kaldor paper' after the group's chairperson Mary Kaldor – makes explicit reference to the concept of human security, which differs markedly from traditional security policy ideas. In particular, the Study Group recommends that foreign and security policy be oriented less towards states and more towards individuals, since territorial defence in the classical sense is no longer effective in the changed global security situation. The Kaldor paper thus calls additionally for a

Table 2.2-1
Main differences between the security strategies of the United States of America and the European Union.
Source: WBGU after Berenskoetter, 2005

	US National Security Strategy	European Security Strategy
Aims and objectives	Global, universal: – liberal world order – human dignity – US hegemony	Predominantly regional: – multilateral world order – credibility of the EU – stability in Europe
Threat scenarios	Grounded in ideology: – terrorism and weapons of mass destruction – 'rogue states' and weapons of mass destruction – tensions in the Muslim world	Defined in a concrete way: – proliferation of weapons of mass destruction – failing states – terrorism and organized crime
Means	Predominantly military: – pre-emptive warfare – coalition of the willing	Predominantly non-military: – conflict prevention – multilateral diplomacy and international regimes
Legitimation	US mandate	UN mandate

Human Security Response Force that is specialized in humanitarian intervention, to complement the EU Military Rapid Response Force proposed in the ISS paper.

In its coalition treaty, the German government also declares its support for a wide-ranging concept of security that links foreign, security and development policy (Große Koalition, 2005). In the face of new and complex risks, such as international terrorism, weapons of mass destruction and highly vulnerable infrastructures, the new security policy of the German government, the White Paper on German Security Policy, also speaks of 'networked security': 'Security cannot be guaranteed by the efforts of any one nation or by armed forces alone. Instead it requires an all-encompassing approach that can only be developed in networked security structures and within the context of a comprehensive national and global security philosophy' (BMVg, 2006). One essential goal of the new security strategy is to strengthen civil crisis management and to improve early warning mechanisms. In order to achieve this, military and non-military approaches need to be better coordinated and crisis prevention measures accorded much greater importance. Security practice so far lacks such an orientation, however, as indicated, for example, by the action plan for 'Civilian Crisis Prevention', adopted in 2004, which was poorly funded and limited in scope (Hauswedell, 2006; Schradi, 2006). Just how difficult and complex it is to implement a new understanding of security is also evident when one looks at the German government's research programme for civil security (BMBF, 2007). This, too, opts for a comprehensive approach and presents a wide-ranging analysis of risks. However, the solution to the problem is seen first and foremost in technical innovations tailored to the needs of the industrialized world.

Security risks that could arise as a result of long-term processes of global environmental change are accorded varying degrees of attention in the strategies put forward by the USA, the European Union and the German government; overall, however, they play a rather subordinate role. Statements about potential linkages between environmental degradation and security often remain vague, unless they are focused on those natural resources whose supply is directly relevant to the interests of the industrialized countries. As another example, the 2005 status report from the Worldwatch Institute, entitled 'Redefining Global Security', stresses the significant role of an oil-based global economy with regard to climate change and calls for a reduction in dependency on crude oil (Box 2.2-1). Yet even the Worldwatch Institute sees the security risks associated with such an economy as being above all a problem of supply shortages and the 'resource curse'; this means that resource richness can often be associated with negative consequences for the development of a country (Worldwatch Institute, 2005).

A similar approach to environmental change also characterizes the security debate in the United Nations context, which has been recently conducted on the basis of reports by UN Secretary-General Annan ('In Larger Freedom', March 2005) and by the High-level Panel on Threats, Challenges and Change commissioned by Annan ('A More Secure World', December 2004) (UN, 2004; UNSG, 2005). While the high-level panel argues for a more wide-ranging concept of security that regards poverty reduction as an indispensable basis for a new collective security, environmental degradation is mentioned only as a marginal issue in the six categories of global threats to security which the panel defined, namely, as a sub-issue under economic and social deprivation.

Given this background, it is remarkable that, in a recent adjustment of its national security strategy in the aftermath of Hurricane Katrina, the United States no longer addresses the scarcity of strategically important resources alone, but also explicitly takes account of environmental degradation (White House, 2006). Germany's White Paper on Security Policy also refers explicitly to the potential risks of global environmental change when it speaks of migration, natural disasters and environmental destruction as being causes of instability (BMVg, 2006).

Independently of these pronouncements, the concepts of 'environmental security' and 'ecological security', which relate to environmentally induced security risks, have already been taken on board by important international actors, such as NATO, OECD and UNEP; these have responded to the policy relevance of the issue by setting up working groups and special departments. Of particular interest is the pan-organizational Environment Security Initiative (ENVSEC), within which the Organization for Security and Cooperation in Europe – OSCE, UNEP, UNDP and NATO exchange information and ideas about the links between environment and security. The main aim of the initiative is to increase the level of cooperation and, increasingly, of security between and within states, not only by examining interactions between the natural environment and human security but also by feeding the issue into concrete project work.

2.3
WBGU's aims and use of terms

Given the various meanings and the political ambiguity of the concept of security, the question as to which concept of security WBGU is referring to is by

> **Box 2.2-1**
>
> **Worldwatch Institute: Changing the oil economy**
>
> (...) Oil saturates virtually every aspect of modern life, and the well-being of every individual, community, and nation on the planet is linked to our oil-based energy culture. Even as oil has become indispensable, however, its continued use has begun to impose unacceptable costs and risks.
> The costs and risks of using oil can be grouped into three broad categories. First, oil threatens global economic security because it is a finite resource for which no clear successor has been developed and because the gap between supply and demand appears to be growing, making the world vulnerable to serious economic shocks. Second, oil's value as a commodity undermines civil security by compromising efforts to achieve peace, civil order, human rights, and democracy in many regions. Third, oil threatens climate stability because its use, which is accelerating, accounts for a major share of global greenhouse gas emissions and because its overwhelming dominance of the transportation fuel market makes it difficult to replace. In short, where oil once helped ensure human security, it now makes us more vulnerable (...).
>
> Source: quoted from Prugh et al., 2005

no means trivial; in fact it is of crucial significance for its analysis of the security threats to be expected in the context of climate change, as well as for the policy recommendations the analysis generates. WBGU supports the comprehensive concept of security as one that does justice to the complexity of the causes of violence and war in a globalized world. Given the kinds of global environmental change that are expected to occur, expanding the traditional concept of security allows the analysis of risks to be extended to include environmentally induced risks. Within this, WBGU's focus is above all on those environmental problems that are generated by advancing climate change and the resulting environmental degradation. The aim of this report is to analyse whether these ecological impacts might lead to security problems – violent conflict in particular – and how this might occur.

WBGU is particularly interested in gaining more knowledge about risks and dynamics that threaten to trigger the destabilization of collective actors, i.e. societies and states. The type of conflict most likely to occur in the context of climate change are 'new wars'. These are characterized by the fact that they combine elements of organized crime and human rights violations with a generally international dimension, and that the distinction between public and private and political and economic actors is increasingly blurred (van Creveld, 1998; Kaldor, 1999). The aim of undertaking an analysis of the complex factors and mechanisms that may lead to conflict in the context of climate change is to generate knowledge on which to base appropriate action, so that future regional and global instabilities can be prevented. A comprehensive concept of security in the sense of human security is of limited usefulness to this end; furthermore, it is hard to capture analytically. However, this by no means implies a departure from the normative aim of placing the protection of people above the integrity of states. Instead, it is the intention of WBGU to take account of scientific possibilities and political circumstances and requirements in the furtherance of a peaceful, global structural policy.

Up to now environmental problems have rarely been the trigger for violent conflict. Even in the future, it will not necessarily be the case that every larger environmental problem will become a security issue. Yet new and ever more compelling scientific knowledge about climate change and the simultaneous lack of effective mitigation and adaptation strategies make it increasingly clear that far-reaching consequences for individuals and society are expected to be seen in the future. This brings with it issues of a highly explosive nature for conflict and security policy that may even lead to warfare. The existing security strategies of the EU, USA, NATO – and Germany as well – pay insufficient attention to these developments. This does not mean that environmentally induced security risks can be eliminated using military means. What it does mean, however, is that, alongside preventive measures in the sphere of environmental, development and research policy, foreign and security policy considerations and options for action will acquire increasing importance. With this analysis of climate-induced environmental change and its destabilizing impacts on international policy, WBGU seeks to draw attention to the essential nature and urgency of this set of issues in the realm of foreign and security policy, and thus to contribute towards shaping a peaceful and sustainable world.

Known conflict impacts of environmental change 3

3.1 State of conflict research at the interface of environment and security

Researching the conditions under which competing political actors either cooperate with one another or engage in conflict has always been one of the core tasks of political science. In the area of security policy, research into the causes of war is one important strand of research that has become established at the crossover between the political science sub-disciplines of comparative politics and international relations. This body of research seeks to determine which factors play a role in triggering and intensifying wars and armed conflicts, making a distinction between conflicts within countries and societies (intrastate conflict) and conflicts between countries (interstate conflict). In the course of such research, large amounts of data have been collated with regard to which countries and societies are particularly prone to armed conflict. Scholars are now able to base their work on a range of basic assumptions that are backed up by empirical evidence.

Although there are historical examples of the link between environmental change and conflict (Box 3.1-1), the environment has not been accorded any great significance as a contributory factor up to now. However, over the last fifteen years a separate field of research has developed outside conflict research that deals specifically with the interactions between environmental degradation and violent conflict. This chapter offers an overview of the research controversies in environment and conflict research along with its major empirical findings, which largely agree with one another (Section 3.1.1). These specific findings are brought together with results from research into the causes of war and from conflict research (Section 3.3), thus enabling relatively reliable statements to be made about the causes and mechanisms through which conflicts arise in the context of massive environmental change. This serves as a means to establish specific conflict constellations (Chapter 6) and as a basis for identifying critical regions (Chapter 7).

3.1.1 Environment and conflict research

Scientific research into so-called environmental conflicts can be traced back to the early 1970s. It was not until the 1990s, however, that scholars entered into serious debate and systematic study of what, up to then, had been vague assumptions about the causal links between environmental degradation and conflict escalation. These efforts acquired an added dynamic at the end of the East-West conflict, and this lasted throughout the 1990s, fuelled additionally by the widespread attention accorded to the concept of human security and expert debate about the general 'securitization' of international politics (Waever, 1995).

Four groups in particular within the community of internationally established conflict researchers have dealt in detail with the relevance of environmental problems regarding the way armed conflicts arise and the course they take. The pioneers include the Toronto group around Homer-Dixon and the Zurich group, which grew out of the Environment and Conflicts Project (ENCOP) set up by Bächler and Spillmann at the Swiss Federal Institute of Technology (ETH) in Zurich. During the early 1990s in particular, both groups undertook empirical studies of the assumed connections between environmental degradation and the escalation of conflict. From the mid-1990s onwards, two further approaches developed out of the critiques made of the work done by the Toronto and Zurich groups. The main representatives of these approaches are, first, the Oslo group around Gleditsch, whose work is based on quantitative studies, and, second, Matthew's Global Environmental Change and Human Security Project (GECHS) based in Irvine, California, which focuses on the adaptive capacity of human societies.

Box 3.1-1

Climate change and environmental change in the past and their impacts on human society

Climate models predict a rise in sea levels, storm surges, flooding and large-scale drought; in the long term it is even conceivable that extreme phenomena, such as a shutdown of the North Atlantic current, will occur. This initially seems far-fetched. However, there have already been comparable events in human history: extreme sea-level changes, periods of severe drought, large storm surges and extensive land losses have been a part of climatic history for the last 100,000 years. With the exception of volcanic eruptions, these changes always took place relatively slowly, so that the societies affected had adequate time to adapt to the changes. Still, archaeological evidence suggests that much environmental change was associated with population movements and the decline of ancient high cultures. No firm conclusions can be drawn as to whether these changes in climate and environment thousands of years ago actually did lead to conflict or war, and to what extent this may have been so, due to a lack of written evidence. The state of knowledge regarding the Middle Ages and modern times is better, though, with evidence indicating the existence of massive societal tensions.

SEA-LEVEL RISE IN THE NORTH SEA

In the Middle Ages a steady rise in sea levels on the German and Dutch North Sea coast led to a considerable loss of valuable cultivated land. The reason for this was not only a global rise in sea levels brought on by the expansion of sea water and the melting of glaciers; it was also due to a geologically induced process of subsidence in the southern North Sea and Baltic Sea area and the settling and erosion of recent layers of sediment. The sea-level rise was always a slow process that did not lead directly to any great loss of land mass. The latter, instead, was caused suddenly by winter storm surges that penetrated deep into the interior of the land and permanently altered the coastline. In the course of such events from 1200 onwards, a series of large storm surges occurred, including the *Julianenflut* in 1164, the First Marcellus Flood in 1219 (approx. 10,000 deaths), the *Luciaflut* in 1287 (approx. 50,000 deaths) and the Second Marcellus Flood in 1362 (approx. 100,000 deaths). Although historians sometimes argue over the exact number of victims, the severe impacts on human populations remain undisputed. The storm tides led to the formation of the Dollard and Jadebusen bays in East Friesland and to the emergence of the 'Halligen' as remnants of a large area of contiguous land in North Friesland (Liedtke and Marcinek, 2002). After the first disasters struck, the population responded by shifting their settlements to artificially created earthen mounds (*Wurten*). This was the beginning of a centuries-long process of adaptation that has lasted until the present day. As the *Sachsenspiegel*, a medieval legal code, reveals, the collective task of coastal protection was organized forcibly through the implementation of drastic measures. Those who were not prepared to engage in community work to improve coastal protection were violently removed from their land and dispossessed. The act of deliberately damaging dyke structures was punishable by death. The rigorous dyke building regulations repeatedly led to tensions in the population and to work stoppages. From about 1500 onwards the population managed to make good the land losses and even to reverse the trend (Kramer, 1989). Nonetheless, after the storm surge of 1634 a quarter of the population was forced to leave the region for good. Storm surges continue to pose a danger to the inhabitants of the German and Dutch North Sea coast. The last larger storm surge disasters occurred in 1953 (Holland flood, approx. 1,300 deaths) and in 1962 (Hamburg flood, approx. 300 deaths).

THE LITTLE ICE AGE (ABOUT 1550–1850)

Since the 14th century there was a slow, irregular cooling of the global climate, which ended a warmer phase with a relatively mild climate in many parts of the northern hemisphere in the Middle Ages. This so-called Little Ice Age became noticeable in Central Europe from the winter of 1564–1565. In the three centuries following this, there were several cold phases with long, snow-filled winters and cool, short summers. Average temperatures in many areas during the Little Ice Age were about 1°C below the 20th century average. The last, rapid-onset cold phase of the Little Ice Age began with the eruption of the Indonesian volcano Tambora in 1815. This largest of volcanic eruptions over the past 2,000 years sent so much volcanic ash into the stratosphere that the global climate changed for several years afterwards. 1816 in particular went down in history as being the 'year without a summer', with summer snowfall in northern Europe and the American Northwest. In many parts of Europe the climatic changes led to several failed harvests in succession, with corresponding impacts on the prices of cereals and other products (Lamb, 1995; Bauernfeind and Woitek, 1999; Landsteiner, 1999). Among the historically observed impacts in the area of food production was also a reduction in fishing yields in the North Sea, as fish such as the cod migrated southwards. In mountainous regions in particular grazing losses were recorded, which brought consequences for food production. The general worsening of harvests led to malnutrition. Apart from hunger, there was also an increase in vulnerability to influenza and plague epidemics. In the Alps especially, and in the mountainous regions of northern Europe, lack of food along with advancing glaciers and the associated loss of land led to outward migration into lower lying areas. The climate-induced deterioration in people's living conditions can also be said to have contributed indirectly to the large-scale migrations to the New World. Alongside other factors such as sudden price rises, a lack of food and epidemics, the climatic conditions of the Little Ice Age led to an intensification of social problems. For example, a connection has been established between climate change and the witch hunts that began in the 15th century, in that it is claimed a collective hysteria broke out in a society that had become highly vulnerable as a result of natural disasters (Behringer, 1999).

3.1.1.1
The Toronto group around Homer-Dixon

The Toronto Project on Environmental Change and Acute Conflict examined the circumstances in which environmentally induced stress causes acute conflicts both within and between states. In order to find out how conflicts induced by environmental problems progress, the Toronto group carried out a number of qualitative case studies on conflicts in developing countries where they assumed an especially close link between environmental stress and acute conflict. Homer-Dixon and his colleagues concentrated on environmental problems that can be put down to the scarcity of renewable resources and environmental services. Six types of environmental change were looked at in the context of the project: climate change, depletion of the stratospheric ozone layer, degradation of agricultural land, deforestation, degradation of water resources and the depletion of fish stocks. The scarcity of renewable resources occupies a central role in the research of the Toronto group. Alongside environmental change, population growth and distributional pressures also influence the availability of resources. Scarcity leads initially to social and economic problems, and these may then cause existing conflicts to escalate.

The Toronto group identified two patterns of conflict that occur especially frequently, arising from the interactions between the above-mentioned causes of resource scarcity. First, resource capture occurs in cases when in a country with a growing population and dwindling resources, powerful groups within society exert influence on the distribution of resources, appropriating them for their own advantage. Second, a process of ecological marginalization occurs in cases where population growth and unequal distribution trigger migration to ecologically fragile regions; this often brings with it environmental degradation and impoverishment. Both patterns pave the way for two types of intrastate conflict. If resource scarcity – resulting, say, from the over-exploitation of utilizable agricultural land – triggers large-scale migration, conflicts may arise when the different group identities of migrants and local inhabitants become mobilized. Uprisings or even civil wars may also break out if resource scarcity leads to economic decline and key social and state institutions are weakened in the process.

No evidence was found of a direct connection between resource scarcity and the violent escalation of conflict. However, the studies from the Toronto group indicate that environmentally induced resource scarcity, in combination with political, economic and social factors, can indeed lead to a destabilization of states and societies likely to cause conflict, and that the destruction or scarcity of environmental resources has already contributed to a dynamic of violent conflict in many developing countries. The Toronto group has also been able to demonstrate that the danger of escalation was relatively higher in the case of intrastate conflicts over fish stocks, forests, water and agricultural land than it is in the context of global problems such as climate change and ozone depletion. There was no evidence to suggest a link between resource scarcity and interstate conflicts, with the exception of conflicts over water (Homer-Dixon, 1990, 1991, 1994, 1999).

3.1.1.2
The Zurich group around Bächler and Spillmann

In 1996 the ENCOP project presented a final report based on qualitative case studies (Bächler et al., 1996; Bächler and Spillmann, 1996a, b). The focus was on developing countries in which both environmental problems and armed conflict are in evidence.

The basic assumption behind ENCOP is that environmental change may lead indirectly to conflict by intensifying the existing potential for socio-economic conflict to the point of violent escalation. According to this view, conflicts are socially or politically motivated in the first instance and are not an irreversible consequence of environmental change. Rather, environmentally induced intensification of conflict is a symptom of the modernization crisis entailed by the transformation from a subsistence economy to a market economy in many countries.

The particular aim of ENCOP was to devise a typology of conflict that links a particular kind of environmental degradation to its socio-economic consequences and the affected parties to the conflict. Drawing on an analysis of 40 environmental conflicts, the following categories were developed: centre-periphery conflicts, ethnoecological conflicts, regional, cross-border and demographically-induced migration conflicts, international water conflicts and conflicts arising from distant sources.

The ENCOP typology shows that contextual factors other than the impacts of resource degradation ultimately determine whether competing actors will seek a peaceful or a violent solution to conflict. Among the most important socio-economic factors identified by the Zurich group as making environmentally induced conflicts more likely are a lack of societal mechanisms for regulating conflict, an instrumentalization of environmental degradation for group-specific interests, group identities, the organization and arming of parties to a conflict, and the influence of past conflict. The typology was applied by the ENCOP group in two follow-on projects with the aim of developing appropriate procedures for cooperation and management in situations of potential conflict in the Horn of Africa (ECOMAN) and in the Nile basin (ECONILE) (Bächler, 1998).

In 2001, building on the ENCOP research, Spillmann and Bächler set up the National Centre of Competence in Research (NCCR) North-South as a broad-based research programme aimed at isolating and identifying syndromes of global change. Alongside systematic analysis of the core problems at the nexus of environment and social development, the main purpose is to identify ways of managing these issues. NCCR North-South defined 30 core problems in all and divided them into five categories. For example, issues such as government failure or the weak geopolitical position of a country come under the category of 'political-institutional core problems'. The categories 'socio-cultural and economic core problems', 'population, habitat and infrastructure', 'services and land use' and 'biophysical and ecological core problems' were also created. These categories are applied in case studies in order to identify the relative importance of individual core problems within a region. In the context of the present report, the NCCR categorization served as a basis for the world map of environmental conflicts (Fig. 3.2-1).

3.1.1.3
The Oslo group around Gleditsch

At the International Peace Research Institute Oslo (PRIO), Gleditsch spearheaded an independent quantitative research approach developed out of a process of critical engagement with the studies from Toronto and Zurich. Its aim was to counter the excessive complexity of the qualitative models and to provide a corrective to their deficiencies regarding the selection of case studies – in particular the tendency to study countries with acute conflicts over resources (Gleditsch, 1998). Robust conclusions regarding the influence of various factors on armed conflict can be reached only when cases in which resource conflicts are conducted violently are compared with those in which there is no escalation of violence. This requires a stronger weighting of political, economic and cultural variables than in the models used by the Toronto and Zurich groups (Hegre et al., 2001). The 'neo-Malthusianism' inherent in the Toronto group approach also comes in for explicit criticism – that is, the assumption that increasing population pressure combined with resource scarcity leads to an escalation of conflict. In contrast to this, the Oslo group argues that an abundance of resources is more likely to lead to violent conflict because rebel groups, for example, draw their funding from the exploitation of natural resources.

Like the Toronto group and ENCOP studies before them, the Oslo group's quantitative studies confirm the basic link between environmental problems and armed conflict (Hauge and Ellingsen, 1998; Diehl and Gleditsch, 2000). The Oslo approach emphasizes much more strongly, however, the circumstance that environmental stress is only one of several variables that may contribute to the escalation of conflict. While environmental factors such as deforestation, soil degradation and water scarcity increase the risk of violent conflict within states, economic and political factors remain crucial as explanations for the outbreak and intensity of such conflicts (Hauge and Ellingsen, 1998). It is also conceivable that so-called environmental conflicts can ultimately be put down to development problems, given that factors such as deforestation, soil degradation and water scarcity are strongly linked to poverty.

3.1.1.4
The Irvine group around Matthew

The Global Environmental Change and Human Security project (GECHS) headed by Matthew was set up at the Center for Unconventional Security Affairs at the University of California in Irvine. The impacts of environmental change on individuals and societies are examined here, taking the concept of human security as a theoretical starting point. The critique levelled by the Irvine group at environment and conflict research to date is aimed at fostering a new theoretical orientation focused more on the long-term adaptability of humans and societies. What research up to now has lacked above all are qualitative frames of access to the issue of environment and conflict. In order to gain a better understanding of the key interconnections and impacts involved, it would be helpful to extend the range of methods and instruments used: this could be done by engaging in interdisciplinary cooperation, making use of research on conflict and cooperation and carrying out microanalyses. What is also lacking is quantifiable empirical research on the relevance of demography as a factor, on the question of whether resource abundance or resource scarcity hold the greater risk of conflict, and on whether environmental degradation might actually promote cooperation rather than stoking conflict.

In addition, the Irvine group stresses the role of 'network threats' for future environmental security research. This is a reference to transnational threats to security arising through an informal, transnational network of individual behaviour, such as decisions about personal energy consumption in the case of climate change (Matthew and Fraser, 2002; Matthew and McDonald, 2004; Matthew et al., 2004). According to Matthew and his colleagues, environment and conflict research can help to provide better policy advice by: identifying environmental protection strategies that promote cooperation, integrating environmental and development policy approaches, taking environmental security into account in the context of town planning and development, enabling better understanding of environmental policy in post-conflict situations, and assessing the effectiveness and sustainability of policy interventions.

3.1.1.5
The German research scene and WBGU's syndrome approach

German conflict research has adopted a critical stance in relation to the international debate about widening the concept of security, particularly with regard to the securitization of environmental policy (Daase, 1991, 1992; Brock, 1992, 1998), and the debate was taken up from the late 1990s onwards by researchers specializing in the field of environmental policy. This is exemplified by the work done by Carius and his colleagues (Carius and Lietzmann, 1998; Carius et al., 1999, 2001; Carius, 2003; Carius and Dabelko, 2004), and in particular by the approach developed mainly by Biermann, Petschel-Held and Rohloff (1998) that is based on an analysis of syndromes and on theories of conflict.

This innovative approach is based on a combination of the syndrome approach, developed mainly at the Potsdam Institute for Climate Impact Research (PIK) under the auspices of WBGU, and the Conflict Simulation Model (COSIMO) of the Heidelberg Institute for International Conflict Research. It involves the use of both quantitative and qualitative methods. The 16 syndromes of global environmental change thus identified by WBGU describe specific dynamic patterns of human-environment interaction, which together capture the main problems of global change (WBGU, 1997, 1998, 2001).

By linking in with the COSIMO approach, these syndromes can be correlated with existing empirical findings as examples of environmental situations likely to lead to conflict. This makes it possible, for example, to confirm the relevance of the Sahel Syndrome in relation to violent intrastate conflict. Evaluation of the COSIMO database in this context points to a clear correlation between violent social conflict and the vicious circle typical of the Sahel Syndrome, namely, increasing rural impoverishment, intensification of low-level agriculture and dwindling natural resources. It is indeed the case that above-average numbers of countries severely affected by the dynamic of the Sahel Syndrome are also affected by intrastate or interstate violent conflict; this linkage is most evident in the case of the countries bordering on the Sahel, that is, Senegal, Niger, Algeria, Burkina Faso and Mali. What is needed as well, though, are studies on Mongolia, for example, where the Sahel Syndrome exists in a state of high criticality but has not led to a violent escalation of conflict. Assumptions regarding the high probability of interstate 'water wars', by contrast, have not been confirmed unequivocally by the syndromes and conflict approach (Carius et al., 2006).

Overall, this approach makes it possible to assess the extent to which specific syndromes are likely to lead to conflict. Unlike the Zurich group's ENCOP project, for example, which works inductively, the environmental situations considered to be critical are established and typologized not on the basis of observed conflicts, but rather independently of conflict. Since this involves including environmental situations that may be considered critical without having led to conflict, it becomes possible to make plausible statements about the likelihood of environmentally induced conflicts (Carius et al., 2006). This represents considerable progress compared to previous environment and conflict research, and underlines the special value of interdisciplinary research in analysing environmental change in the context of global political processes.

3.1.1.6
Fundamental critique of environment and conflict research

As the empirical research described above developed further, critical voices began to be raised in the scientific debate, calling into question the very notion of dealing with ecological issues in the context of security discourse (Deudney, 1990, 1991; Brock, 1992; Levy, 1995). Both Daase and Brock, for example, argue that overloaded concepts such as environmental security are suitable neither as a means of scientific description nor as a way of explaining new critical developments. The lack of high analytical resolution that characterizes such approaches makes it difficult to make empirical distinctions; it levels the differences between relevant fields of policy and suggests that the interests of one party coincide with the interests of all others (Daase, 1992, 1996; Brock, 1992, 1997, 2004). In the view of these authors, environmental security is less a theoretical innovation than an empty formula that serves different political agendas. Discourse about ecological security can therefore serve to legitimate new areas of military deployment, while lessening the problem of public acceptance of the armed forces and promoting repressive tendencies in the sphere of internal security policy. In addition to preventive conflict resolution, then, efforts aimed at achieving environmental security are also a means of legitimating violence. Critics appeal instead for peace and conflict research that avoids the concept of security, in order to pre-empt political instrumentalization of the concept and to do greater justice to the complexity of environmental change.

A fundamental critique of the idea of environmental security has also been expressed in the context of North-South discourse, with critics highlighting the inappropriate 'colonization of environmental problems' by security discourse (Barnett, 2000; Dalby, 2002). According to these critics, the literature on environmental security suggests that the underdeveloped South poses a physical threat to the prosperous North, in that population explosion, migration and resource scarcity necessarily lead to disputes over distribution and conflicts of interest that can be solved only by military means. The industrialized countries, it is said, are under suspicion of exploiting such scenarios of threat in order to attack the 'uncivilized South' and to close off their own borders. It becomes clear in this view that environmental security is committed less to the security of people on the ground than to the national interests of the industrialized world. The actual causes – and causers – of environmental problems, as well as the large-scale injustices that exist in the global use and distribution of natural resources, are hidden from view, it is claimed, in favour of shoring up the global political status quo.

Last but not least, the critics question the premise that environmental change necessarily leads to conflict at all. This one-sided focus prevents a more precise examination of the processes and adaptive mechanisms involved and therefore leads inevitably to a dramatization of the problems. What also comes in for particular criticism in this context is the determinism inherent in the debate. This suggests that the environment as such is capable of triggering such conflicts, whereas they can only be understood as the products of complex social relations (Hagmann, 2005). What is required, therefore, is a fundamentally different research orientation – one that is concerned more with peace than with war, with the concept of sustainability rather than that of security, and with holistic analysis rather than one-sided, deterministic perspectives.

3.1.1.7
Key findings from environment and conflict research

The literature that fundamentally questions the security-related approach to environmental issues draws explicit attention to the ambiguity of the conceptual linkage between environment and security and to the potential for political instrumentalization. In doing so, it makes an important contribution to the critical self-reflection of policy-relevant research and can provide a helpful addition to the theoretical approaches of empirical research into environment and conflict. Despite these fundamental objections towards environment and conflict research and the sometimes pointed critique of its premises and methods, it is nonetheless evident that a considerable degree of agreement exists as far as the main research findings are concerned:

- *Multicausality*: All approaches emphasize the multicausality of the conflicts observed. There is a consensus that environmental degradation is always only one of several complexly connected causes of conflict and that environmental degradation rarely seems to be the decisive factor.
- *Locality:* There is also a consensus regarding the locality of the conflicts believed to involve an environmental element. They are predominantly intrastate conflicts; even when they can be categorized as cross-border conflicts they are generally not classical interstate conflicts in the sense of large-scale wars between countries but rather regionally limited clashes at the sub-national level, such as between states that border on the same rivers and lakes.
- *Problem-solving capacity:* Finally, all the approaches emphasize the central role of a state's or society's problem-solving capacity with regard to the emergence and management of conflicts: in places where political and societal institutions are weak, there is a proportionally higher probability of conflict occurring. Future crisis hotspots are therefore assumed to be located in countries and regions considered problematic in terms of their problem-solving capacity.

So far, there has been no evidence that environmental problems are the direct cause of war – that is, there have been no 'environmental wars' manifesting the most extreme form of interstate conflict. At least no evidence exists to date to suggest any unambiguous causal links between environmental change and violent interstate conflict. Indeed there are some striking examples in which efforts to solve environmental problems have led to constructive and cooperative engagement between fundamentally hostile parties (e.g. water use between Israel and Palestine or Egypt-Israeli cooperation in the context of the Mediterranean Action Plan). However, it certainly cannot be ruled out that environmental degradation can have destabilizing impacts that may lead to conflict – this remains a plausible possibility, as can be seen from various conflicts in the recent past.

3.2
World map of past environmental conflicts

3.2.1
Resource conflicts over land, soil, water and biodiversity

A world map of environmental conflicts that have occurred in the recent past is presented below (Fig. 3.2-1). It is based on close empirical observation of 73 conflicts that occurred in the period between 1980 and 2005, in which environmental problems regarding water use, land use, biological diversity and fish resources played a crucial role. It shows which environmental problem was relevant to which conflict, along with the geographical distribution of the 73 conflicts and their degree of intensity. For purposes of simplification, four different levels of intensity have been identified: (1) diplomatic crises; (2) protests that may entail an element of violence; (3) violent conflicts that affect the entire nation; (4) conflicts characterized by systematic, collective use of violence. The map should be interpreted with caution, as it is difficult to draw any conclusions that can be generalized. However, it does show, for example, that Europe has manifested diplomatic crises only – in other words, none of the recorded conflicts of interest over environmental assets escalated into violence (Carius et al., 2006).

With regard to the environmental problems that caused conflict, it is clear that soil degradation and water scarcity are closely interconnected (Houdret and Tänzler, 2006). In order to make it easier to visualize this phenomenon, the constellations in which soil degradation or conflicts over land use were a factor were combined into a single category. Dam projects were also assigned to this category if they were accompanied by resettlement measures or loss of land ownership. Even though dam projects frequently lead to water scarcity, they count first and foremost as land use conflicts (Carius et al., 2006).

A number of examples from Central America, the Caribbean and Africa will be mentioned in this section for illustrative purposes. With regard to the resource problems underlying the escalation of conflict, it is noticeable that in Central America these are overwhelmingly conflicts over land use (Fig. 3.2-2). In some cases, they involved the mass expulsion of local populations and deforestation. In three of the seven conflicts recorded in the region it is possible to speak of systematic, collective violence, leading to the deaths of 70,000 people in El Salvador and about 200,000 people in Guatemala. The conflict over land use rights in El Salvador needs to be seen against the backdrop of a history of conflict between El Salvador and Honduras, providing an exemplary illustration of the multicausality of conflict. The way the conflict developed also shows clearly how conflicts over a certain environmental resource can spread geographically and turn into an international issue (Carius et al., 2006).

The South American examples are also frequently about conflicts over land, and these are closely connected to the problem of soil degradation, sometimes being further compounded by the loss of biodiversity. This affects indigenous population groups in particular (Carius et al., 2006).

In Africa, by contrast, most of the 22 cases recorded in total show clear evidence of a connection between soil degradation and water scarcity (Fig. 3.2-3). Both forms of environmental change operate as a driving force for internal migration, which in some cases has led to conflicts being 'exported' to neighbouring countries and regions. Conflicts of this sort are found across the entire Sahel region, in many cases involving the use of systematic and collective violence. One important factor in the conflicts typical of this arid region were shown to be issues of land distribution, or rather the unjust distribution of land (Carius et al., 2006).

3.2.2
Conflict-related impacts of storm and flood disasters

In view of the growing risks from storm and flood disasters, a study was undertaken to determine in which historical cases such disasters had a positive or negative impact on the course of conflict. This was done using the database of the World Health Organization's (WHO) Collaborating Centre for Research on the Epidemiology of Disasters (CRED, 2006), which has recorded all the storm and flood disasters since 1950 that involved at least 1,000 victims. 171 cases were identified and their impact on conflict examined using newspaper reports, disaster research literature and historical documents. In spite of the sketchy state of information about disasters up until the mid-1990s, a clear connection was established in 12 cases between storm and flood disasters and an intensification of conflict, violent unrest and/or political crisis. A 13th case that does not appear in the pre-given analytical framework (because it involved fewer than 1,000 victims) was an event that occurred on the island of Haiti in 1954; it represents an important case study nonetheless (Fig. 3.2-4).

In several cases storm and flood disasters triggered domestic political crises. The most serious of such events in this regard was a typhoon that raged in the Bay of Bengal in 1970, causing an estimated

3 Known conflict impacts of environmental change

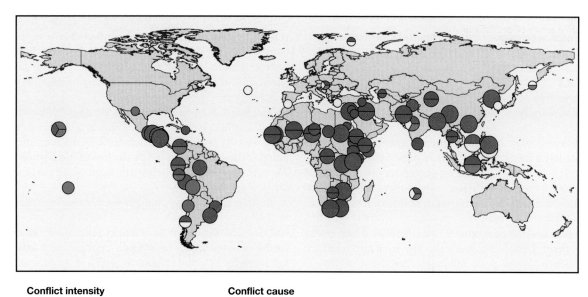

Figure 3.2-1
World map of environmental conflicts (1980–2005): Causes and intensity.
Source: Carius et al., 2006

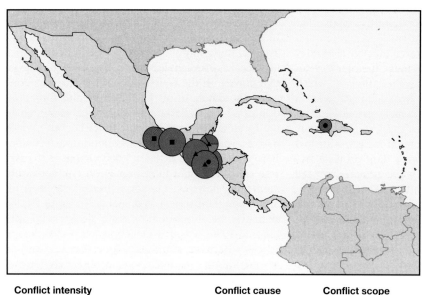

Figure 3.2-2
Environmental conflicts in Central America and the Caribbean (1980–2005).
Source: Carius et al., 2006

Figure 3.2-3
Environmental conflicts in Africa (1980–2005).
Source: Carius et al., 2006

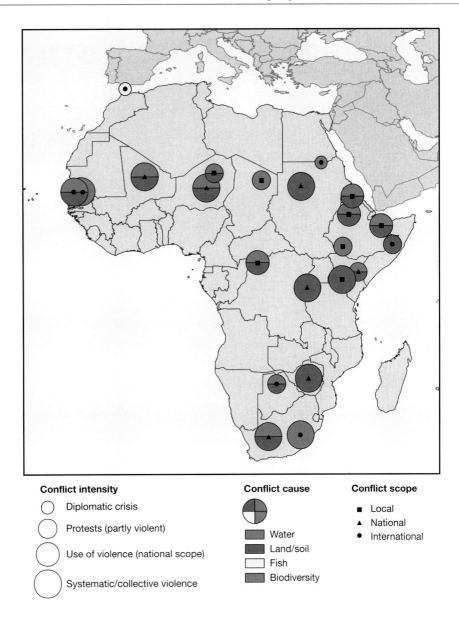

300,000–500,000 deaths. Most seriously effected was today's Bangladesh, which at that time was part of Pakistani state territory (East Pakistan) and was dominated politically and militarily by the western part of the country (West Pakistan). Dissatisfaction over the paltry aid measures provided by the Pakistani central government as well as the apparent indifference of political leaders towards the human suffering led to a strengthening of the opposition separatist movement and an open declaration of struggle for independence. When the separatist Awami League emerged as the clear winners of the 1970 elections to the National Assembly in East Pakistan, the central government responded with draconian measures involving repression and violence. The civil war that followed claimed an estimated 3 million victims and led to the independence of Bangladesh in 1971 (Drury and Olson, 1998; Heitzman and Worden, 1989; Jones, 2002). During the years that followed, the country went through a series of domestic political crises, which were intensified to such a great extent by storm and flood disasters in 1974 and 1988 that they led to the violent overthrow of the government of the day (Choudhury, 1994). Similar events took place on Haiti in 1954, after Hurricane Hazel destroyed large parts of the island's agriculture as well as its infrastructure. Here, too, domestic political crisis led to a forced change of government (Drury and Olson, 1998; Metz, 2001).

Many large storm and flood disasters have been followed by looting. While in most cases this was a case of people getting hold of the basic necessities of

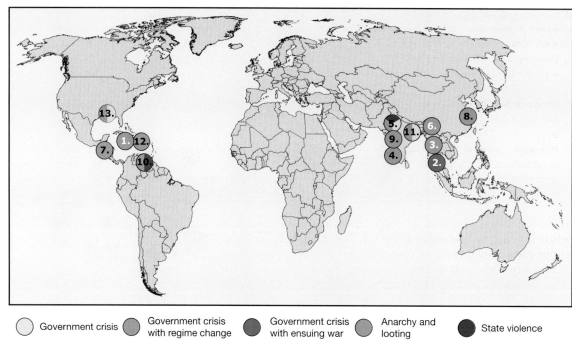

Figure 3.2-4
Storm and flood disasters with destabilizing and conflict-inducing consequences.
* In these cases, disasters led to an intensification of existing tensions.
Source: WBGU

1.* **Hurricane Hazel in Haiti, 1954:** The misappropriation of international financial aid led to widespread resentment within the population. When the President attempted to extend his period in office in the midst of this situation, a general strike was called and martial law introduced. The President was finally forced to leave the country. A year of political chaos ensued.

2.* **Typhoon in East Pakistan, 1970:** 300,000 people fell victim to a typhoon in East Pakistan (today's Bangladesh). Dissatisfaction over the government's insufficient aid measures led to a strengthening of the separatist opposition. The government responded with repression and violence. The civil war that followed claimed about 3 million lives. Bangladesh gained independence in 1971.

3.* **Flooding and typhoon in Bangladesh, 1974:** In addition to claiming 30,000 victims, the ensuing destruction of a large part of the rice crop triggered a famine. In a political situation that was already tense the government called a state of emergency and established a presidential dictatorship. In the same year the President was murdered by the military. A transitional military government took over.

4. **Flooding in Orissa (India), 1980:** In the course of collecting donations for flood victims, a conflict flared up between students and business owners. Severe rioting followed in which at least 34 people were injured and several hundred arrested.

5. **Flooding in Bihar (India), 1987:** When survivors began looting aid supplies, the police responded with force. Batons were again used against looters. In one case, the police fired shots into the crowd. The government was accused of gross failure.

6.* **Flooding in Bangladesh, 1988:** Anti-government resistance intensified in the aftermath of the disaster. Civil unrest grew, headed by the oppositional parties. Two years of political chaos followed, eventually leading to the overthrow of the President.

7. **Hurricane Mitch in Nicaragua and Honduras, 1998:** Food was looted in Nicaragua as a result of poor provision in the disaster areas. An armed group forced its way into a storage depot containing international aid supplies. The police responded by making arrests. In Honduras the government imposed a curfew and instructed the army to deploy all necessary measures against looters.

8. **Flooding of the Yangtze in Anhui (China), 1998:** Official reports reveal that there were fears of the situation sliding towards anarchy. The government instructed local authorities to heighten the presence of security forces in order to maintain public order and to punish all crime severely.

9. **Typhoon in Orissa and West Bengal (India), 1999:** Starving survivors looted aid convoys in many places. A group of politicians trying to get an overview of the situation was attacked by survivors. The group was only just able to escape in their helicopter.

10. **Flooding and landslides in Venezuela, 1999:** After one of the most severe natural disasters in Latin America, looting was widespread. Soldiers fired warning shots in order to protect the delivery of food supplies. According to reports by human rights organizations, in the process of re-establishing public order alleged looters were subject to summary executions.
11. **Flooding in West Bengal (India), 2000:** Due to delays in the distribution of aid supplies, trains and aid convoys were looted, aid workers were attacked and aid trucks were stolen. In order to protect one aid convoy, police fired warning shots. At the political level, blame was attributed to the regional state government.
12. **Hurricane Ivan in Haiti, 2004:** The distribution of international aid supplies was accompanied by violence; convoys were looted and lorries stolen by force. Armed gangs posed a considerable security problem during the entire emergency aid operation.
13. **Hurricane Katrina in New Orleans (USA), 2005:** Disaster victims went into stores in many parts of the city and stole vital supplies. There were reports of gangland activity and violent crime. Public order was re-established only with the help of armed members of the National Guard. Inadequate disaster management plunged the government into a crisis of public confidence.

daily living, such as food and clothing, there were also various cases of outbreaks of violence when groups of survivors attempted to appropriate aid deliveries for themselves. After Hurricane Mitch (1998) in Nicaragua, for example, an armed group forced its way into a storage depot containing international aid. In the Indian states of Orissa and West Bengal, several aid convoys were looted after a typhoon in 1999 (ACT, 2000). Just one year later similar scenes were re-enacted in West Bengal after a severe flood. In the course of these occurrences aid staff were set upon and disaster aid vehicles stolen (AFP, 2000; Reuters Foundation, 2000b).

In some cases, looting escalated due to the particular severity with which the security forces responded. After a flood in the Indian state of Bihar in 1987, the police took violent action against looters on several occasions, shooting into the crowd (Sahay, 1987). A quasi-anarchic situation also arose in Venezuela in 1999 after severe flooding and landslides. When public order was re-established by the military in the disaster area, human rights organizations reported that there had been on-the-spot executions of alleged criminals (Reuters Foundation, 2000a; Amnesty International, 2001).

To summarize, there has frequently been violent unrest, political destabilization and an intensification of existing conflicts due to storm and flood disasters in the past. Thus, storm and flood disasters are different from other environmental influences in which the links to conflict are often disputed. The analysis also demonstrates that storm and flood disasters are never the sole cause of conflict, but rather give rise to conflict situations in interaction with other factors (such as already existing domestic political tensions). Section 6.4 below examines the factors, mechanisms and typical sequence of events involved in such cases.

3.3
War and conflict research

The main question addressed by this report is whether global climate change and its consequences can lead to security problems and, if so, under what circumstances. More specifically: when do climate change and environmental change lead to violent conflict within and between states and societies? In order to be able to determine the potential for conflict entailed by environmental change more precisely, the major factors that are generally considered to be decisive in the escalation or de-escalation of conflict will be presented below. These general insights from research on conflict and the causes of war together with the findings from environment and conflict research serve WBGU as anchor points for developing both conflict constellations (Chapter 6) and concrete recommendations for action (Chapter 10).

3.3.1
Regime type, political stability and governance structures

In conflict research it is widely agreed that a link exists between the type of regimes and their vulnerability to armed conflict and war: democracies and autocracies are far less prone to internal violent conflict than 'anocratic', that is, partially democratic states (Muller and Weede, 1990; Hegre et al., 2001). Violent conflict rarely occurs in democracies because the political opposition is permitted to express divergent opinions. In dictatorships, violent conflict is rare because the state apparatus usually suppresses uprisings and is thereby able to prevent an escalation of violence. Anocracies do not permit the expression of oppositional opinion yet are not in a position to suppress dissidents effectively (Fearon and Laitin, 2003). They therefore seem to be more prone to conflict than authoritarian states.

Despite the consensus that exists regarding the high level of vulnerability to conflict in anocracies, there is lively debate among researchers in this field as to whether democracies display a fundamentally smaller tendency towards conflict than autocracies. On the one hand, political liberalization appears to lower the risk of war – at least, democracies do not wage wars against one another and are considered, for this reason among others, to be relatively less prone to conflict in comparison with autocracies (Hauge and Ellingsen, 1998; Gurr, 2000; Elbadawi and Sambanis, 2000). On the other hand, it can be demonstrated that democracies are affected just as frequently as autocracies by violent internal conflict (Fearon and Laitin, 2003; Collier and Hoeffler, 2004). Furthermore, Reynal-Querol (2002) finds that democracies constituted on the basis of a 'winner-takes-all' form of election law are relatively more prone to conflict than democracies that manage better – at least in part – to take account of the interests of minorities and to protect minority rights.

Regardless of political constitution, states with weakly developed governance structures are considered to be fundamentally more vulnerable to conflict in comparison with states that have strong governance structures and consolidated forms of government (Fearon and Laitin, 2003; Lacina, 2004). This has to do, above all, with the effective maintenance of the state monopoly on violence (Section 4.2.1). However, strong governance structures and governance capacity do not necessarily correlate with a democratically constituted state. These findings have also been confirmed by environment and conflict research, with Homer-Dixon (1999) and Bächler (1998) establishing a correlation between political stability, legitimacy and state capacity and vulnerability to conflict.

3.3.2
Economic factors

3.3.2.1
Economic performance and distributive justice

A general link is universally acknowledged to exist between the economic development of a country and its vulnerability to conflict. In particular, a low level of economic development increases the risk of conflict within societies (Collier and Hoeffler, 2004; Fearon and Laitin, 2003; Fig. 3.2-5). Some studies even identify the level of economic development as the key explanatory factor in the violent escalation of conflict (Hauge and Ellingsen, 1998; Smith, 2004). The level of economic development can be ascertained in a number of ways, including economic per-

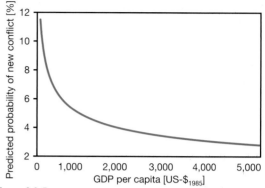

Figure 3.2-5
Predicted likelihood of the emergence of new conflicts within five years in relation to per capita income. The figure shows an average function over countries and points in time.
Source: UN Millennium Project, 2005

formance, the size of the agricultural sector or the vulnerability of a country to price fluctuations on the international markets (Avery and Rapkin, 1986). Some scholars assume a linear connection, so that the higher a country's per capita economic performance, the lower is its vulnerability to conflict (Hauge and Ellingsen, 1998). Others, by contrast, assume a non-linear relationship and stress that economic growth in very poor countries can give rise to political instability and an increased likelihood of conflict, while in more affluent countries it tends to reduce the risk of war (Gates, 2002).

Various lines of causality are discussed in this regard as a means of explaining the links between the level of economic development and the likelihood of conflict. Fearon and Laitin (2003), for example, point out that a low per capita GDP is generally accompanied by low-level governance capacity in the country concerned. The greater the capacity of a state, however, the more effectively it is able to solve problems, suppress potential unrest and integrate dissatisfied groups. Collier and Hoeffler (2004) focus on the extent to which oppositional groups prepared to use violence are able to organize an uprising at all, which depends crucially on their access to resources. Avery and Rapkin (1986), by contrast, emphasize that political destabilization is caused above all by the dynamic of international markets – by price shocks, for example – whose impacts are mediated through a country's level of development. Thus, countries that are dependent on the export of a single commodity, or a few commodities, are affected particularly severely by falling world market prices and are therefore likely to experience a greater degree of vulnerability to conflict.

In addition to economic development, people's share in public goods and the distribution of economic prosperity are also relevant to conflict. Countries in which the gap between the most affluent and the poorest sections of the population is especially large are considered to be fundamentally more prone to conflict. One specific aspect of poverty and distributive justice that deserves particular attention in the predominantly agriculture-based developing countries is the issue of land rights and the relevance of different sorts of land use systems to conflict (Carius et al., 2006).

3.3.2.2
Natural resources

The existence of natural resources in a country and the way these are used also plays a crucial role in economic development. It continues to be a topic of lively debate whether the existence and export of natural resources influences a country's vulnerability to internal conflict, and under what conditions this is the case. De Soysa (2000), for example, takes the view that the occurrence of abundant renewable resources in otherwise poor countries and of non-renewable resources in every country serve to increase the likelihood of armed conflict. Collier and Hoeffler (2004) assume a non-linear relation between the proportion of raw materials exports and agricultural products as part of a country's GDP on the one hand and the risk of civil war on the other. From this perspective, the risk of conflict is greatest when exports of raw materials and agricultural products amount to roughly one third of GDP. If this proportion is lower, the risk falls. However, if it is higher, the state generally has sufficient capacity to either fend off rebellion or to make it an unattractive option. Le Billon (2001, 2002) assumes further that states with a large degree of dependence on natural resources are prone to a greater risk of conflict because some natural mineral resources, such as diamonds, are vulnerable to theft. Additionally, dependence on natural resources can have a distorting influence on economic and political life. The way violent conflict is conducted in such circumstances depends on who has what access to the relevant resources.

A perspective differentiated according to natural resources shows that the export of crude oil in particular appears to increase a country's vulnerability to conflict. Studies by Ross (2004a, b), for example, show that oil, minerals and drugs increase the likelihood of conflict, whereas no linkage is apparent between various agricultural products and armed conflict. One possible explanation for this could be that countries with large oil reserves frequently have weak governance structures and, as 'rentier economies', tend to neglect other economic sectors.

3.3.3
Societal stability and demography

3.3.3.1
Population trends

A recurring theme of conflict research in general and of environment and conflict research in particular are the assumed linkages between socio-demographic dynamics, population density and the absolute size of a country's population on the one hand, and its vulnerability to conflict on the other. Hauge and Ellingsen (1998) and de Soysa (2002), for example, concentrate on the relevance of population density to conflict and find that countries with a high population density seem to be more prone to violent conflict than countries with a lower population density. Fearon and Laitin (2003), by contrast, look at the influence of the size of a country's population on its vulnerability to conflict and come to the conclusion that countries with large populations face an increased risk of conflict. Collier and Hoeffler (2004), working on the same set of issues, have calculated that the risk of conflict increases in proportion with the size of the population, but that a high population density correlates to a lesser extent with vulnerability to conflict. In addition, findings from PRIO show that high population growth per se is not likely to cause conflict, and does so only in combination with a scarcity of utilizable land (Urdal, 2005).

3.3.3.2
Socio-cultural composition of the population

It is unclear what the precise connection is between the heterogeneity of a country's population and the country's vulnerability to internal societal conflict involving violence. Montalvo and Reynal-Querol (2003) and Reynal-Querol (2002) hold the view that certain forms of population heterogeneity increase vulnerability to conflict. They suggest that a situation of societal polarization in which an ethnic, linguistic, cultural or religious majority exists over against an almost equally large minority holds the greatest potential for conflict escalation. This is especially the case when low incomes coincide with social exclusion. Steward (2004) makes a similar argument when she points out the danger of 'horizontal inequality' between regional, ethnic, economic or religious groups running parallel with political and economic dimensions. Whether a high degree of 'horizontal inequality' between various groups ultimately leads to conflict is influenced, among other things, by the status of these groups within the society.

Other research work, however, stresses that the socio-cultural diversity of a society has no influence on the likelihood of conflict within a country and may possibly even reduce the likelihood of conflict. Fearon and Laitin (2003), for example, reach the conclusion that a larger ethnic, religious or other socio-cultural heterogeneity has no correlation with a greater likelihood of conflict. The likelihood of conflict, according to them, does not grow even when groups become polarized or individual groups are disadvantaged. Collier and Hoeffler (2004) argue that it is not polarization itself that increases vulnerability to conflict, but rather that the dominance of one ethnic group is the deciding factor. Collier (quoted in Smith, 2004) even holds the view that the existence of multiple ethnic and religious fractions lowers the risk of conflict. In sum, the various views converge on the assessment that cultural and ethnic differences do not necessarily increase the risk of conflict but can, in principle, be instrumentalized for political purposes. The real danger thus exists in the deliberate political exploitation of such differences (Lohrmann, 2000). The study by Carius et al. (2006) also underscores the fact that cultural and ethnic tensions frequently represent an important contextual factor, particularly in environmental conflicts.

3.3.3.3
History of conflict

A broad-based consensus exists regarding the fact that countries that have experienced war or armed conflict on their territory in the recent past bear a considerably higher risk of conflict than countries where this does not apply (Gates, 2002; Walter, 2004). Hegre et al. (2001) explain this by referring to the special difficulty of resolving the security dilemma typical in post-conflict situations. Collier et al. (2003) make the additional point that a past conflict probably intensifies other risk factors, which in turn increase the likelihood of conflict. They name factors such as political instability, frequent regime change, low-level economic development and a large, potentially violent diaspora supported from abroad.

Analysis of interstate wars reaches comparable conclusions, namely, that states that wage war against one another have usually already experienced frequent violent conflict with one another in the past. Thus, the likelihood that two warring states will wage war against one another again in the future seems to grow significantly (Bremer, 2000). Social constructivist research explains this link in part by showing that rivalries between warring parties are reproduced through hostile actions (Adler and Barnett, 1998).

3.3.4
Geographical factors

The influence of geographical factors on the degree of risk of both violent intrastate conflict and interstate war, as well as on the course such conflicts take, is also undisputed. As far as 'internal' geographical factors are concerned, it is assumed that countries whose territory extends over large areas of rough terrain (e.g. mountainous terrain in Afghanistan and Pakistan), are more prone to conflict than countries with more accessible regions, because rough terrain provides ideal defensive cover for rebel groups (Fearon and Laitin, 2003).

With regard to 'external' geographical factors, countries with a neighbouring state in which there is armed conflict are generally more at risk than countries where this is not the case (Ward and Gleditsch, 2002; Buhaug and Gleditsch, 2005), due to the concrete danger of such conflicts 'spilling over'. The reasons that have been identified for this are military infiltration, provision of rebel cover in the neighbouring country, refugee flows, political and ideological 'infection' ('demonstration effect') and group solidarity (Buhaug and Gleditsch, 2005). However, not all types of conflict are equally likely to spill over into neighbouring states. While separatist conflicts (Buhaug and Gleditsch, 2005) and ethnic and other identity-based conflicts (Sambanis, 2001) increase the risk of conflict in neighbouring countries, those that are waged over access to central government power do not generally spill over. Furthermore, not all states are considered to be equally at risk. Small, democratically constituted and rich states seem to be largely immune to spillover conflicts, while states that have a high risk of conflict in any case are considerably more vulnerable (Buhaug and Gleditsch, 2005).

Wars between countries are also waged predominantly by states that share a common border (Vasquez, 2000). Indeed, the fact of being neighbouring states itself may even be the most important factor for predicting interstate wars (Raknerud and Hegre, 1997). States whose territories border one another generally have a high potential for conflict with regard to disputes over natural resources, cross-border migration and other typical interstate conflict issues. The large number of smaller, less powerful states generally do not have the means to wage wars over large distances (Boulding, 1962; Gleditsch, 1995).

3.3.5
International distribution of power and interdependencies

Of particular significance for violent interstate conflict are the distribution of power between states and the degree of their interdependence. It is widely recognized that the risk of an armed encounter between states increases when the difference in power between them is reduced. Stable asymmetries of power between two states correlate far less with interstate violence than balances of power or decreasing imbalances of power. This empirically robust finding is explained by reference to the fact that weaker states tend to fall in with stronger states when conflicts of interest are at issue. Moreover, stronger states generally begin wars against weaker states only when they assume that avoiding war in the present will have serious consequences for the future (Geller and Singer, 1998; Russett and Oneal, 2001). In addition, it is evident that large powers are more likely to become involved in interstate wars than smaller, weaker countries.

A growing number of studies put forward the view that interstate interdependencies reduce the risk of war. There is growing agreement about the fact that economic interdependence inhibits the potential for conflict through trade and international capital flows (Russett and Oneal, 2001; Gartzke et al., 2001). What remains doubtful is the extent to which growing interstate economic interdependence also inhibits conflicts within societies. In cases where transnational economic relations and interests dominate and constrain national development efforts, it is also plausible that conflicts may be induced. Wolf (2006), for example, assumes that, in view of the trend towards transnational privatization in the water sector, it is far more likely that local revolts over water directed at transnational corporations will escalate than that violent cross-border conflicts over water courses will occur.

With regard to political interdependencies, the empirical findings clearly support institutionalist assumptions, according to which states that are members of the same international organizations will wage war less frequently against one another than states that share no common membership in an international organization or that coincide in only a few (Russett et al., 1998; Russett and Oneal, 2001). The main reason given for this is that international organizations mitigate the security dilemma posed by the 'anarchic' world of states. Moreover, as cooperative relationships increase and particular fields of policy become subject to greater regulation, the costs of using violence increase, which in turn reduces the likelihood of conflicts being dealt with through violence.

3.3.6
Main findings of conflict research

Research on war and conflict has gathered extensive data concerning which states and societies are particularly prone to armed conflict, so that its findings are now supported by a range of basic assumptions that are empirically robust. Widespread unanimity exists about the fact that countries are especially prone to violent conflict within their own borders when at least one – though generally several – of the following factors apply: they are anocratic (that is, neither clearly democratic nor clearly autocratic) in constitution, have weak state structures and capacities, are at a low level of economic development, have a large population and/or a high population density, are characterized by rough terrain, border on a neighbouring country in which a violent conflict is being waged, and/or have themselves experienced violent conflict in the very recent past on their own state territory.

Research shows that interstate wars are likely to occur above all when conflicts of interest between democratically and autocratically constituted states escalate, when territorial conflicts become virulent, when a balance of power exists between states (or an existing imbalance of power decreases), when no or only few interdependencies exist between the conflicting parties, when the countries are neighbours and when the states waging war against one another have already done so in the past.

3.4
Conclusions

As is clear from this overview, the interconnections between environmental conditions, society and conflict are extremely complex. This is due to the diversity of mutual dependencies between political, societal and economic factors on the one hand and ecosystem factors such as availability of water, soil quality and climate change on the other. Nonetheless, overall it is possible to record the following key findings of environmental and conflict research, which counter the partially alarmist views that prevail in the media and caution against any rash policy moves:

- An escalation of violent conflicts that might be regarded as genuinely 'environmental conflicts' is not currently likely to occur.
- Environmental degradation may be one factor of conflict among many, but it is socio-economic factors and governance problems above all that are decisive.

- Conflicts that display a marked environmental dimension are generally limited to a local area, while some are also of a cross-border nature.

Little can be said about future longer-term developments on the basis of these findings, however. While there is a growing body of scientific evidence about climate change and its potential consequences, knowledge about the resulting tendency towards violence both within and between countries remains hazy.

In order to reach a more accurate assessment of the relevance of climate change and environmental change to security, WBGU seeks to open up the debate by incorporating the following considerations:

- Studies on environment and security to date have focused on environmental change that is limited to a locality or region, such as soil degradation, water scarcity and conflicts over resources. In contrast to this, WBGU seeks to focus its analysis on the destabilizing influences of climate change and its potential to trigger conflict around the world.
- Empirical research to date has been concerned with the period from 1980 up to the turn of the millennium. Taking a new look through the lens of climate change requires an extension of the analytical time horizon because the security-relevant disruption that is to be expected as a result of climate change is only likely to occur in the coming decades.
- Previous studies in the context of environment and security have looked above all at the area of development cooperation practice. As the findings from research into the causes of war suggest, however, avoidance of conflict requires action on many levels and in a variety of contexts. In the context of climate change, this applies not least to the international centres of power, given that future development policy success depends crucially on resolute action on climate policy from these quarters.
- Research on environment and conflict has been dominated largely by political science. Transdisciplinary – or even interdisciplinary – cooperation with the natural sciences and other social sciences has so far occurred only tentatively or not at all. The WBGU report deliberately approaches the theme from the interdisciplinary perspective of global change research.

Rising conflict risks due to state fragility and a changing world order 4

4.1
Introduction

Environmental changes pose major challenges to the problem-solving capacities of states and societies. It must be assumed that all the ensuing threats to security will be greatly exacerbated by global climate change. Never before has a world population of more than six thousand million people been confronted with such rapid and profound change in its natural environment. The link between stable and effective political institutions and the risks associated with climate change is therefore of major significance.

WBGU anticipates that two structural trends affecting the constitution of the nation states and the global political order will fundamentally impair their capacities to adapt to and mitigate climate change:

- *Weak and fragile states:* It is becoming apparent that states which, based on current information, are classed as weak or fragile (Box 4.2-1), are poorly equipped to protect their societies from the impacts of global climate change. These impacts have the potential to exacerbate the destabilization of these states by intensifying the problems they face, and in extreme cases, may even be a contributory factor in their collapse (Section 4.2). The concern, then, is that the number of weak and fragile states could increase further as a result of unabated climate change.
- *Unstable multipolarity:* A structural shift in global politics can currently be observed, driven primarily by the economic and political rise of China and India. In the new multipolar world order which is likely to emerge, the cooperation that is essential for the effective management of major global problems such as climate change and poverty could be seriously impaired or even deadlocked completely (Section 4.3). The global challenge in the next two decades is to master two Herculean tasks simultaneously: achieving a peaceful transition of power in the international system, and developing and implementing effective multilateral policies to mitigate dangerous climate change.

It is extremely uncertain, at this stage, whether both will succeed.

4.2
State fragility and the limits of governance

Not all states and regions are equally exposed to the risk of destabilization due to the effects of climate change. This risk depends on how severely they are impacted by climate change and their capacity to respond effectively. However, many countries which are already characterized by state fragility will be affected in both ways: they will be exposed to a relatively severe degree to the impacts of climate change, and their problem-solving capacities are poor at best. However, this does not mean that developed or newly industrializing countries will necessarily remain unscathed. As Hurricane Katrina demonstrated in the USA in 2005, flood disasters, for example, undoubtedly have the potential to trigger a breakdown in public order, at least for a brief period, even in highly developed societies. Given that such disasters could occur more frequently, with greater severity and in several places simultaneously in future, this could become a problem for developed countries as well over the long term (Sections 7.2.3 and 7.8.2). Coastal regions are typically the hub of a dense supply infrastructure (e.g. energy and water supply pipelines, regional concentration of energy infrastructure or ports of supply) or international trade structures (e.g. dependency on just-in-time production and high mobility through transportation networks), resulting in significant vulnerabilities to the risks posed by climate change.

Nonetheless, it seems unlikely at present that distortions will occur in the developed countries with such severity that massive security problems will ensue. It may be assumed that the developed countries will mainly be affected as target countries for migration and as financial donors (humanitarian assistance). In its analysis, WBGU therefore focuses primarily on the group of countries which can already

> **Box 4.2-1**
>
> **Qualitative categorization of state stability**
>
> - *Consolidated states:* Ideal-typical modern nation states in which all three core functions are safeguarded in the long term, represented primarily by the OECD countries. May also include consolidating states which are recognizably undergoing a sustainable transformation process towards a democratically constituted state with market economic structures (e.g. Costa Rica, Chile, Estonia, Latvia, Lithuania, Slovenia, South Africa).
> - *Weak states:* States in which the monopoly on the use of force is still largely safeguarded but which display grave deficiencies in fulfilling the welfare and rule of law functions. Examples include Eritrea, Uganda, Venezuela, Macedonia and Albania. Authoritarian or semi-authoritarian Islamic/Arabic regimes often fall into this category: despite appearing strong with regard to stability and basic public service provision, they perform poorly in fulfilling the rule of law and welfare functions (e.g. Saudi Arabia, Egypt, Iran).
> - *Fragile states:* Here, the state continues to perform in essence welfare and rule of law functions. . However, its monopoly of the use of force is either severely restricted or entirely absent, and it does not completely control its territory or external borders. Many states which are formally democratic but which are challenged by separatist forces fit in this category (e.g. Colombia, Sri Lanka, Indonesia, Georgia), but others are authoritarian states (e.g. Sudan, Nepal).
> - *Failed states / state collapse:* A state may be regarded as failed if none of the three state functions is effectively performed. As the example of Somalia/Republic of Somaliland shows, the 'failed state' category does not necessarily imply chaos or anarchy, generally because relatively powerful non-state actors emerge in place of the failed state and perform or substitute key regulatory functions. Besides Somalia, Afghanistan, Iraq and the Democratic Republic of Congo can currently be regarded as failed states. However, this category does not apply to processes in which several new states are formed from a predecessor state, either relatively peacefully (e.g. the Soviet Union in 1991, Ethiopia/Eritrea in 1991, Czechoslovakia 1993) or by violence (e.g. Pakistan/Bangladesh in 1971, Yugoslavia in 1995), which must be considered separately.
>
> Source: WBGU, based on Schneckener, 2004

be classed as weak or fragile and whose problem-solving capacities are poor.

4.2.1
Characteristics of state fragility

Countries' governance capacity is a key factor for the prevention or management of inter and intrastate armed conflicts. This applies irrespective of the cause of the conflict: in other words, it initially makes no difference whether the conflict dynamic is driven mainly by socio-economic, environmental or other factors. Individual countries' governance capacities vary, sometimes very considerably, and there is a close correlation between these capacities and the vulnerability of the states and societies concerned.

Against this background, the 'state failure' phenomenon plays a key role in political theory and practice. It first emerged as a theme in social science discourse in the late 1980s and has acquired major significance and topicality in the security debate in the wake of the 9/11 attacks.

The scientific conceptualization of the phenomenon termed 'state failure' is difficult and contentious (Schlichte, 2005), as is apparent from the multitude of analogous terms used. Authors refer, for example, to weak states, quasi-states and shadow states; to fragile, eroding and precarious states; to para-states; to failing, collapsing and collapsed states; and to state implosion, state collapse and states at risk, to cite just a few examples (Jackson, 1990; Herbst, 1996; Tetzlaff, 1999; von Trotha, 2000; Spanger, 2002; Milliken and Krause, 2003; Rotberg, 2003; Ottaway and Mair, 2004; Roehder, 2004; Schneckener, 2004). Despite sometimes considerable differences of opinion on matters of detail, there is a broad consensus that states characterized by state fragility lack the capacity to fulfil three core functions:

1. to guarantee the state's monopoly on the use of force, internally as well as externally;
2. to safeguard socio-economic welfare through the provision of basic public services such as infrastructure, health and education;
3. to maintain institutions that are essential for the rule of law and to establish and enforce legal and social norms and public order.

The establishment of institutional arrangements to safeguard social participation is often regarded as the fourth core function in terms of a democratic state ideal (Roehder, 2004). However, as a criterion for state stability, this is contentious because empirically, it can be shown that numerous states under autocratic regimes have proved to be extremely stable, whereas democratization processes tend to have a destabilizing effect, at least in some phases. WBGU therefore bases its analysis mainly on the three core functions listed above. According to this definition, states which perform one or more of these functions poorly should be regarded as weak and fragile states. States which have lost all their capacity to fulfil these core functions are failed states and in extreme cases, their continued existence may be at risk. So the quest for state stability is not simply the pursuit of an abstract constitutional theory and an end in itself; it is a political necessity, enabling the state to perform the core functions without which peaceful social relations in the complex interaction of our modern societies are well-nigh impossible.

In the analysis of states, it has proved helpful to establish differentiated categories based on an ideal-typical continuum of state stability. At the positive end of the spectrum, there is the fully-functioning modern nation state which meets all three criteria (e.g. Norway); at the negative end, there is the failed state which, when measured against legal and formal criteria, is barely recognizable as a state at all (e.g. Somalia). Based on Schneckener (2004), four categories representing diminishing quality of state stability are presented, with distinctions being drawn between consolidated, weak, fragile and failed states (Box 4.2-1). However, the classification of the various individual states can only ever be transitory, presenting a snapshot of current conditions and trends, and it is not always possible to draw a clear distinction between weak and fragile states, for example (Roehder, 2004; Schneckener, 2004).

Numerous examples bear witness to the 'state fragility' phenomenon in its various facets and forms. From a regional perspective, sub-Saharan Africa is particularly striking; many countries here can be classified as weak (Mehler, 2002; Grimm and Klingebiel, 2007). Indeed, as many as one-third of them are considered to be at acute risk of state failure (Roehder, 2004). However, examples of weak states also exist in Asia (e.g. Afghanistan), South-East Europe (e.g. the Serbian province of Kosovo) and South America (e.g. Colombia).

CAUSES
It should be noted, first and foremost, that there is still a lack of reliable empirical research about failed/failing states, especially as regards the causes of state failure processes. According to one plausible causal model, state failure is triggered when the state first ceases to deliver essential public goods and thus forfeits its legitimacy. Only then are the state's own institutions affected: the administrative apparatus's capacity to act gradually diminishes and the state loses its tax sovereignty and monopoly on the use of force, etc. (Lambach, 2005).

What is undisputed is that the erosion of the state's monopoly on the use of force often triggers intrastate spirals of violence which in turn have a destabilizing effect. Many security problems result from the breakdown of law and order which typically occurs first in peripheral border regions or urban slums where the state has a low level of penetration. The spread of organized crime and criminal violence are then almost inevitable (Ottaway and Mair, 2004). The threat is compounded by corruption and lack of capacity of the state's law enforcement agencies (especially the police), which leave a vacuum of power that is then gradually filled by non-state actors such as private security companies, vigilantes, militias and warlords. To move closer to an analytical evaluation of the causes of state fragility, a distinction can be made between structural, process and trigger factors, according to Schneckener (Box 4.2-2).

MANIFESTATIONS
The weakening of state structures to the point of possible state collapse by definition affects individual states, so analysis must focus on the national level. However, the phenomenon of state fragility becomes a matter for international politics once the impacts of such failure spill over the borders of the directly affected state. Weak and fragile states have a destabilizing effect on neighbouring states and regions, e.g. through cross-border migration or black markets. Furthermore, fragile states are considered to be vulnerable to 'new wars' (Kaldor, 1999) and, it is assumed, can become safe havens for terrorist organizations and centres for the trade of drugs and arms. For that reason too, it is in the international community's interest not to ignore weak and fragile states

---

Box 4.2-2

State fragility: Destabilizing factors

- *Structural factors:* Conditions which relate to natural features of a country, e.g. minerals or climate, and long-term political, cultural and socio-economic characteristics, e.g. ethnic diversity, demographic development, 'colonial legacy', regional power constellations.
- *Process factors:* Conditions which in the medium term trigger or drive the erosion of states, with the response of the actors involved (especially elites) to internal or eternal crises playing a key role. Includes political instrumentalization of social discontent or of ethnic/cultural differences, political or religious extremism, separatist tendencies, repression by the state, corruption and mismanagement, privatization of violence, economic crises, etc.
- *Trigger factors:* Conditions which trigger abrupt change, including factors which may result from longer-term developments with a catalytic effect. Examples are military intervention, refugee flows, military coups and revolution, massive violent repression of the opposition (e.g. massacres), social unrest, famine, civil war, etc.

Source: Schneckener, 2004

but to utilize the available opportunities to consolidate these states' institutions and structures (UN, 2004; for a critical appraisal, see Schlichte, 2005).

POLICY OPTIONS

In light of the risks posed by weak and fragile states, the question which arises is this: in extreme cases, who is prepared to intervene in collapsing states or acute trouble spots, and when, under which circumstances and on which political basis? Various policy options have been discussed in this context, e.g. in the framework of the Democracy and Rule of Law Project established by the Carnegie Endowment for International Peace and the German Institute for International and Security Affairs (SWP), which identify three basic approaches that put security and stability first (Ottaway and Mair, 2004):

- Long-term strategies to promote economic development and thus achieve the desired reduction in fragility and vulnerability;
- Military interventions by international troops in acute crises in order to restore order at domestic level in the short term;
- Non-military interventions (external assistance) aimed at the targeted short-term stabilization of specific state functions within a country at risk.

These approaches involve sometimes considerable risks both for the directly affected states and societies and for the intervening external actors. They are discussed critically in the literature, with a primary focus on the interfaces between security and development policy (Faust and Messner, 2004; Klingebiel and Roehder, 2005; Debiel et al., 2005).

Within the framework of the Fragile States Group set up by the OECD Development Assistance Committee (DAC), a short list of working principles, the Principles for Good International Engagement in Fragile States and Situations (Box 4.2-3), was adopted in April 2007. As part of the DAC's work programme for the next two years, the practical implementation of these principles will be a key focus of the international consultations.

4.2.2
Destabilizing effects of environmental degradation

The role of environmental degradation in the process of state failure has not yet been studied explicitly. Instead, attention has focussed primarily on natural resources and issues of political economy rather than on the environment as a factor in the ecological sense. For example, the exploitation of natural resource deposits, especially oil but also diamonds or precious metals, has been discussed by many authors (Berdal and Malone, 2000; Collier et al., 2003).

In essence, however, various regional and local environmental changes whose intensity is increased by global influences can operate as destabilizing factors if they lead to genuine impairment of the state's capacity to perform its core functions. For example,

---

Box 4.2-3

OECD-DAC: Principles for Good International Engagement in Fragile States and Situations

- *Take context as a starting point:* It is essential to understand the specific context in each country – blue-print approaches for the stabilization of fragile states should be avoided.
- *Do no harm:* Negative impacts through international interventions must be carefully avoided.
- *Focus on state-building as central objective:* International engagement should focus on strengthening the capability of states to fulfil their core functions.
- *Prioritize prevention:* An emphasis on prevention will reduce fragility, and lower the risk of future crisis and violent conflict.
- *Recognize the links between political, security, and development objectives:* The challenges faced by fragile states are multi-dimensional. Tensions and trade-offs between objectives need to be addressed ('Whole of Government Approach').
- *Promote non-discrimination as a basis for inclusive and stable societies:* Real or perceived discrimination needs to be avoided as it is associated with fragility and conflict and can lead to service failure.
- *Align with local priorities in different ways in different contexts*: Where governments demonstrate political will to foster development, international actors should seek to align assistance. Functioning systems within existing local institutions should be strengthened and activities avoided which undermine nation-building, such as developing parallel structures.
- *Agree on practical coordination mechanisms between international actors*: Close coordination of engagement between international actors is important to avoid inconsistencies and thus undermine stabilization efforts.
- *Act fast, but stay long enough to give success a chance*: Assistance to fragile states must be flexible enough to take advantage of windows of opportunity at the same time as being prepared for longer-duration engagement. Volatility of engagement is often potentially destabilizing.
- *Avoid pockets of exclusion*: International actors need to address states where international engagement and aid volume is low to avoid destabilization.

Source: OECD Document DCD/DAC(2006)62 of 24 November 2006

storm and flood disasters can be a typical trigger for destabilization (Box 4.2-2). Depending on the specific context and perspective, famine due to soil degradation and water scarcity can be classed as a structural factor (e.g. long-term climate change) or a process factor (e.g. political instrumentalization). Such impacts of environmental change have the potential to destabilize states and societies and in extreme cases may even lead to state collapse.

It is likely that increased environmental stress resulting from progressive environmental degradation itself operates as a structural factor which can take a variety of forms (e.g. air pollution, water contamination, toxic waste). Environmental stress often occurs where the external costs of humans' use of nature are not internalized. The precarious effects of the ensuing environmental degradation are often only visible over the long term, but then may reach a critical level which can also trigger destabilization and violent conflicts (the pollution of the Niger Delta by the oil industry is a case in point). Such destabilization can affect the domestic constitutional structure of individual states as well as interstate relations. Weak and fragile states can thus have an indirect 'spillover' effect on the wider region and on the international community as a whole.

At present, there is little empirical evidence to suggest that the high conflict relevance of weak and fragile states is significantly increased by environmental degradation. Wherever destabilization and state failure processes have been described to date, a variety of factors has always come into play. One issue which should be considered, however, is whether and to what extent this might change in future if environmental stress and possible conflict constellations arise on a hitherto unprecedented scale.

Empirically, the number of states that must be considered potentially at risk of further destabilization is very large. The potential risk is greatest in the low-income countries. However, even some middle-income countries, such as Venezuela, Jordan, and Kazakhstan, qualify as weak or fragile states (Ottaway and Mair, 2004). Over the long term – i.e. the period to 2050 or beyond – destabilization of states which at present are still characterized by consolidated statehood could also occur. Environmental changes on an unprecedented scale could be a driving force behind this development. Hurricane Katrina which hit the US coast in August 2005 and devastated the city of New Orleans shows that even well-performing industrialized countries can face major problems as a result of environmentally induced extreme events.

In view of the regional and global environmental changes which may occur in future, the environment-security nexus may have to be re-evaluated for some countries and regions. Depending on the extent to which weak and fragile states are integrated into world trade, multilateral cooperation and transnational political processes, critical developments in the affected regions and countries are relevant to the international community as well. A key task in future is therefore to limit the number of at-risk countries by drawing on current information to identify the specific causes of their fragility and develop appropriate policy responses.

In WBGU's view, these are likely to be states which are classed as critical in various crisis and governance indices (Foreign Policy, 2006; Freedom House, 2006; World Bank, 2006b; AKUF, 2007). According to the indices drawn up by the World Bank (Bad Governance Index), Freedom House (Level of Freedom Index), the Hamburg-based Working Group on the Causes of War (AKUF; Prevalence of Armed Conflict) and the Failed States Index compiled by the Fund for Peace and *Foreign Policy* journal, the following states must be judged as critical against at least two of these four indices (the number of indices in which the country is cited is stated in brackets; Fig. 4.2-1), even though the assessment bases used in these indices vary in some cases very widely:

- Sub-Saharan Africa: Burundi, Chad, Central African Republic, Côte d'Ivoire, Democratic Republic of Congo, Liberia, Nigeria, Sudan (four each), Angola, Ethiopia, Guinea, Sierra Leone, Somalia, Zimbabwe (three each), Cameroon, Guinea-Bissau and Malawi (two each).
- Central America and the Caribbean: Colombia and Haiti (three each).
- Middle East: Iraq (four), Yemen (three), Lebanon, Saudi Arabia and Syria (two each).
- Eastern Europe and Central Asia: Afghanistan (three), Russia and Uzbekistan (two each).
- South and South-East Asia: Myanmar (four), Bangladesh, Indonesia, Laos, North Korea, Pakistan and Sri Lanka (two each).

4.3
Unstable multipolarity: The political setting of global change

The scenarios developed in global change research, especially in climate research, provide a fairly clear picture of how the Earth will change over the coming decades. Biogeophysical changes resulting from climate change can be predicted with relative certainty, within a certain spectrum, to around 2050. Even changes up to 2100 can be estimated for different development pathways, which vary according to the assumptions made about anthropogenic greenhouse gas emissions. Global political conditions or the

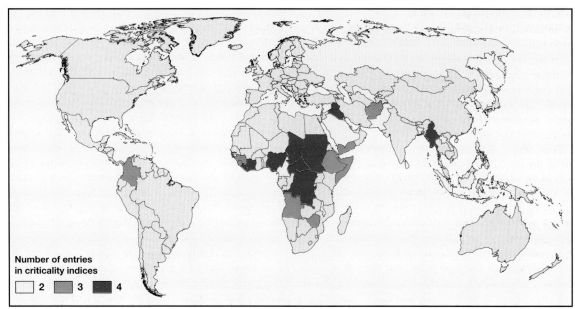

Figure 4.2-1
Weak and fragile states: A global overview. The colours reflect the number of times the individual countries are mentioned in various crisis and governance indices (see text).
Source: WBGU

development of the world economy are more difficult to predict reliably, however. Nonetheless, empirical trends can be identified both in the world economy and international politics as a basis on which to extrapolate and predict probable future trends with some degree of plausibility. For example, it is becoming increasingly apparent that the world's two most populous countries, the People's Republic of China and India, will in future play a far more important role than in previous decades (Worldwatch Institute, 2006). This is likely to have far-reaching and unprecedented effects on the world economy and global political order.

WBGU assumes that China and India in particular, due to their population size and newly acquired economic dynamism, will substantially influence the world economy and therefore also gain in global political significance and capacity in the near future. Besides the two Asian giants (Winters and Yusuf, 2007a), countries such as Brazil, Indonesia and Russia could also become relevant actors in some areas of global governance. The United States of America – since the end of the East-West conflict, the world's only superpower – is likely to experience a relative loss of power at the same time. The rise of China and India therefore marks a major shift in world order, which will move from a unipolar to a multipolar system. The ensuing political, institutional and socio-economic adaptation requirements may trigger numerous conflicts of interest within the international community, potentially increasing countries' vulnerability to violent conflict and thus making classic power struggles – rather than the task of mastering the impending global climate crisis – the focus of international attention. However, in an ideal scenario, climate change could also become a catalyst which unites the international community, provided that the key political actors recognize climate change as a threat to humankind and take resolute counteraction by adopting a globally coordinated approach. Unless this happens, however, climate change will in future draw ever-deeper lines of division and conflict between the Western industrialized countries and the rising Asian nations.

4.3.1
Conflict or cooperation through the transformation of the world order?

This does not mean that the anticipated global adaptation processes will necessarily be conflictive. Unlike the major international conflicts of the 19th and 20th century, no serious territorial conflicts can be identified aside from the permanently contentious issues of the status of Taiwan and the Palestinian territories and the continued division of Korea. Notwithstanding the stylized discord between the West and Islam, there are no fundamental ideological conflicts between the existing and the rising major powers that could be compared with those during the Cold War. Rather, as a result of the highly visible economic interdependencies, there is a strong mutual interest in international stability and a rules-based world economy.

However, as Kupchan et al. (2001) point out sceptically, a glance back at history shows that transitions from one type of world order to another – in other words, the supplanting of a world power by one or more ascendant nations – have rarely taken place peacefully. Indeed, there are many historical precedents which indicate that the 'rise and fall of the great powers' (Kennedy, 1988) were invariably turbulent turning points in world politics which often escalated into violence (Münkler, 2005). The shift of power between Great Britain and the United States at the start of the 20th century was one of the rare exceptions in that it did not involve armed conflict. As a rule, however, hegemonic powers are rarely prepared to break with a strategy of global dominance and move towards a system of global or even shared global leadership. This is currently apparent from the example of the United States of America (Kupchan, 2003; Brzezinski, 2004).

Given that mitigation of the major problems facing the world, especially climate change, is most likely to be achieved through a generally stable, cooperative and effective global governance architecture, WBGU is a firm advocate of multilateral solutions. Against this background, the present report aims to explore how much scope is available to Germany and Europe to influence the processes of change which can be anticipated during the coming decades.

4.3.2
Global trends: China, India and the path towards multipolarity

In the globalization debate, China's development is currently attracting considerable attention, and India's significance is also increasingly being recognized. The gain in economic power and anticipated global political influence of China and India – known as 'Asian drivers' (Humphrey and Messner, 2006c) or 'Asian giants' (Winters and Yusuf, 2007a) – are clearly reflected in statistics and projections about trends that are anticipated in the coming decades.

Strong economic development in China and India
Since it initiated its reform and liberalization policies in 1978/1979, China's economic performance has been spectacular, with GDP growth rates averaging more than 9 per cent per annum in real terms over the last ten years. With GDP at around US$2300 thousand million, China's share of world nominal GDP was around 5 per cent in 2005. For comparison, Germany's share was around 6.3 per cent. Continued growth in the Chinese economy is forecast, with average annual growth rates of 5.5 per cent predicted to 2020, meaning that China's share of world GDP is likely to continue to rise substantially (World Bank, 2006a; Winters and Yusuf, 2007b). By contrast, India is still at an early stage of economic growth with very promising prospects for the future. Here, real growth averaging around 6 per cent per annum is predicted for the coming decades (DBR, 2006; Winters and Yusuf, 2007b). India's GDP currently stands at around US$800 thousand million, i.e. around 1.7 per cent of world GDP, and is also expected to rise substantially (Goldman Sachs, 2003; Winters and Yusuf, 2007b). An even more impressive picture emerges if a country comparison is undertaken on the basis of purchasing power parity, i.e. if nominal GDP values are adjusted for differences in purchasing power, inflation and exchange rate effects. Based on this adjustment, China accounted for 14 per cent of world GDP in 2005, with India's share estimated at 6 per cent, while Germany's share falls to 4 per cent (IMF, 2007).

Shift of power relations in the world markets
The increasingly important role being played by China and India in the world markets is apparent primarily from their growing share of world trade. Between 1990 and 2005, China's share of world exports of goods (excluding Hong Kong) rose from 1.8 per cent to 7.5 per cent, while its share of world exports of services rose from 0.7 to 3 per cent (WTO, 2006). Today, the People's Republic is the world's third largest exporter and importer of goods and is also one of the top ten trading nations in the service sector. China's shares of world trade are expected to rise further: long-term export growth averaging 8 per cent per annum is regarded as realistic, especially given that China's export growth has exceeded 20 per cent per annum in recent years (Winters and Yusuf, 2007b).

Although India has a different economic and export structure (Dimaranan et al., 2007), a similar trend can be observed here as well. India's share of world exports of goods almost doubled between 1990 and 2005 from 0.5 to 0.9 per cent, standing at around US$100 thousand million in 2005; exports of services amounted to around US$56 thousand million in 2005. India thus moved up the ranking to become the 11th largest exporter of services in the world, behind China and Hong Kong (WTO, 2006). India is also becoming increasingly significant as an importing nation; India's share of world imports also almost doubled between 1990 and 2005, from 0.7 to 1.3 per cent. Average growth of 7–8 per cent per annum in the Indian export sector seems entirely feasible for the coming decades (IMF, 2006; Winters and Yusuf, 2007b). Trade liberalization is also likely to result in further substantial growth in imports, enabling India to further increase its overall world market share.

In sum, China and India, along with the dynamic newly industrializing countries of South-East Asia, will become ever more important for the industrialized countries in particular, both as increasingly productive competitors and suppliers and as export markets. However, the substantially increased and still rising demand for raw materials (oil, ores, timber) and food in the Asian growth economies is pushing up prices. Exporting countries will benefit from this: simulations for the period to 2020 show that these will mainly be countries in Africa, the Middle East, the former Soviet Union and Latin America, along with Canada and Australia, while some purchaser countries are likely to be negatively impacted (Dimaranan et al., 2007). Furthermore, China is already playing a key role on the international capital and financial markets; in 2005, for example, around US$70 thousand million in direct investment – 8 per cent of the world's total – flowed into China (UNCTAD, 2006). India, by contrast, attracted just US$6 thousand million in direct investment in 2005, but this has recently risen substantially (to US$10 thousand million in 2006), and in view of the growth forecasts for India and the removal of investment barriers, this trend is likely to continue. China and, increasingly, India are therefore not only important as target countries but are becoming increasingly important as countries of origin for direct investment as well.

In view of the very high currency reserves – around US$1000 thousand million – accumulated by China as a result of its foreign trade surpluses and exchange rate policy, the People's Republic is viewed as a key actor in the international financial markets, capable of exerting significant influence over interest and exchange rates through the deployment of its currency reserves. India's currency reserves presently stand at around US$170 thousand million (2006) – a relatively small share of the reserves held in the Asian region, which (excluding Japan) exceed US$2000 thousand million.

More complex conflict patterns in world trade policy

In view of the importance of international trade and capital movement for their own growth processes, China and India have benefited from their integration into the world economy. Nevertheless, at the recent GATT negotiations on the liberalization of trade in goods, India positioned itself above all as a powerful representative of the group of developing countries and attempted to assert traditional developing country interests vis-à-vis the industrialized nations.

In the WTO's current Doha Round, however, it is becoming apparent that the supposedly clear North-South line of conflict over the removal of trade barriers is becoming increasingly blurred as interests diverge. The major exporters of agricultural goods among the newly industrializing and developing countries, especially Brazil, have joined forces with a number of industrialized countries to demand the swift and unequivocal abolition of US and EU farm subsidies. However, moving beyond agricultural trade issues, these countries have established themselves in the G20 as a powerful counterforce to the group of industrialized countries in the trade negotiations. The other developing countries are hoping that the impetus emanating from the G20 will lead to the liberalization of agricultural markets in the industrialized countries and therefore indirectly to more open agricultural markets in the newly industrializing countries as well. At the same time, however, they see their special treatment in the WTO system coming under threat from the interest-led actions of the newly industrializing countries (Wiggerthale, 2004).

Both India and China are key actors within the G20. Whereas India, with Brazil, has recently actively sought a compromise with the industrialized countries, China's position has been relatively restrained. Unlike the other members of the G20, most of which are net exporters of farm products and would therefore benefit from rising world market prices resulting from trade liberalization, China is a net importer of farm products (Wiggerthale, 2004; Langhammer, 2005). Similarly complex differences in interests also arose in relation to the liberalization of the textile trade and are still causing conflicts between developing countries (Kaplinsky, 2005; Langhammer, 2005). Overall, China and India's roles in the world trade regime are far from identical, fluctuating between advocacy for the interests of the developing countries, on the one hand, and the assertion of their own national interests, on the other – against the interests of other developing countries.

Rivalry over resources: Signs of a renaissance of geoeconomics

The growth processes taking place in China and India would not have been possible without massive inputs of raw materials which, for the most part, had to be purchased on the world market. In order to maintain the growth momentum, demand for resources in both countries is likely to soar in the coming years as well. For example, China's share of world demand for key base metals has risen from 5–7 per cent in the early 1990s to 15–30 per cent in 2005 (Humphrey and Messner, 2006c; Winters and Yusuf, 2007b). The competition with other industrialized and developing countries which are also reliant on supplies of raw materials via the world markets will therefore intensify further. This competition focuses strongly on fossil fuels and other mineral resources. These scarce and strategically important goods are not only distributed via the markets, however. Political and military flanking measures are being taken to help secure exploration and trade routes (pipelines and seaways) in and through crisis regions – an activity which, in the past, was primarily undertaken by the USA, with a number of EU countries also being involved.

In view of the growing scarcity of raw materials, Chinese and, indeed, Indian foreign policy is increasingly geared towards securing long-term access to energy and resource supplies. Both countries are thus entering into competition with other major powers, especially the USA but also Japan and the EU. For example, China is already the world's second largest oil importer after the USA. Against this background, some observers are predicting a 'renaissance of geoeconomics' (Klare and Volman, 2006; Wesner and Braun, 2006), implying that the competition for energy reserves and other raw materials will exert a very substantial influence over world politics in future. The more conflictive this competition becomes, the more likely it is that cooperative multilateral approaches to the management of global problems such as climate change or poverty reduction will be weakened and fall victim to the major powers' unilateral, resource-oriented regional strategies (Humphrey and Messner, 2006b, 2006c; Section 4.3.3).

China's current engagement in Africa can be regarded as symptomatic of this trend. In the EU and the USA, which – especially in the 1990s – claimed to promote conditionality-based development cooperation in the interests of democracy-building, there are concerns that China's massive investments in the African extractive industries not only undermine efforts to foster economic diversification and industrialization in Africa but may also consolidate autocratic structures of governance or become a contributory factor in the emergence of rentier economies with unstable social institutions (Goldstein et al., 2006; Tull, 2006). In line with this argument, efforts to promote political liberalization in Africa are thus being undermined, especially as resource-dependent countries have often proved resistant to democracy (Ross, 2001; van de Walle, 2001, 2005). Admittedly, neither American nor European development cooperation has always lived up to its own claims. For example, the conditionality associated with 'Western' values such as democracy and human rights is by no means enforced consistently, and in practice, development cooperation is invariably influenced by donor countries' own resource supply needs, the securing of export markets, and overarching geostrategic considerations (e.g. the 'war on terror'). Overall, it is apparent that China's expanding trade relations and direct investments in developing countries are weakening the industrialized countries' political position in the competition for the remaining energy and raw material supplies and eroding the basic parameters for the use of conditional aid in development cooperation. What's more, both these trends run counter to the Western countries' key security policy interests.

CHINA AND INDIA AS RELEVANT CLIMATE POLICY ACTORS

The growth dynamic in China and India is linked with a high demand for energy. For example, China's share of global primary energy consumption increased from 10 per cent to 14.5 per cent between 1990 and 2005, while India's share increased from 4.1 to 5.1 per cent over the same period (Fig. 4.3-1a). In order to maintain growth, demand for energy – despite decreasing energy intensity – is likely to remain high and may

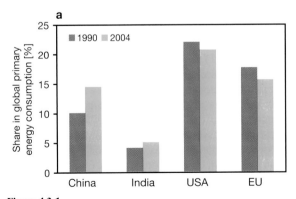
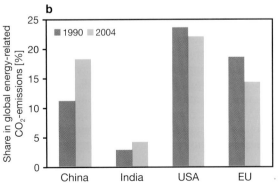

Figure 4.3-1
China's, India's, US and EU shares of: a) global primary energy consumption and b) global energy-related CO_2 emissions.
Source: WBGU, 2007; data from IEA, 2006c, CAIT WRI, 2007

even rise further. China's primary energy consumption is expected to double again between 2004 and 2015, while India's is predicted to increase by more than one-third over the same period (World Energy Outlook Reference Scenario; IEA, 2006c).

With the energy supply structures that are likely to be in place over the medium term, the two countries' growing energy needs will mainly be met from fossil fuels, resulting in a strong increase in CO_2 emissions. In 1990, China accounted for around 11 per cent of the world's CO_2 emissions, but this had already risen to 18 per cent in 2004. Within the next few years, China will take over from the USA as the world's largest CO_2 emitter. India's share has also increased from just 3 per cent to more than 4 per cent since 1990 (Fig. 4.3-1b).

As two of the world's main emitters of CO_2, China and India are also becoming increasingly important in international climate policy. Without their constructive participation in the further development and implementation of international climate agreements under the United Nations Framework Convention on Climate Change (UNFCCC), the aim of preventing dangerous anthropogenic climate change will not be achieved (Scholz, 2006; Richerzhagen, 2007).

Both countries have signed and ratified the Kyoto Protocol, but as non-Annex I countries, they have not entered into any reduction commitments for the period to 2012. China and India are also target regions for CDM projects, by means of which Annex I countries, i.e. the industrialized nations, undertake to fulfil some of their own reduction commitments abroad. The key issue is whether over the long term, China and India will commit to emissions reductions of their own and have the capacity to implement them in time. At present, neither of these two countries is showing any willingness to enter into commitments (even voluntary commitments) to limit their greenhouse gas emissions. Nonetheless, China and India occupy divergent positions in the international climate policy forums. The industrialized countries are exerting very little pressure on India to take on commitments in the near future. India itself is at present categorically rejecting any prospect of differentiated commitments in future on the grounds that its per capita emissions are still very low, and is drawing attention to its right to pursue 'catch-up' economic development (Hörig, 2006). Furthermore, India is insistent that the responsibility for climate change mitigation lies primarily with the industrialized countries (Narain, 2006).

By contrast, far greater pressure is being exerted by the industrialized countries on China to commit, at the least, to progressive reductions or 'no-lose' targets in the near future. As a result of its rapid economic growth, China faces massive environmental problems. According to official figures, China is now spending 1.8 per cent of its GDP on environmental protection, but environmental damage is costing China 3 per cent of GDP. Indeed, other sources calculate that 8-12% of China's annual GDP is being lost because of the severe consequences of environmental pollution (Sternfeld, 2006; Heberer and Senz, 2006a).

For these reasons too, China currently has a key interest in mitigation measures, not least in order to be able to solve its own problems of local air pollution. However, China is expecting active support from the developed nations, in the form of international research and technology transfer. As China sees itself as a high-quality partner aiming to achieve rapid progress in the development of its own technologies, there are concerns in the industrialized countries that this type of technology transfer could result in the loss of their own competitive advantages, especially in the field of renewable energies, also in view of China's uncertain legal position with regard to intellectual property. Notwithstanding this basic willingness to engage on climate change mitigation, China has emphasized that while it is willing to adopt measures on a voluntary basis, under no circumstances will it enter into voluntary commitments within the climate regime. It remains to be seen to what extent the possibility of more far-reaching steps, as tentatively implied in Chinese politics recently, will be pursued further.

GROWING GLOBAL POLICY ENGAGEMENT BY CHINA AND INDIA

As a result of the economic 'catch-up' processes in China and India, these countries are intensifying their global political engagement: China's growing presence in international politics can no longer be ignored (Amineh, 2006; Gill, 2007). China's engagement in international organizations and its participation in multilateral treaty systems has also increased significantly since the 1970s and its previous strategy of isolating itself internationally has given way to a pro-active role in world affairs (Johnston, 2003; Medeiros and Fravel, 2003; Heberer and Senz, 2006a). The new diplomacy being pursued by the People's Republic is most apparent in Asia, where China was a driving force in the establishment of the ASEAN+3 mechanism, a dialogue forum involving the South-East Asian countries, Japan and South Korea. The People's Republic has also taken on a leading position in the Shanghai Cooperation Organisation, an intergovernmental organization whose membership includes besides Russia and China various Central Asian countries. Not least, as a permanent member of the Security Council and a participant in various UN peace missions, the People's Republic is now playing a far more important role than at the time of the Cold War (Fravel, 1996; Medeiros and Fravel, 2003; Berdal, 2003).

Unlike China, India has always been involved in international organizations and has always seen itself, in this context, as a mouthpiece for the developing world, e.g. in international trade policy. In 2003, conscious of its own global political strength and in line with its claim to represent the interests of the developing countries, India joined forces with two other newly industrializing countries, Brazil and South Africa, to form a 'trilateralist' diplomatic partnership and forge common positions for multilateral negotiations and political processes on key global issues (Alden, 2005). In order to expand its influence on the world stage, India is also lobbying intensively for a permanent seat on the UN Security Council. In sum, there are increasing signs on the subcontinent that India is on the verge of becoming a great power (Raja, 2006).

In parallel to their economic catch-up processes, China and India have also expanded their military capacities: in 2005, China and India accounted for 4 per cent and 2 per cent of global military spending respectively (compared with 3 per cent for Germany). This puts China and India among the ten countries with the highest military budget (SIPRI, 2006). They are also nuclear powers, and India has even become a geostrategic partner for the USA in this context (Mitra, 2006; Wagner, 2006). By contrast, the military modernization of the People's Republic is viewed with scepticism in the West (Cordesman and Kleiber, 2007). Regardless of their security policy motives, their growing military power flanks China's and India's increasing engagement on the world stage, underscoring these two countries' regional and global weight.

4.3.3
Global governance in the context of Chinese and Indian ascendancy

The transformation of China and India into world economic heavyweights is significantly changing the conditions in the global competition for markets, resources and pollution rights. In tandem with their global political engagement, it is likely that this will also lead to power shifts in the political world order which in turn could have profound implications for international relations.

In light of these developments, it is astonishing that leading analysts of international politics ignored the de facto increase in these two countries' significance for so long (Humphrey and Messner, 2006a, c; Box 4.3-1). As late as 2004, Brzezinski, for example, was still arguing that China was a developing country which posed no real challenge to the USA, while India featured only peripherally in his analysis (Brzezinski, 2004). Nye (2002) also emphasizes that China has a very long way to go before it can play a genuinely global role. Katzenstein (2005) urged readers to think of the world as regions which would be strong but would nonetheless be organized by America's imperium. Other authors focussed intensively on the future of transatlantic relations between the USA and the EU, without considering whether China and India could change the overall global panorama (Daalder and Lindsay, 2003; Kupchan, 2003). Only recently have leading analysts and think-tanks in the British- and American-dominated debate begun to

---

Box 4.3-1

Interpretations of the post-1990 world order

- Mearsheimer (1990) focussed on the re-emergence of conflicts between nation states after the collapse of the bipolar regime of the Cold War. From his point of view, proliferation of weapons of mass destruction was the most important challenge.
- Fukuyama's (1992) analysis of the 'End of History' perceived a strong trend towards global democratization; the major conflict line of his map of global politics is the struggle between democratic and non-democratic states.
- Huntington (1993, 1996) postulated the 'Clash of Civilizations' between Western and other cultures as the central line of conflict in the future.
- Kaplan (1994) as well as Kennedy and Connelly (1994) discussed the socio-economic asymmetries between 'the West and the Rest' as the challenge most likely to cause conflicts.
- In the global governance discourse, many authors have analysed the effects of globalization on the nation states' capacities to act. They argue that without new forms of multilateral cooperation, the major problems facing the world will remain unresolved, causing turbulence in international relations (Rosenau and Czempiel, 1992; Messner, 1998; Young, 1999; Donahue and Nye, 2000; Kennedy et al., 2002).
- Robert Kagan (2002, 2003) predicted a lengthy period of unilateral dominance by the USA, with transnational terrorism being the major conflict in the coming decades. Both Mearsheimer and Kagan put 'security' at the top of their global political agenda, with the diffuse 'war on terror' taking the place of classic confrontations between nation states.

Source: WBGU, based on Humphrey and Messner, 2006b

include China and India in their frame of reference when analysing global power relations (e.g. CSIS and IEE, 2006; Kaplinsky, 2006; Worldwatch Institute, 2006; Gill, 2007). In the debate about the options for the development of a system of global governance in a rapidly changing, post-Cold War world too, the emerging ascendancy of China and India was barely considered at first. Instead, analysts focussed primarily on issues relating to the future of the classic, territorially organized nation state and its capacity to act and exert influence (Messner, 1998; Zürn, 1998b) and the role of new actors 'beyond the state' (Risse, 2002; Jachtenfuchs, 2003; Dingwerth and Pattberg, 2006).

4.3.3.1
Multipolarity as a threat to multilateralism?

Against this background, the question which arises is whether, and to what extent, China and India, but also the United States, will be willing to participate promptly and constructively in the development and expansion of multilateral institutions as the framework for effective and legitimate global governance. If China or India opts in favour of unilateralist foreign policy strategies which are oriented towards their own national interests, the international community could face the renaissance of a 'balance of power' politics last witnessed during the Cold War. Unlike a cooperative, multilateral approach, this type of power constellation based on rivalry, would absorb substantial capacities and resources which are urgently needed to tackle major global challenges – from poverty reduction to climate change.

It is axiomatic that the responsibility for setting an appropriate policy course does not, and cannot, lie solely with China and India. The much-lamented crisis besetting multilateralism in recent years is due in no small part to the foreign policy being pursued by the United States of America, with its very limited focus on cooperation (Menzel, 2003; von Winter, 2004; Hummel, 2006). The European countries' efforts to counteract this approach are ubiquitous but often lack political impact. It remains to be seen whether the foreign policy model which is strictly geared towards national concerns will gain ground, or whether international cooperation can prevail. The options for future international relations are becoming clear: 'An effective multilateralism and either a gradual return to a world of great power competition or a world overwhelmed by disruptive forces or both' (Haass, 2005).

4.3.3.2
General dynamics of global political change

In the analysis of the global political processes which can be observed at present, two parallel and inherently contradictory dynamics can be identified which have the potential to influence the architecture of the world order on a permanent basis. They are, firstly, the growing and increasingly complex interdependencies between states, economies and societies, and secondly, the trend towards unilateral foreign policy action by governments, which is typical of phases of a shifting balance of power in international relations.

COOPERATIVE MULTILATERALISM AS A REACTION TO COMPLEX INTERDEPENDENCIES

The globalization that has taken place in the 20th and 21st century is characterized by complex interdependencies between states, economies and societies (Held et al., 1999; Keohane and Nye, 2000). As a result of diverse political, economic, cultural, military and, not least, environmental linkages, it is becoming increasingly difficult for nation states to solve problems, even those falling within their own scope and competences. The complex interdependencies in the world therefore also raise the threshold for violent conflict (because the opportunity costs of such conflicts increase with growing integration) and offer a greater incentive to pursue institutionalized, rules-based cooperation (which, among other things, promotes confidence-building by increasing the reliability of mutual expectations; Keohane and Nye, 1977, 1987; Zürn, 2002).

The rapid growth of international organizations and regimes in recent decades bears out this institutionalist view and reflects countries' needs to address complex global problems through a stable system of international institutions. Greater interdependence encourages cooperation, as is apparent from successful examples of regional integration such as the EU and other regional organizations like ASEAN. This shows that dependable international agreements based on international law can be more important than military power in safeguarding peace, security and development. Climate change amply demonstrates the multilayered complexity of interdependencies, not only regional but global, and the great and mutual vulnerability which it will create, such as the potential impairment of global economic development and the possible increase in the number of destabilized states as a result of climate change (Section 8.3).

EROSION OF MULTILATERALISM THROUGH A
RELATIVE POWER SHIFT

Running counter to the approach which sees international relations in terms of complex interdependencies, there is the so-called 'realistic' view which fixates on states, their power potential and the primacy of security policy. In line with this latter approach, relative shifts in power potential among the key actors in an anarchic international system generates a dynamic of its own which undermines cooperative behaviour and thus triggers the erosion of multilateralist politics. Unilateral action then becomes increasingly attractive for powerful states, and is viewed as a rational alternative to cooperation with less powerful countries.

If the major powers' foreign policy action aims to safeguard peace – or their own security – through classic balance-of-power politics, this will lead to less global governance and more traditional politics between sovereign nation states. This is especially likely wherever distrust of the intentions of countries which are perceived to be rivals outweighs the trust in cooperative solutions. This type of premise conforms with the type of logic which dominated European history in the 19th and early 20th century. The conduct of the expected conflicts between the USA, China, India and other states over the basic pillars of the world order in the 21st century will therefore be determined, among other things, by the extent to which the major powers' foreign ministries subscribe to this logic. The relativization of both international law and the United Nations' scope for policy-making and action which has been observed in recent years could be the prelude to a renaissance of aggressive power politics (Messner et al., 2003).

The foreign policies which have emerged as a result are unilateralist in focus and based on classic interpretations of sovereignty, security and the nation state which conflict with the widely proclaimed desire for global solutions to typical transnational problems such as terrorism, HIV/AIDS or global warming. It must be borne in mind, in this context, that rising powers such as China and India can hardly be expected to jettison the concept of the sovereign nation state without good reason. The conceptof the state which has developed in Europe during 50 years of European integration cannot be assumed here. What's more, given that the current superpower, the USA, is playing out the classic model of sovereignty in all its contradictions, it is hardly surprising if the governments of China and India invoke the words of Robert Kagan, the US Government's foreign policy adviser, to the effect that 'multilateralism is a concept for weak actors' (Humphrey and Messner, 2006b).

Against this background, it is apparent that the development of a multipolar new world order is, in essence, open-ended. A reversion to a world dominated by great power politics is just as possible and plausible as the world's progressive development towards a network of international cooperation (Messner et al., 2003). The debate makes it clear that multipolarity and multilateralism are independent from each other, so a trend towards multipolarity should not, per se, be viewed as a problem or as a worsening of the situation compared with the unipolar status quo. By the same token, the current unipolar world order dominated by the USA cannot be regarded as the apogee of multilateralism (Hummel, 2006). As things stand, the development of a cooperative system of global governance cannot be taken for granted, and the conduct of the new great powers in the anarchic international system is by no means clear and predetermined (Brzezinski and Mearsheimer, 2005).

Nonetheless, the parallelism between both dynamics described here will crucially determine the development of global political structures and economic parameters. Ultimately, the political efforts and interests of all the relevant global actors will influence the interaction and thus determine whether the adaptation of the world order takes place primarily by peaceful and stable means, or whether it is fraught with conflicts and instability.

4.3.3.3
China and India as the driving forces of global political change

With a combined population of more than two thousand million people, China and India – the world's most populous countries – are, compared with the Asian tigers in the 1980s and 1990s, not only ascendant economies but are also emerging as driving forces in global political processes (Humphrey and Messner, 2006a, c). These dynamics of change affect international institutions such as the United Nations, the World Bank and the International Monetary Fund as well as the international 'clubs' such as the G8 and the G20. So it is likely that these power shifts will have a tangible influence on a range of multilateral treaties, e.g. the Kyoto Protocol in climate policy or the world trade agreements established under the auspices of the World Trade Organization (WTO) and on a range of transnationally operating non-state actors and the many different governance mechanisms through which global players interact.

All these elements of the current global governance architecture will be subject to powerful pressure to adapt as soon as China and India begin to wield their new-gained power, while at the same time, the quasi-hegemonic dominance of the USA will subside. A key, if not the primary, issue in this context is whether the two Asian countries will acquire global political significance as swiftly as they have come to dominate the world markets (Heberer and Senz, 2006a).

It is likely that as a result, not only will the long difficult relationship between the industrialized and the developing countries change radically; so too will South-South relations. It is possible, for example, that India will in future further intensify its efforts to do justice to its traditional claim to be a mouthpiece for the developing world by stepping up its demands and wielding power more assertively than before. It is more likely, however, that China and India, as rising economic powers, will in the medium term pursue their own interests to a greater extent than before, and that these interests will no longer accord with those of most other developing countries. This trend is already becoming apparent from the asymmetrical power relations between China and India and the majority of African and Latin American economies. If the two countries become relevant donor countries in the coming years, the developing countries could choose between the assistance available from the OECD countries, on the one hand, and that offered by the two Asian giants, on the other, which would undermine the industrialized countries' development and geopolitical dominance.

Another factor which should be considered is that with the People's Republic of China, an authoritarian state is emerging as a global power, raising concerns that calls for respect for human rights and good governance, for example, will in future be even more difficult to defend via the international system than before (Kurlantzick, 2006). Due to religious and social tensions, India too barely meets the expectations placed on a leading global power at present. It is unclear how this development will impact on the legitimacy of global governance processes which largely depend on recognition of the actors involved in them.

With the rise of China and India, the unipolar world order which has developed since the end of the Cold War as a result of the USA's dominance is likely to remain a brief moment in history. If the ascendancy of China and India continues at such a rapid pace in the coming decades, it will no longer be possible to refer to a Western-dominated world order.

The competition for power and global influence triggered by a shift in the world order could, on the one hand, become a key line of conflict in the global governance architecture in the coming decades and potentially cause a situation which would be comparable to the system conflict of the Cold War. On the other hand, the transformation of international relations could also take positive shape and result in global cooperation, based on equality, between the world's regions, which would then be able to address global problems far more effectively and vigorously than before. Ultimately, the future interaction between old and new global players will crucially determine whether, and how, the global challenges and risks arising in the 21st century can be resolved and which role the 'rest of the world' can play in this context.

4.4
Conclusions

As the outline on state fragility and governance and the expected changes in the international system shows, national and international institutions and actors are poorly prepared to address the challenges described. In relation to fragile states, there is currently no evidence that environmental degradation substantially increases conflict relevance. However, it can be assumed that as the pressure of the problems increases, such states – which even under current conditions barely have the capacity to maintain a functioning polity – will be overwhelmed by the growing environmental stress. So it is likely that climate change will lead to further destabilization in weak and fragile states, potentially leading to distortions in international politics as well.

Furthermore, looking at the current situation of global governance architecture, it is apparent that for the foreseeable future, the international community will lack the requisite capacities to respond effectively to the problems described. As the section on unstable multipolarity shows, the political world order is at the start of a radical and probably turbulent transformation. It is becoming apparent that new global actors, especially China and India, could change the rules of international politics on a lasting basis. This new competition for power and influence does not necessarily have to be conflictive. However, it will absorb valuable time and resources which will then no longer be available for other purposes. The question of the development and establishment of a functioning multilateral system which effectively counters the impending risks of climate change will arise in any event. Whether and how rising and declining world powers cooperate in future will, not least, be a key factor determining the success of international climate policy (Section 8.2 and 8.3).

Impacts of climate change on the biosphere and human society 5

This section discusses expected climatic changes and their consequences based on data from two sources: observed trends in the period 1975–2004, and evaluation of the results of a series of model scenarios. This then serves as the basis for identifying regions in which particularly critical developments are to be expected (Chapter 7).

5.1
Changes in climatic parameters

WBGU assumes that climate change will become the most important driver of global environmental changes in the coming decades. Added to this are the effects of direct human interventions in the environment, which vary from region to region, e.g. river channel alignment that exacerbates flood risk, and unsustainable land-use practices that lead to soil degradation.

Simulations of regional climatic changes are fundamentally more uncertain than global mean values, because regional climate is heavily influenced by atmospheric and oceanic circulation (e.g. by the prevailing weather situations and wind directions). Changes in these have scarcely any impact on global mean values, but at regional level they represent a factor whose influence is highly unpredictable. Furthermore, there is greater uncertainty when it comes to calculating precipitation than when calculating temperature, because the processes that determine rainfall involve complex physics, while temperature is the outcome of a relatively straightforward thermal balance. This is why there is considerable uncertainty regarding regional precipitation scenarios in particular, and scenarios may vary quite markedly depending on the model used.

Similarly, extreme events such as tropical cyclones are also captured less well by climate models because the spatial resolution of global climate models is mostly inadequate for describing them, and the relevant processes are strongly non-linear.

This section discusses developments that can be considered robust and reliable. The discussion also touches on other risks about which there is still considerable uncertainty given the current state of knowledge. However, 'one-off' developments that arise in only one or other of the models are not discussed here. WBGU has opted to focus instead on phenomena occurring in at least several of the climate change models. Reference is also made to observational data and physical contexts in order to underpin the plausibility of the risks under discussion. This does not make it possible to draw up definite forecasts regarding where and when particular critical developments will occur, but plausible threats can be outlined.

A distinction is made between two types of climate risk that are loosely associated with different time periods. A first is climatic changes in the period up to 2050, which are relatively easy to predict and have a high likelihood of actually occurring. In many cases these are developments that can already be observed today, trends such as increasing intensity of hurricanes, vanishing mountain glaciers, and drought-related problems in some regions of the world. Second, Section 5.3 examines risks that may be associated with the possible occurrence of non-linear and qualitative changes in the climatic system – for example far-reaching change in the Asian monsoon. This refers to eventualities that are difficult to assess and have very serious consequences, and which may be more likely to occur in the second half of the century, although it is not always possible to rule out their occurring earlier.

5.1.1
Temperature

Past and possible future global temperature dynamics are depicted in Figure 5.1-1. The development of the global temperature up to 2030 is foreseeable: the rise in temperature compared to 2005 will very likely be in the range of 0.4–0.6 °C, irrespective of the emissions scenario assumed. This temperature range refers to the trend that emerges after adjustment to eliminate small year-on-year fluctuations overlying the long-term trend. Simple extrapolation of the cur-

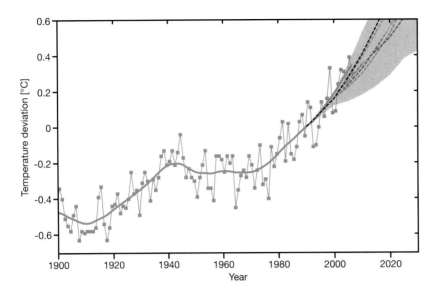

Figure 5.1-1
Global-mean temperature development over land and sea up to 2006 based on the NASA data set, shown as deviations from 1990 (annual data points; the line drawn through the points shows the course of temperature dynamics smoothed over 11 years). 2005 was the warmest year since records began. The grey area and the dotted lines show the future scenarios of the IPCC (2001), which take 1990 as the baseline year. Observed warming is currently in the upper reaches of the IPCC scenarios.
Source: Rahmstorf et al., 2007

rent trend would also be within this range. The narrow bandwidth of the various developments concerning temperature is due on the one hand to the fact that emissions scenarios assume gradual changes rather than drastic leaps in emissions, and on the other to the inertia of the climate system, notably the thermal inertia of the world's oceans. Only very rapid and radical changes in emissions of greenhouse gases or aerosols, or unforeseeable events such as a major volcanic eruption or meteorite impact would bring about any marked deviation from the forecast outlined above in the next 25 years. Gradual emissions reductions over the coming decades, which would most likely be implemented as part of an effective climate protection policy, would not bring about any notable slowdown in the global warming trend until after 2030.

There will be considerable regional variation in warming compared to the global trend, as indeed is already the case. Figure 5.1-2 shows observed warming trends for recent decades over land. While the global temperature has risen by 0.6 °C in the given period, some regions are already registering a rise of 2 °C or more. In this context, two effects merit particular mention. First, the continents are warming up faster than the global mean (in some regions up to twice as fast, due to the moderating effect of the oceans). Second, higher-latitude regions are warming up especially quickly because shrinking snow and ice cover leads to increased absorption of solar radiation. As a result, people in two regions in particular are starkly affected by the direct increase in mean temperatures: the northern Polar regions (Alaska, Siberia), due to the particularly rapid rise in temperatures and its impact on permafrost soils and infrastructure; and the tropical climate zones, due to the

fact that temperatures there are already high to start with.

Heatwaves have special significance. The European heatwave in the summer of 2003 cost the lives of between 30,000 and 50,000 people and, according to figures published by the Munich Re insurance group, is the biggest natural disaster to have occurred in central Europe in living memory (Münchener Rück, 2004). The example shows that heatwaves can have devastating consequences even in temperate zones and affluent countries, if society is ill-prepared to deal with them. In Germany, 2003 saw the biggest losses in terms of wheat yield since at least 1960 (Sterzel, 2004). This heatwave occurred suddenly and was well above the long-term temperature trend (in Switzerland, temperatures for June were around 3 °C higher than in the previous record year, 2002); it could not have been anticipated, therefore, even taking global warming trends into account. The physics underlying this heatwave is not yet fully understood; various attempts are currently being made to explain it. For this reason, it is also impossible to predict where and when in the future such extreme heatwaves may be anticipated. Around the middle of the century, however, it is expected that high summer temperatures such as those experienced in 2003 will no longer represent a heatwave, but will have become the norm (Schär et al., 2004; Seneviratne et al., 2006).

In the longer term (in the second half of the 21st century), temperature dynamics depend to a large extent on which emissions scenario is applied. With effective climate protection measures (stabilizing the greenhouse gas concentration at below 450ppm CO_2eq), it is anticipated that global warming can be restricted to a maximum of 2 °C above pre-industrial levels. In the absence of effective climate pro-

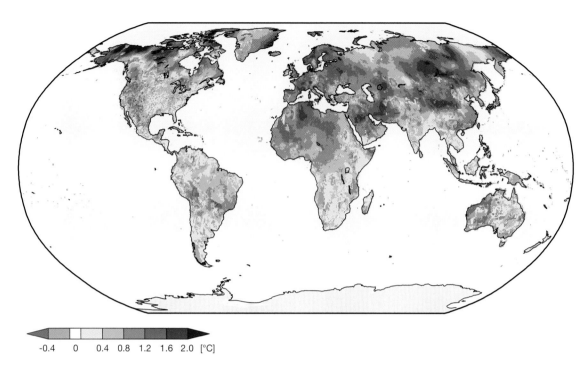

Figure 5.1-2
Linear temperature trends from measurements on land in the period 1975–2004.
Source: WBGU; data: Potsdam Institute for Climate Impact Research (PIK) climate database

tection, however, a temperature rise of between 2°C and 7°C above pre-industrial levels may be expected (IPCC, 2007a). The higher levels may be generated by a positive feedback from the carbon cycle occurring as a result of more pronounced warming. Recent studies based on both simulations and historical climate data show this kind of positive feedback mechanism (Friedlingstein et al., 2006; Scheffer et al., 2006), where oceans and biosphere are no longer able to absorb the same fraction of anthropogenic emissions as before, with the result that the CO_2 concentration begins to rise disproportionately.

As regards the impact of warming, it must be borne in mind that some regions (notably continents) are likely to experience much greater effects than the global mean. In the event of major warming, i.e. in the upper reaches of the range described above, the increase in mean temperature and increased temperature variability (Seneviratne et al., 2006) would trigger considerable changes in the biosphere, because, for example, the distribution of most plants and animals and patterns of competition in their current habitats would shift. Glaciers would rapidly disappear, the Arctic Ocean would be ice-free in summer, and it is very likely that complete melting of the Greenland ice sheet would be induced. These effects begin to occur even in 'moderate' scenarios where the global mean temperature rises by 3°C; if warming is more marked, these phenomena would occur all the more rapidly and with greater certainty.

5.1.2
Precipitation

Globally averaged, rainfall would increase in a warmer climate. Depending on the model used, there would be a 1–2 per cent increase for every degree of warming (IPCC, 2007a). The reason for this is the increase in evaporation. After spending a few days, on average, in the atmosphere, evaporated water must fall again as rain. Moreover, the amount of water vapour in the atmosphere increases because warmer air can hold more water before becoming saturated. Given the same relative atmospheric humidity (measured as a percentage of the saturation humidity), this increase is 7 per cent more water for every 1°C increase in temperature. The average relative atmospheric humidity remains almost constant because an increase leads to more precipitation.

While everywhere in the world water evaporates and precipitation falls, the difference between these two magnitudes varies considerably from one region to another. Figure 5.1-3a shows the regional distribution of the climatic water balance, i.e. the difference between precipitation and potential evaporation (evapotranspiration) in the reference period 1961–

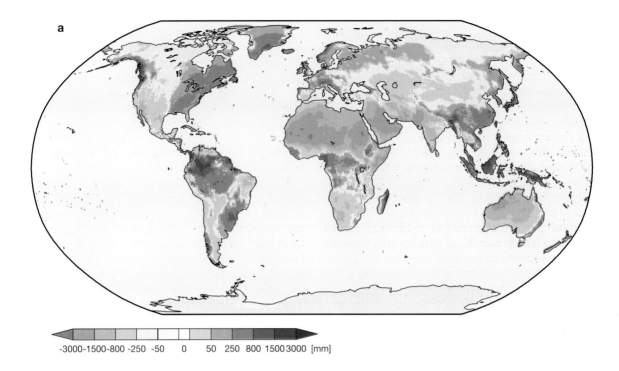

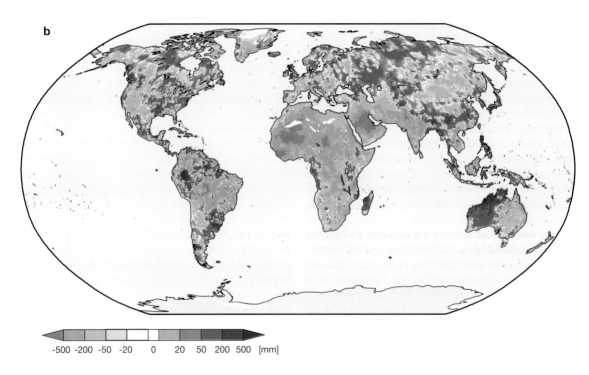

Figure 5.1-3
Climatic water balance: a) mean for the period 1961–1990. b) Change in climatic water balance based on observations in the period 1975–2004. Climatic water balance is the result of the difference between precipitation and potential evapotranspiration.
Source: WBGU; data: Potsdam Institute for Climate Impact Research (PIK) climate database

1990. The basic pattern of the global water cycle is as follows: in the tropics, plenty of water evaporates due to the high temperatures, but due to the rising air mass near the equator (the Inter-Tropical Convergence Zone) it soon falls again as rain; these are the tropical cloud and rain-belts (e.g. Brazil, Congo, Indonesia). Evaporation of water is considerable in the subtropics, too, but precipitation is scant because sinking air masses mean that the air is dry and clouds are few. This is why the desert regions of the world are located in the subtropics: in the northern subtropics lies the Sahara, for example, while the Namibian and Australian deserts are in the southern subtropics. A large proportion of the moisture that evaporates in the subtropics is transported to middle and high latitudes and falls as rain there, so in these regions precipitation exceeds evaporation.

An important aspect of the water cycle is the monsoon circulation, which causes highly seasonal precipitation in some regions of the world (e.g. southeast Asia); in some countries such as India and China, agriculture and food security depend on the monsoon rains. The monsoon winds and rains are driven by seasonally varying differences in temperature between land and sea. It is a strongly non-linear phenomenon that can be influenced both by aerosol pollution of the atmosphere and by global warming.

The effects of global warming amplify this water cycle: evaporation in the subtropics increases, and this on average leads to heavier precipitation in the medium and high latitudes. There is already evidence from observations that this is indeed happening. In the subtropical areas of the Atlantic, salinity has been on the increase for decades, while in the high latitudes it is declining. In addition, long time series for Eurasian rivers flowing into the Arctic Ocean show that river flow has been increasing constantly for several decades, as is to be expected with an increase in rainfall in the high latitudes (Peterson et al., 2002). As a result, subtropical regions are tending to become more arid, while high-latitude regions are becoming more humid, although there are many regional and micro-regional idiosyncrasies in this regard (Fig. 5.1-3b).

In this way, many arid areas have become even drier in recent decades, especially in the Saharan and Mediterranean regions, southern Africa, the southwestern United States of America, north-eastern Brazil, India and the northern half of China. The reasons for this, however, are not always related to macro-climatic factors alone (e.g. Sahel: Zeng, 2003; Hutchinson et al., 2005), but also to human interventions in the environment at regional level, as mentioned above, some of which will be discussed in more detail later (Chapter 7 and Sections 6.2 and 6.3).

Model scenarios depict a clear pattern for the future (Fig. 5.1-4): the high-rainfall regions in the tropical rain belt and the middle to high latitudes will become even more humid overall, while the arid regions of the subtropics will become even drier. The models also single out the Mediterranean region and southern Africa in particular in terms of increasing aridity. For other regions, the projections are less robust; for the Sahel region, for example, different models project major changes, but with contradicting signs (IPCC, 2007a, b).

With regard to precipitation, it is not just annual or seasonal amounts that are of concern, but also extreme weather events such as periods of drought and unusually intense rainfall. Both the data and the model calculations reveal a trend where an increasing share of annual rainfall is concentrated in such intense precipitation events (in other words occurring over a few days), accompanied by a simultaneous increase in the duration of periods without rainfall. This tendency increases both the risk of floods and the frequency of periods of drought.

Based on one model scenario, Figure 5.1-5 shows the regions where a major increase in the risk of drought due to longer periods of dry weather is likely. Again, the regions most implicated in this are the Mediterranean area, southern Africa and Brazil, where drought also extends into the Amazon region. Another aspect of the drought is revealed (based on another model) by the projected dynamics of the climatic water balance (Fig. 5.1-6.). In most continental regions, even in areas where precipitation increases in absolute terms, the climatic water balance declines. There is thus less water available for human use, because the increase in evaporation exceeds the increase in precipitation.

Figure 5.1-7, based on the same model simulations as those used in Figure 5.1-5, shows a rise in the incidence of extreme precipitation in most areas of the world. This increase is probably largely attributable to basic physics, as mentioned above, in other words to the fact that warmer air can hold more water vapour.

5.1.3
Tropical cyclones

Since the extreme hurricane season of 2005, if not before, public interest has focussed increasingly on the risk posed by tropical cyclones. In recent decades a distinct increase in the intensity of tropical cyclones has been observed, probably due primarily to the increase in tropical ocean temperatures (Emanuel, 2005; Webster et al., 2005; Hoyos et al., 2006). The rise in tropical ocean temperatures has so far followed a

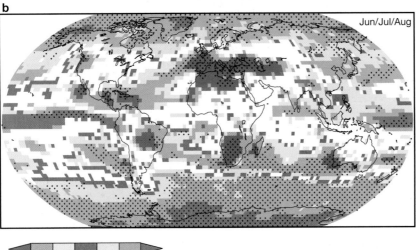

Figure 5.1-4
Relative changes in precipitation towards the end of the century as compared to 1990 under SRES scenario A1B. The figure shows the multi-model averages for two seasons. In the dotted areas, the sign of the projected changes tallies in more than 90 per cent of the models examined; in the white areas, the sign tallies in fewer than 60 per cent of the models. It is therefore difficult to predict the direction of the changes in precipitation in these regions, although considerable changes are nonetheless possible.
Source: IPCC, 2007a

similar pattern to that of the global mean temperature (Fig. 5.1-1) and this will probably continue to be the case.

This means, first of all, that the threat to regions already endangered by hurricanes, such as the Caribbean and coastal areas of China (Fig. 5.1-8), will greatly increase. In these regions, destruction resulting from tropical storms could become so frequent and so severe as to be a massive obstacle to the socio-economic development of the affected countries, particularly the poorer island states.

Second, the areas at risk from tropical storms could expand towards the poles, since the warm temperatures that help to generate hurricanes will be present in an ever-greater area of the oceans. Expansion of the risk zone in the direction of the equator is not expected, because the Coriolis force disappears in the immediate vicinity of the equator and hence cyclones are unable to develop. In recent years, there have already been signs of a possible expansion of the cyclone risk area in the direction of the poles. The year 2004 saw the first ever hurricane to occur in the South Atlantic, namely Hurricane Catarina, off the coast of Brazil. In 2005, two tropical storms took an unusual path, heading towards Europe: Vince, which had weakened by the time it reached the Spanish mainland, and Delta, which caused considerable damage on the Canary Islands.

5.1.4
Sea-level rise

Sea-level rise was discussed in depth in the WBGU Special Report 'The Future Oceans' (WBGU, 2006). For this reason, only the most important aspects are summarized here.

In the 20th century, the sea level rose by 15–20cm. This is for the most part a modern phenomenon, as no sea-level rise that was in any way comparable had occurred in the thousand years prior to this (IPCC, 2007a). This means that the observed rise in sea level must be ascribed primarily to global warming. Other non-climatic, anthropogenic effects also play a lesser

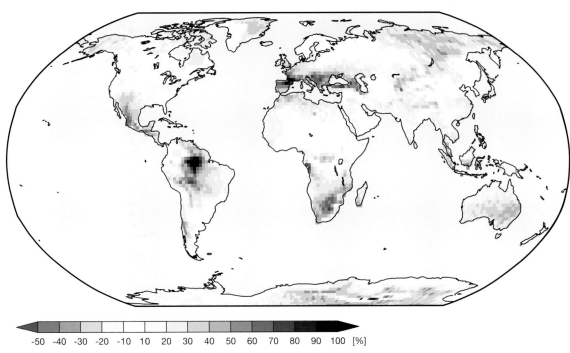

Figure 5.1-5
Percentage change in maximum dry periods under scenario A1B in a simulation by the Max Planck Institute (MPI), Hamburg. A maximum dry period is defined here as the maximum number of consecutive days within a year with a daily precipitation amount below the threshold of 1mm. The figure shows the percentage change in the 30-year mean values for the period 2071–2100 relative to the means for 1961–1990.
Source: MPI, 2006

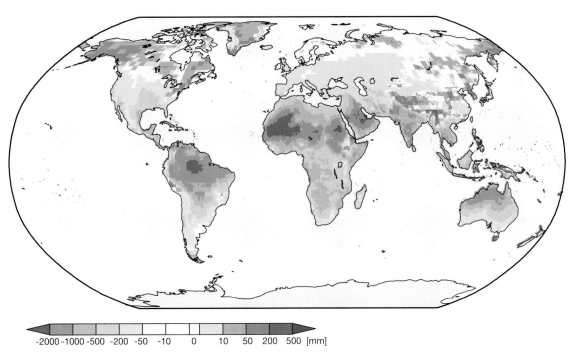

Figure 5.1-6
Future dynamics of drought risk. The figure shows the absolute changes in the climatic water balance from a simulation by the Hadley Centre for the period 2041–2070 compared to similarly simulated data for the period 1961–1990.
Source: WBGU; data: Hadley Centre

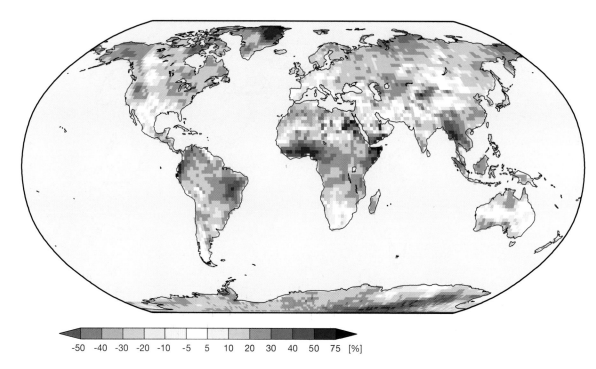

Figure 5.1-7
Percentage changes in annual extreme precipitation under scenario A1B. Annual extreme precipitation is defined here as the maximum amount of precipitation in a 5-day period in a given year. The figure shows the percentage change in the 30-year means for the period 2071–2100 relative to the means for 1961–1990.
Source: MPI, 2006

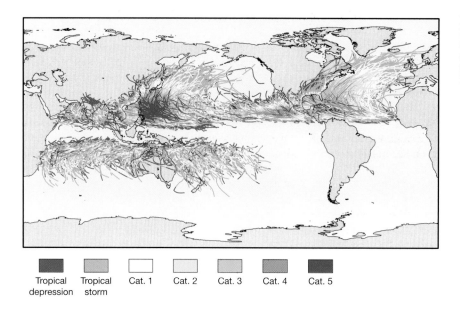

Figure 5.1-8
Risk from tropical storms: storm tracks and intensities over the past 150 years.
Source: WBGU, based on Rohde, 2006

Figure 5.1-9
Global sea-level rise as measured by tide gauge (brown) and satellite (black). Values are given as deviations relative to the 1990 sea level. The dotted lines show the various sea-level projections from the IPCC Third Assessment Report 2001 for comparison, together with the stated range of uncertainty.
Source: Rahmstorf et al., 2007

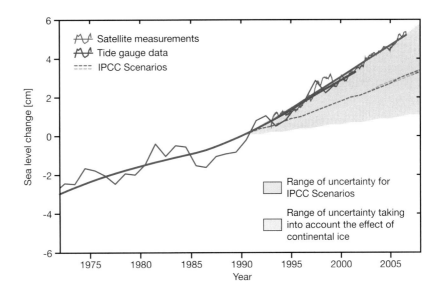

role, such as water storage in artificial lakes and the use of 'fossil' water from underground aquifers for irrigation purposes. Since 1993, accurate satellite data relating to the sea level have become available, and these indicate a rise of around 3mm per year since that date (Fig. 5.1-9). In contrast to coastal tide gauge measurements, satellites provide global coverage and show interesting regional variations. There are some regions, for example, where the sea level is falling, e.g. in the area around the Maldives. However, given that the quantity of water in the Earth's oceans is increasing due to meltwater input and the global average seawater temperature is becoming warmer, the global trend in the longer term is likely to override regional fluctuations and lead to a sea-level rise everywhere. The rise in sea level relative to coasts is a result of the absolute (as seen from satellites) local sea-level rise plus a contribution from uplift or subsidence of land masses. The latter phenomenon is largely an ongoing reaction to the retreat of the ice from the last ice age on the northern continents, and is therefore negligible in the tropics.

The current rise in sea level is due in roughly equal measure to warming of the oceans (thermal expansion) and to meltwater input, although the exact contribution of each remains uncertain, and it also depends on the period under consideration. As regards future sea-level rise, there is particular uncertainty surrounding the contribution of melting continental ice masses. At the same time, however, this presents the greatest potential risk in the long term, because the quantities of water contained in the continental ice masses could cause the sea level to rise by up to 70m. Anthropogenic warming poses a particular threat to the Greenland ice sheet (complete melting of the Greenland ice sheet equates to a sea-level rise of approx. 7m; see Section 5.3) and the West Antarctic ice sheet (sea-level equivalent approx. 6 m). The larger East Antarctic ice sheet is considered stable on the whole, at least for many centuries, even in the event of more dramatic warming.

Without mitigation, a global sea-level rise of around half a metre by the year 2100 can be expected (IPCC, 2007a), and possibly significantly more (Rahmstorf, 2007). Over several centuries into the future, a sea-level rise of several metres is likely.

In terms of pinpointing 'hotspots' of threat relating to sea-level rise, these would include low-lying islands and coastal regions (especially river deltas) on the one hand, and areas where sea-level rise and subsidence of the land mass have an additive effect. To some extent this also relates to anthropogenic subsidence of cities due to physical pressure from buildings and infrastructure and groundwater withdrawal. If the ocean currents change, additional dynamic sea-level changes may occur in some regions. In the event of a breakdown of the North Atlantic Current, the sea level along the northern Atlantic coast could rise by up to 1m.

5.2
Climate impacts upon human well-being and society

The greater the changes in temperature, precipitation and sea level in a given region, the more keenly the impacts of climate change will be felt in terms of human well-being and society. Societies' adaptive capacity also plays a major role, however. The less developed the country, the lower its adaptive capacity. Climate change will thus exacerbate existing poverty due to its impact on ecosystem services, health, incomes, and future growth prospects (WBGU, 2005). The impacts highlighted in Sections 5.2.1 to 5.2.3 below have been identified by WBGU as having particular potential for triggering conflicts; these are discussed in greater depth in the chapter on conflict constellations (Chapter 6).

5.2.1
Impacts upon freshwater availability

Climate change is having an increasing impact on the amount of freshwater available for human use. Freshwater makes up only about 2.5 per cent of worldwide water resources, and more than two-thirds of it is not in readily accessible form, such as glaciers and permanent snow pack; the ice of Antarctica alone stores 60 per cent of the world's freshwater. Groundwater represents a further 30 per cent, while the remainder is to be found in freshwater lakes, soil moisture, in the atmosphere, in moors and wetlands, rivers and living organisms (WBGU, 1998; UNESCO, 2003). In terms of human use, the static amount of freshwater available at a given time is of less relevance than the quantity of sustainably useable, renewable freshwater resources. For example, the amount of water contained in rivers worldwide totals only around 2,000km^3, while annual run-off amounts to more than 40,000km^3. The amount of water withdrawn by humans, for comparison, totals about 3,800km^3 (Oki and Kanae, 2006).

Freshwater is a renewable resource whose renewal rate (and thereby the maximum quantity that can be used sustainably over a set period of time) is determined by the speed at which water circulates in the global hydrological cycle.

As outlined in Section 5.1.2, climatic change influences virtually every element of the global hydrological cycle and thus also the availability of renewable water resources for human use. The most important parameters in this context are changes in precipitation, evaporation and snowmelt. Although climate change intensifies the hydrological cycle, thereby increasing the quantity of water available globally, a significant reduction in available resources may nevertheless arise in particular localities or regions. As a rule, regional water shortages cannot be balanced out at supraregional level; local solutions are required, as transporting water over long distances from regions with abundant water to arid regions (for example by means of channels) is generally only feasible if there is a suitable gradient (Oki and Kanae, 2006).

A variety of indices are used to measure water shortage, water scarcity and water stress. Box 5.2-1 gives an overview of these. Figure 5.2-2 shows the regions where water scarcity, defined by the relationship between water withdrawal and available renewable water resources, occurs at the present time (Oki and Kanae, 2006).

Almost one-third of the world's people currently live in regions where the absolute quantity of water available is less than 1,000m^3 per capita per annum (approx. 1.4 thousand million people, Arnell, 2006; approx. 2 thousand million people, Oki and Kanae, 2006). Projections based on the SRES scenarios of the IPCC show that the number of people living under water stress, as defined above, could increase to 2.9–3.3 thousand million by 2025, rising to 3.4–5.6 thousand million by the year 2055 (Arnell, 2004) as a result of the postulated population increase alone – in other words, without taking into account the consequences of climate change. Climate change increases water stress for the inhabitants of certain regions (such as the Mediterranean region, Central and South America and southern Africa). As a result, depending on which scenario is used, an additional 60–1,000 million people could be affected by 2050. For 700–2,800 million people already affected by water stress, the situation would worsen. In other regions, meanwhile, the number of people affected may fall as a result of climate change. Again depending on the scenario chosen, this could be the case for 90–1,400 million people, and water stress would lessen for 1.8–4.3 thousand million people. This reduction relates almost exclusively to South Asia and the north-west Pacific region. However, as the increase in precipitation predicted by the simulations occurs in the rainy season, and therefore may not produce an increase in available water during the dry season, the benefits of this additional precipitation are limited (Arnell, 2004).

Regional projections of this sort relating to water availability are very imprecise, however, due to uncertainties surrounding precipitation projections. These uncertainties, meanwhile, also depend significantly on the region in question. In some regions, the modelling results deviate so markedly from each other that it is sometimes unclear even whether precipitation is likely to increase or decrease (Section 5.1-2). This does not mean that changes in precipitation will not occur in these regions; it simply means that they are difficult to forecast. This forecasting difficulty may in fact present one of the most major challenges for water management in these regions.

Around one-sixth of the world's population lives in regions where seasonal variability in run-off is determined by snowmelt (Barnett et al., 2005). On the one hand the increase in temperature resulting from climate change increases evaporation, especially in summer, while on the other it means that less of the precipitation falls as snow, and the spring

Box 5.2-1

Water shortages affecting people: Indices

Various indices are used to measure water shortage, water scarcity and water stress affecting people. For the two indices below, comprehensive projections have been carried out that take climate change into account:

- *Per capita water availability (Falkenmark indicator):* This index describes the long-term mean renewable water resources available relative to the population. If available water resources are less than 1,000m³ per capita per year, this is classified as a high level of water stress for the population (Fig. 5.2-1a).
- *Quotient between water withdrawal and available resources:* This index uses quotients between water withdrawal excluding desalinated seawater on the one hand (although current global capacity for seawater desalination is only around 13km³/year (Gleick, 2006), in some regions it is a vitally important resource) and renewable water resources on the other. Water stress is indicated when water withdrawal exceeds 40 per cent of available resources, in other words, when the index is greater than 0.4 (Fig. 5.2-1b).

Using both indices for comparison, Figure 5.2-1 shows the number of people that could face water stress by the end of this century according to the IPCC's various SRES emissions scenarios.

Neither of these indicators can be considered more than a very broad marker of water shortage. They fail to take account of seasonal water availability, and they do not allow representation of other impacts of climate change, such as impacts on water quality.

Besides physical water scarcity, where water supply is so inadequate that current water requirements cannot be met sustainably even using efficient technology, there are other concepts of water scarcity that capture the social and ecological dimensions of the issue as well.

Economic water scarcity affects regions in which there is no hydrological water scarcity, but lack of investment in water infrastructure means that people's water requirements cannot be adequately met, or marginalized population groups lack access to water (e.g. in sub-Saharan Africa and parts of Asia; IWMI, 2007; Fig. 6.2-2).

Ecological water scarcity – in other words, water scarcity as seen from an environmental perspective – is present when withdrawal of water resources for human use is so great that it threatens the integrity of ecosystems, and people who depend on the services of these ecosystems suffer damage (Smakhtin et al., 2004).

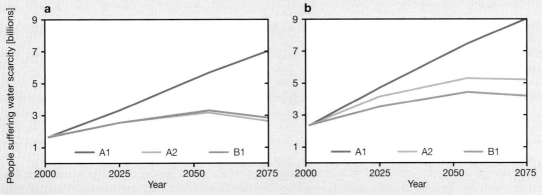

Figure 5.2-1
Projections of populations suffering severe water stress under three of the SRES scenarios. a) Falkenmark indicator: available water amount < 1,000m³/capita; b) Quotient between water withdrawal and available amount > 0.4.
Source: Oki and Kanae, 2006

melt takes place earlier in the year. Consequently, in these regions the amount of run-off increases in winter, but decreases in spring and summer. The consequences of rising temperatures are even more serious in regions where water supply in the dry season is dependent on meltwater from glaciers. In these cases, increased melting of the glaciers initially leads to an increase in run-off. Once the glaciers have melted away, however, the additional water source and the seasonal water storage they represent disappear (Section 6.2.2.1). Besides the impact of climate change on freshwater quantity and temporal availability described here, it can also impair water quality. Higher water temperatures, storm rainfall events or periods of drought can all exert a negative impact by different mechanisms, and increase contamination of water resources with sediments, nutrients, pathogens, etc. (IPCC, 2007b).

Overall, in many places, climate change brings about a situation that presents overwhelming challenges for both existing water infrastructure and current management practice (IPCC, 2007b). Empirical knowledge from the past concerning the natural availability and variability of renewable water resources is no longer applicable to the future (IPCC, 2001; 2007b). This applies especially to the frequency and intensity of extreme events such as droughts and intense precipitation. Empirical knowledge thus no longer provides a meaningful basis for water management planning. Instead, planning may have to rely on

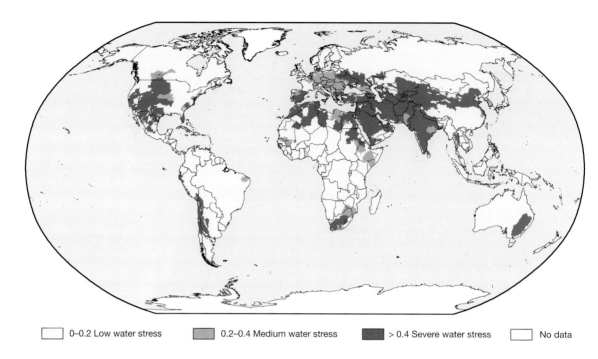

Figure 5.2-2
Current global distribution of the water scarcity indicator (Box 5.2-1). If more than 40 per cent of the renewable water resources in a region are withdrawn by people (indicator > 0.4), then the region is deemed to be suffering from water stress.
Source: Alcamo et al, 2007

model-based regional projections of water yield that take account of climate change. As projections of this sort are fraught with a level of uncertainty that may, as already mentioned, be considerable in the case of some regions, data used for water management planning will become less reliable in many cases.

The climatic impacts described above could lead to a dramatic scarcity of freshwater in some regions. This crisis may in turn fuel existing internal or interstate conflicts and social conflict and heighten competition among different users of the scarce water resources. In certain circumstances the conflict could culminate in violent clashes. This is the subject of the conflict constellation: 'Climate-induced degradation of freshwater resources' (Section 6.2).

5.2.2
Climate change impacts upon vegetation and land use

Climate change and the increasing concentration of CO_2 in the atmosphere affect terrestrial vegetation in a variety of very different ways. Increasing atmospheric CO_2 concentrations can directly impact on the productivity and water use of vegetation (Körner, 2006). The rise in air temperature, too, affects productivity, but it also has an impact on biological diversity and species distribution. Changes in precipitation, and thereby in water availability, influence both productivity and species distribution (Kaiser, 2001). Vegetation responds to extreme events, e.g. extreme temperatures, major or rapid fluctuations in temperature, or extreme wind speeds (Potter et al., 2005; reviews: see e.g. Peñuelas and Filella, 2001; Walther et al., 2002).

As plants require CO_2 for photosynthesis, it was initially assumed that, physiologically, an increase in atmospheric CO_2 concentration would have a major, direct positive effect on carbon fixation (the CO_2 fertilization effect; see e.g. Cure and Acock, 1986; Stockle et al., 1992). While numerous greenhouse-based experiments in the 1980s confirmed the fertilizing effect of increased CO_2 concentrations on crop growth (Cure and Acock, 1986), in the open field, plants adapt to the higher CO_2 concentrations. The opening of leaf stomata, plants' mechanism for gas exchange of CO_2 and water with the atmosphere, is restricted to reduce water loss. Production increases thus tend to be lower, generally below 30 per cent (Körner, 2006). Experimental application of CO_2 gas to crops in open-field production has shown, for example, that grain yields increase by a mere 11 per cent on average instead of the expected 23–25 per cent (Long et al., 2006). The fertilizer effect appears to be much weaker than hitherto assumed in the case of rice, wheat and soybeans, and there is little or no effect on millet and maize. These findings are

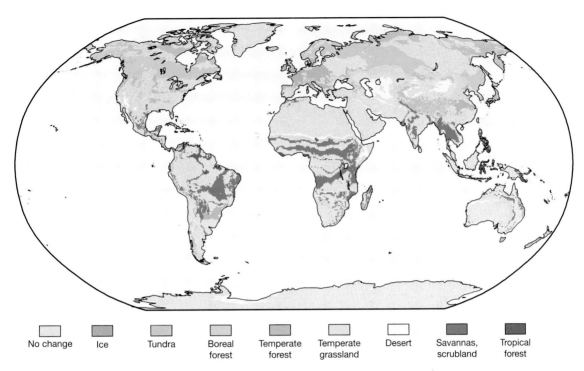

Figure 5.2-3
Terrestrial ecosystems that will be affected by changes in the event of a global average temperature increase of 3 °C (based on HADCM-GCM).
Source: Leemans and Eickhout, 2004

borne out in the case of woodlands too (Norby et al., 2005; Asshoff et al., 2006). Consequently, in global terms, the CO_2 fertilizer effect will compensate to a much lesser extent than previously assumed for the declines in crop yields that are expected as a result of rising temperatures (and concomitant increase in transpiration losses) and decreasing soil moisture (Parry et al., 2004). The fertilizer effect, moreover, may be completely absent in the event of major climate change.

Vegetation distribution also changes in response to changes in temperature and precipitation. An in-situ shift in vegetation may occur if previously non-dominant or subdominant species in a plant community become dominant as a result of the altered environmental conditions, or if species from other ecosystems migrate into the plant community. These two mechanisms presumably also overlap and interact; first, a shift in the species composition within a plant community takes place, permitting alien species to take on a significant functional role in this plant community (Neilson et al., 2005). Bio-geographical vegetation modelling can demonstrate, for example, that exotic alien species may come to dominate the ecosystems of Mediterranean islands in the future (Gritti et al., 2006). Vulnerable species could thus be suppressed or even face extinction (Kienast et al., 1998; Thuiller et al., 2005). This was the mechanism behind the disappearance of some southern ecotypes of widespread Euro-Siberian species from the Mediterranean flora in the period from 1886–2001. Thomas et al. (2004) conjecture that by 2050 some 15–28 per cent of plant species could face risk of extinction due to climate change, but also due to changes in land use.

Based on modelling calculations, it seems unlikely that biomes as a whole will shift as a consequence of climatic changes (IPCC, 2001). Vegetation models show, however, that in the event of an average temperature rise of 3 °C, around one-fifth of the Earth's ecosystems will change, although major variations are to be expected from region to region (Fig. 5.2-3). Expansion of tropical forests and savannahs will remain relatively constant in the event of an increase of 1–3 °C in the air temperature (Leemans and Eickhout, 2004). The negative impact of climate change is more likely to be a result of the increasing water deficit. African tropical forests may respond more sensitively than savannahs to changes in precipitation, because not only do they depend more heavily on the amount of precipitation, but also on the time of year that the precipitation occurs (Hély et al., 2006). Simulations relating to biodiversity in the Amazon rainforest, too, show that by 2095 climate change-induced habitat modification could be so extensive as to threaten the survival of 43 per cent of rainforest

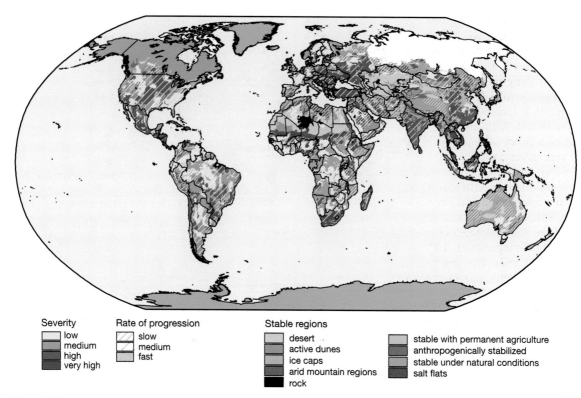

Figure 5.2-4
Global soil degradation by severity and rate of progression.
Source: Oldemann et al., 1991

plant species (Miles et al., 2004). The hardest-hit area in this regard is likely to be north-western Amazonia, because it is here that the most significant changes in precipitation and its seasonal distribution are anticipated (Section 5.3.4).

Species in temperate and boreal forests face a greater degree of change if the atmospheric temperature increases, and they will shift towards the poles. It is expected that the surface area of temperate forests will increase on average, while the tundra is likely to shrink considerably as a result of boreal forest expansion (Leemans and Eickhout, 2004). A meta-analysis of data from various studies on species distribution shows that, on average, the occurrence of the bird, insect and plant species investigated will shift more than 6km polewards and 6m upwards in altitude per decade (Parmesan and Yohe, 2003). Climate change also affects seasonal events such as plant flowering and bird breeding. Even today, the phenological onset of spring in temperate latitudes has already shifted forwards by around 5 days per decade (Root et al., 2003). In Europe, spring is starting earlier by an average of 2.5 days every decade, while the rise in temperature is having a greater impact on the phenological onset of spring in warmer countries (Menzel et al., 2006).

In addition to the impact of climate change on vegetation, environmental changes also occur as a result of land use, and these are no less relevant than direct climate-induced changes (MA, 2005a). Alteration of physical and chemical processes in the soil due to human activities, whether in the past or ongoing, may be significant and sometimes irreversible, often resulting in a deterioration in soil characteristics (Fig. 5.2-4). Inadequate data are available for estimating future soil degradation. Baseline data at global level are patchy or out of date. The best global data on soil degradation are from the GLASOD study carried out between 1990 and 1992 (Oldeman et al., 1991; Oldeman, 1992; WBGU, 1995). Many of the processes that are connected to or triggered by changes in land use are not yet (adequately) represented in global models linking materials cycles to global climate (e.g. agricultural use, urbanization, disruption of natural and semi-natural ecosystems and soils). As a result of this flawed information base, it is almost impossible to make any reliable forecasts regarding future land cover and use. The sections below therefore rely on status-quo data (e.g. Oldeman et al., 1991; Oldeman, 1992; MA, 2005a) or only consider demographic developments up to approximately 2025 (Wodinski, 2006).

Erosion by wind and water is already a major problem, and one that will be aggravated by rising temperatures and a resulting increase in aridity. Where land use has eliminated the dense vegetation cover, thus exposing the soil directly to the effects of the weather, e.g. following forest clearing or overgrazing, desiccation facilitates rapid erosion of the surface layer of soil. With loss of the fertile topsoil, including as a result of landslides that occur due to lack of vegetation cover, nutrients vital for plant growth are lost, and this in turn leads to reduced agricultural productivity. Often, this process of erosion also entails compaction and encrustation of the soil surface. This further reduces the water-carrying capacity of the soil and accelerates soil degradation even more (Schlesinger et al., 1990; Oldeman et al., 1991; Oldeman, 1992). For this reason, erosion is given particular attention as a factor when considering climate change. Wind erosion is already a typical phenomenon in semi-arid and arid zones. It occurs on the fringes of most deserts, where the sparse plant cover is highly sensitive to overgrazing and arable soils are particularly prone to desiccation (e.g. the Sahel region). But the west coast of South America, the Mediterranean region, the Middle East, India and north-eastern China are also affected by wind erosion (USDA, 1998). It can therefore be assumed that susceptibility to soil degradation due to erosion will increase further in future in tandem with higher air temperatures worldwide.

Alongside these physical factors, chemical degradation of soils due to inappropriate agricultural practices also causes loss of organic matter and nutrients, especially in soils where fertility is poor, which in turn is exacerbated by climatic changes. Inappropriate intensive farming often results in soil impoverishment and thereby to a decline in crop yields. A particularly critical problem is posed by soil salinization, which occurs only very rarely in nature. In agriculture, soil salinization occurs primarily as a result of artificial irrigation in semi-arid and arid zones, where the evaporation rate is very high. These are the very regions where even higher or extreme temperatures are to be expected in future, along with highly variable and generally scant annual precipitation. This category includes for example North Africa, the Arabian Peninsula, Central Asia, northern India and parts of Pakistan. In China, Australia and Argentina, Mexico and the west coast of South America, there are also large areas with salinized soils (Oldeman et al., 1991; Oldeman, 1992), and the trend is increasing. In the medium term, the decline in yield can be reined in for a time by switching to more salt-tolerant crops (e.g. maize, wheat), but the scope of agriculture is tightly constrained by plant metabolism. In the long term, however, salinized soils will be irreversibly lost to agriculture and therefore also to food production. This development is particularly relevant in terms of world food supply; even today around 40 per cent of the world's food is produced on irrigated land. With changing climatic conditions, the massive expansion of irrigation agriculture predicted by many observers will entail serious risks.

Soil salinization can also, however, occur in coastal regions and regions with saline aquifers if there is intrusion of seawater or fossil brine into the groundwater. The reason for this, likewise, is inappropriate and excessive use of water (Oldeman et al. 1991).

Whether and how the expected decline in food production due to climate change might bring about food crises and conflicts is discussed in the conflict constellation 'Climate-induced decline in food production' (Section 6.3).

5.2.3
Climate change impacts upon storm and flood events

Even today, storms and floods are responsible for nearly 60 per cent of all natural disasters. Between 2000 and 2006, a total of 1,885 storm and flood disasters were recorded around the globe, causing the deaths of more than 57,000 people. Just under one thousand million people were directly affected by these disasters, and the economic damage totalled nearly US$390 thousand million (CRED, 2006).

In many regions, climate change will increase the risk of floods and storm damage. Reasons for this include the anticipated increase in intense precipitation events (Section 5.1.2) and in higher-intensity tropical cyclones (Section 5.1.3), and sea-level rise (Section 5.1.4).

These changes attributable to climate change are further exacerbated by a number of other factors. For example, deforestation in the upper reaches of rivers leads to a reduction in water storage capacity and thereby to increased surface run-off. In the event of intense precipitation, this can lead to floods (MA, 2005a). The severity of the risk posed to coastal towns by sea-level rise and increasing tropical storm intensity grows in direct proportion to rising population density and concentration of assets and critical infrastructure in many coastal regions. At the present time, between 600 and 1,200 thousand million people, in other words 10–23 per cent of the world's population, live in the vicinity of the coast, i.e. within 100km of the coast and less than 100m above sea level. In the future, the number of people living in coastal areas is expected to rise even further (IPCC, 2007b). Moreover, the subsidence of coastal cities that has been observed as a result of groundwater abstraction and

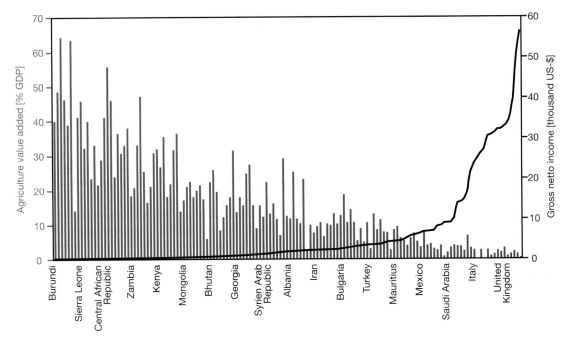

Figure 5.2-5
Share of agriculture in GDP and per capita income (2004).
Source: Stern, 2006

the physical pressure of buildings and infrastructure is contributing to exacerbate the associated problems (Nicholls, 1995).

Overall, the factors described here could, in conjunction with the changes in precipitation, tropical storm intensity and sea level induced by climate change, lead to an increase in storm and flood events, especially on islands and in coastal regions. If disaster management on the part of government institutions fails in a situation of this sort, collapse of the infrastructure and a growing humanitarian disaster could bring about a crisis. This crisis in turn could exacerbate existing internal and social conflicts, culminating in violent clashes. This is the subject of the conflict constellation relating to storm and flood disasters, which is discussed in Section 6.4.

5.2.4
Indirect economic and social impacts of climate change

Over and above the direct impacts of climate change – e.g. freshwater availability, vegetation, soil degradation, frequency and intensity of storm and flood events – there are other ways in which it can affect the economy and society.

5.2.4.1
Climate change impacts in selected economic sectors

Climate change affects many important sectors of the economy by influencing the supply of and demand for goods and services. One of the sectors most severely affected is agriculture (Section 5.2.2). Since agriculture accounts for a large share of the economy in many developing countries, these countries will be particularly hard hit (Fig. 5.2-5).

However, forestry and fisheries, industrial sectors (e.g. energy and construction sectors) and services (e.g. insurance sector, tourism) are also affected. The aggregate effects on many of these sectors have not been widely researched and are therefore poorly understood (IPCC, 2007b).

In the event of a temperature rise of between 1 and 3 °C (compared to 1990), global agricultural production is initially likely to increase overall, because higher yields in higher-latitude regions (e.g. Canada, Russia and Scandinavia) will more than compensate for declines in production in many developing countries. Significant declines will be noted particularly in Africa, because less land will be available for agricultural uses in arid and semi-arid regions, the length of the crop production period will be reduced and potential yields will decline. In some sub-Saharan countries, yields from rain-fed crop production could fall by up to 50 per cent by 2020 (IPCC, 2007b). If the global mean temperature rise exceeds 2–4 °C, it is likely that agricultural production will decline worldwide. If the temperature increase exceeds 4 °C, serious impairment of global agriculture may be anticipated (IPCC, 2007b). Hardest hit by these changes will be the millions of smallholders in developing countries whose agricultural output is geared largely towards subsistence. While more affluent income groups in developing countries may be in a position to compensate for the decline in domestic food production by means of imports, production losses affecting subsistence farmers bring about direct food problems. Almost all the projections assume that world market prices for grain will rise at the latest when warming reaches the 2 °C threshold (see e.g. Adams et al., 1995; Fischer et al., 2002). If warming reaches 5.5 °C or more, prices may rise by as much as 30 per cent on average (IPCC, 2007b). Climate change exerts direct negative effects on livestock production, forestry and fisheries by impinging on the quality and surface area of pasture, soil (Section 5.2.2) and limnic and marine ecosystems (IPCC, 2007b).

How the dynamics of global net energy use will develop with increasing global warming remains unclear (IPCC, 2007b). For example, increasing use of air-conditioning systems will cause a rise in demand for energy, while there will be a decrease due to falling heating requirements. The higher temperatures rise, the more likely it is on balance that global energy demand will increase, because the increasing demand for cooling systems is likely to exceed the decline in demand for heating (Hitz and Smith, 2004). Climate change will probably also influence energy production and thereby energy supply. For the time being, however, research findings relating to this are thin on the ground. Possible influencing factors include for example the increase in extreme weather events and dependency of some regions on hydropower for generating energy (negative influence), or changes in conditions for energy derived from biomass, wind or solar sources (positive or negative influence; IPCC, 2007b). One concrete example illustrating the immediate relevance of extreme weather events for energy production was Hurricane Katrina in 2005: in the wake of Katrina, loss of refinery capacity led to a global rise in fuel prices. Increased use of biomass for energy production, moreover, could lead to a rise in food prices.

As climate change brings with it an increase in the frequency of extreme weather events, the risk of damage to property and infrastructure also rises. Insurance companies will need to significantly increase the amount of capital they hold to be able to provide insurance cover at a level comparable to today. For example, given an increase in storm intensity of 6 per cent, as predicted by many climate models for a rise in temperature of around 3 °C, the capital requirement of insurers for hurricanes in the USA would have to increase by more than 90 per cent (Association of British Insurers, cited in Stern, 2006). It is therefore foreseeable that the insurance market will grow, whereby premiums for insuring against climate-related losses are likely to rise and certain risks will increasingly be classed as no longer insurable (IPCC, 2007b).

In the tourism sector, too, some shifts will occur as a direct or indirect consequence of climate change (see e.g. Hamilton et al., 2005). As a direct effect of climate change, tourism in the higher-latitude regions will become more attractive. In regions where it is already hot and dry, however, visitor numbers may decline. Islands and coasts may become less attractive as a result of sea-level rise and the increase in extreme weather events, while at higher altitudes areas that are ideal for skiing-related tourism may suffer from lack of snow. In addition, the tourist industry may be subject to indirect climate change effects relating to drinking water availability, the cost of ambient cooling systems, and climate-induced landscape changes (IPCC, 2007b).

5.2.4.2
Climate change impacts on the global economy

Estimations of the overall economic costs of climate change vary considerably depending on the assumptions regarding the future and the methodological parameters used (e.g. climate sensitivity, discount rate and regional aggregation). Most models nevertheless come to the conclusion that an increase of just a few degrees Celsius could result in a global loss of welfare in the order of several per cent of global GDP (IPCC, 2007b).

According to the calculations of Stern (2006), in the event of unabated climate change these losses will reach at least 5 per cent, and in all probability will far exceed this figure. Welfare losses, according to Stern, could reach as much as 20 per cent of GDP if high climate sensitivity is assumed, if the calculation takes greater account of damage to non-market goods (such as human health and biodiversity) and the risk of disasters due to positive biophysical feedbacks in the climate system, and if the relatively higher welfare losses of poorer countries are given greater weight in absolute terms. Of course the procedure adopted by Stern, like those employed in other cost estimates, has not been methodologically proven, and the normative assumptions may always be queried. Estimates like these nonetheless give reference points regarding the quantitative extent of the potential impact of unabated climate change on the economy and development. Although Stern's figures tend to be at the upper end of the scale compared to other estimates currently circulating, even his quantitative estimates fail to include the economic upheavals that would arise as a consequence of climate-induced conflicts or might be triggered by climate-induced migration.

Global aggregation of climate-related welfare losses does not show how variable these losses may be from one region to another. Regions in the lower latitudes will tend to be worse affected. All of the regional studies concur in their estimations that Africa will suffer GDP losses of several percentage points in the event of a temperature increase of just 2 °C compared to 1990; losses in some smaller countries may reach as much as 30 per cent (IPCC, 2007b).

5.2.4.3
Climate change impacts on society

There is a danger that unabated climate change could cause an additional 30–170 million people to suffer from malnutrition or undernutrition by 2080. Three-quarters of these people will live in sub-Saharan Africa (IPCC, 2007b), because in this region the impacts of declining agricultural output, a high dependency on agriculture and a low purchasing power act together (Stern, 2006). According to Stern (2006), in South Asia and sub-Saharan Africa alone, an additional 145–220 million people could fall below the US$2-a-day poverty line. This figures illustrate the extent to which climate change can jeopardize chances of meeting the Millennium Development Goals.

This is also particularly clear with regard to the impact of climate change on human health. Already in the year 2000, climate change was estimated to be responsible for 0.3 per cent of deaths (150,000 people) worldwide (IPCC, 2007b). In 2100, some 165,000–250,000 more children may die than would be the case in a world without climate change (Stern, 2006). At global level, moreover, it is projected that, as temperatures rise, so too will detrimental impacts on health resulting from undernourishment, disease and injury. Causes for this include heatwaves, floods, storms, fires and drought. Higher atmospheric concentrations of ozone and allergens, for example, could give rise to cardiovascular diseases (IPCC, 2007b). In addition, climate change will bring about an expansion of malaria into previously uninfected areas. It is possible that the incidence of malaria will increase overall. Scientists are still discussing the quantitative evidence in this regard, because the spread of malaria also depends on a large number of factors unrelated to climate (IPCC, 2007b).

The distribution of these health risks will exacerbate existing regional disparities. Although fewer people will die of cold in the temperate latitudes, the negative health impacts of the rising temperatures will predominate, especially in developing countries. In East, South and South-East Asia, for example, there will be an increase in diarrhoea-related mortality connected with floods and droughts. In general, the hardest hit groups in all countries will be the urban poor, older people and children, traditional societies, subsistence farmers and residents of coastal regions. Education level, extent of health care provision and infrastructure, and level of economic development will be critical factors in terms of the health situation (IPCC, 2007b). Moreover, millions of people will face the risk of migration due to undernourishment, lack of access to drinking water and settlements in areas prone to flooding. Twenty-two of the 50 largest cities in the world are located in coastal regions, where there is a direct increase in the risk of floods resulting from climate change, especially if, as in New Orleans, they lie below sea level. Around the globe, nearly 200 million people currently live in endangered areas close to the coast that are prone to flooding, with 60 million living in South Asia alone (Stern, 2006).

Research on the direct impacts of climate change on human society has already advanced. Investigation into its impacts on institutions, governance structures and processes, on patterns of social and economic distribution, and on capacity for action in policy terms, however, is still in its infancy. It is vitally important to increase understanding of these contexts in order to adequately comprehend the impacts of climate change on the stability, security and welfare of societies and regions.

5.3
Non-linear effects and tipping points

So far this chapter has discussed climate system developments that are largely continuous and predictable, and that have a high or extremely high likelihood of occurring. In the event of high global warming, however, beyond 2–3 °C the risk of additional, qualitative changes occurring in the climate system increases. Such strongly non-linear responses by system components are often referred to as 'tipping points' in the climate system. This term is used to refer to the behaviour of the system when a critical threshold has been crossed, triggering runaway changes that are then very difficult to bring under control again. Broad-scale parts of the Earth System capable of triggering such instability have been termed 'tipping elements' (Lenton et al., 2007). Examples include the Green-

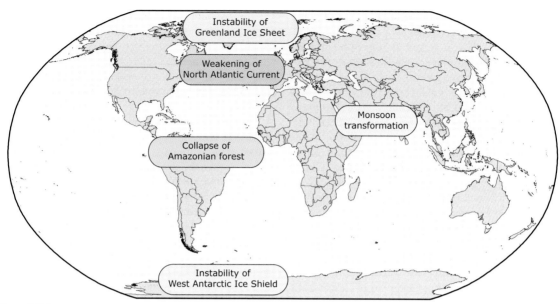

Figure 5.3-1
Map with some of the tipping elements in the climate system discussed in the text.
Source: WBGU, based on Schellnhuber et al., 2005

land ice sheet, which, via positive feedback mechanisms, could potentially start to slide and disintegrate if warming exceeds a critical threshold, and the Amazon rainforest, which could die back irreversibly if a critical point is crossed (Fig. 5.3-1). Climate history shows that the climate system is indeed capable of drastic and rapid changes (see e.g. Rahmstorf, 2002).

Non-linear phenomena like these are generally much less easy to predict than smooth trends. This is why there are no robust projections or systematic risk assessments available to date, just one-off studies that generally only convey a qualitative impression of the risks involved. The considerable uncertainties do not, however, mean that these are negligible risks that society can choose to ignore.

5.3.1
Weakening of the North Atlantic Current

The best-known example is the Atlantic thermohaline circulation, which is likely to be significantly weakened both by warming itself and by the increase in freshwater flowing into the ocean as a result; if the freshwater inflow is sufficiently large, the thermohaline circulation in the Atlantic could break down completely. It is not yet possible to quantify objectively the likelihood of such an eventuality, since climate models are not yet able to capture some of the relevant processes, e.g. the danger of a massive inflow of freshwater from the Greenland ice sheet. Risk assessment is therefore based on expert opinion, taking into account all modelling studies, current trends in the data from measurements and information from Earth's history. The most recent report of the IPCC (2007a) describes the likelihood of the thermohaline circulation weakening in this century as 'very high' in the event of unabated climate change, while the probability of its complete collapse is estimated at up to 10 per cent. In a detailed survey of experts carried out recently (Zickfeld et al., 2007), subjects were asked to assess the likelihood of a complete breakdown of the thermohaline circulation being triggered in this century as a result of various levels of warming by 2100. Given warming of up to 3 °C, most of the 12 experts questioned agreed with the assessment of the IPCC (a risk of up to 10 per cent); two of the experts consider the risk to be higher, at 30 per cent and 40 per cent respectively. Many experts predict a sharp increase in the risk if warming is higher: with warming of 5 °C, half of the subjects predict a risk of 20 per cent or more, and one-third estimate the risk at 50 per cent or more.

At the present time it is only possible to assess in broad outlines what the consequences of such far-reaching reorganization of the global system of ocean circulation might be. There appears little reason to fear a marked cooling below present-day temperatures (the 'little ice age scenario' investigated in the Pentagon report; Schwartz and Randall, 2003), because warming resulting from greenhouse gases will probably more than compensate for the effect of diminishing heat transport in the ocean. There are, however, two circumstances in which regional cooling could occur, especially in north-western Europe: first, if, contrary to expectations, the current changes very early on, before warming due to the greenhouse effect has reached several degrees; and second, centuries into the future, if warming due to greenhouse gases abates but the current has broken down permanently. Other consequences include for example an additional sea-level rise of up to 1m in the North Atlantic region (Levermann et al., 2005), a shift in the tropical rain belts (Peterson and Haug, 2006; Zhang and Delworth, 2005), and a possible collapse of the North Atlantic marine ecosystem (Schmittner, 2005), currently one of the most productive marine regions on Earth. Even in the absence of regional cooling, therefore, the consequences of a breakdown in the Atlantic thermohaline circulation could be so severe that the livelihoods and food situation of many people in the North Atlantic region and in the tropics would deteriorate significantly.

5.3.2
Monsoon transformation

Akin to the shifts occurring in the ocean circulation, massive shifts could also take place in the atmospheric circulation. In this context, the discussion focuses primarily on the monsoon circulation. As a consequence of the regularity of the monsoon rains and their vital contribution to total annual precipitation, the agricultural sector in some regions is highly dependent on the monsoon, making these regions highly vulnerable to any changes affecting it (Webster et al., 1998). Large parts of the Indian subcontinent, for example, depend on the monsoon rains. The summer monsoon accounts for as much as 90 per cent of annual precipitation in some regions (Lal et al., 2001). It is known from paleoclimatological research that phases of weaker monsoon activity have occurred in the past (Gupta et al., 2003).

As is the case with ocean circulation, atmospheric circulation patterns such as the monsoon circulation respond in a markedly non-linear manner to external disruptions. In the case of the monsoon, contamination of the atmosphere with aerosol particles is probably an even more important control parameter than increased greenhouse gas concentrations (Zickfeld et al., 2005). These two parameters are currently acting in opposite directions: increasing aerosol pollution is weakening the monsoon, while the rising CO_2 concentration is tending to make it stronger. Any significant change in the monsoon – weakening or strengthening, or more marked fluctuations and thereby a loss of predictability – can have major consequences for agriculture and thus also for the food supply of people in Asia. There are signs, moreover, that the various circulation systems in the oceans and the atmosphere have mutual influences; research in this area, however, is still in its infancy.

5.3.3
Instability of the continental ice sheets

A third example is the behaviour of the continental ice sheets, melting of which would lead to a sea-level rise of several metres. Via a process called 'ice elevation feedback', ice sheets have a well known critical threshold: within certain limits, an ice mass such as the Greenland ice sheet maintains its own equilibrium, because its thickness of approximately 3km ensures that a large part of the surface of the ice is located in higher and therefore cooler atmospheric layers. Atmospheric temperature decreases by an average of 6.5°C with every additional kilometre of altitude. When the ice sheet shrinks, its surface is exposed to increasingly warmer air, thereby triggering a positive feedback that ultimately culminates in the ice sheet's complete disappearance.

There is a whole series of other positive feedback mechanisms that can lead to acceleration of ice sheet disintegration if warming goes beyond a critical point (WBGU, 2006). Measurement data have shown, for example, that glacial flow in Greenland is speeding up significantly, one reason being that meltwater from the surface of the ice is percolating through to the glacier bed via holes (called moulins) and acting as a lubricant (Zwally et al., 2002). In Antarctica, too, there are increasing indications that the ice may be responding dynamically, especially the smaller West Antarctic ice sheet. In February 2002, the several thousand-year-old Larsen B ice shelf collapsed and disintegrated off the Antarctic Peninsula. As ice floats on the sea, a shelf's disintegration initially has no impact on the sea level. However, it apparently does affect the continental ice sheet. The ice streams discharging from the ice sheet in the wake of the Larsen B ice shelf have since accelerated considerably, reaching up to eightfold velocity (Rignot et al., 2004; Scambos et al., 2004).

The risk associated with the collapse of continental ice sheets is due especially to the fact that it entails a major sea-level rise. Melting of the Greenland ice sheet alone would increase the global sea level by around 7m. This could already happen with global warming of 2–3°C compared to 1990 temperatures (Gregory et al., 2004; Ridley et al., 2005). There is still considerable uncertainty regarding the timeframe in which melting of this nature might take place. Up until a few years ago, it was assumed that it would take a few thousand years. Recently discovered mechanisms, however, raise the possibility that the ice sheet could melt completely within centuries (Hansen, 2005).

5.3.4
Collapse of the Amazon rainforest

Strongly non-linear effects could arise in the biosphere, too, in response to climate change. The net primary production of vegetation falls by 10–20 per cent in periods of drought in El Niño years compared to the long-term average (Potter et al., 2001). As a result, the ecosystem turns into a source of carbon dioxide in the short term. A modelling study by Cox et al. (2004) that has been the subject of heated debate among the experts sets out a scenario in which 65 per cent of the Amazon rainforest dies back by 2090 as a consequence of global warming. Favourable conditions for these droughts are created by warm ocean temperatures in neighbouring regions of the Atlantic and the Pacific (El Niño conditions), and this is also true of the droughts observed in the Amazon region. Debate on this scenario has sharpened further because in 2005 the increasing droughts experienced by the Amazon region reached their highest level so far. In this scenario, positive feedback between biosphere and atmosphere plays a crucial role, because the high level of evaporation resulting from the presence of the Amazon rainforest causes increased rainfall in the region. This in turn sustains the rainforest. More than half of the precipitation in the Amazon region is 'recycled' in this sense. An experimental study in which water was diverted from the roots of the trees by means of plastic panels has recently demonstrated that trees respond highly sensitively to even just a few years of drought (Nepstad et al., 2002). However, there are other model calculations that show the sensitivity of rainforest vegetation to changes in precipitation in more relative terms. Cowling and Shin (2006) go as far as to suggest that a decline in precipitation of 60 per cent in the tropical rainforest of Amazonia would only have a minimal effect on net primary production and heterotrophic respiration. According to their model calculations, the stability of the 'recycling' process between vegetation and precipitation would not be jeopardized unless there was an 80 per cent reduction in precipitation, causing a major decline in net primary production.

Changes in land use (e.g. forest clearance) also need to be taken into consideration in the discussion on feedback mechanisms. Deforested areas have a major feedback effect on local climatic conditions due to their reduced evaporation rate, diminished groundwater storage capacity and increased reflection of solar irradiation (Morton et al., 2006).

In addition, droughts in open landscapes in the Amazon region (on the forest perimeter, clearings, savannahs) intensify competition between grassland and forest in favour of grasses, and this produces vegetation that has less capacity for carbon storage (Botta and Foley, 2002). Grasses also favour the spread of wildfires, which in turn release fixed carbon into the atmosphere.

A massive transition in the carbon balance of the Amazon rainforest system that converted it from carbon sink to source would not only entail consequences for the region as a whole and cause the loss of animal and plant species; it would also result in considerable carbon dioxide being released into the atmosphere by the vegetation and soil (including as a result of fire), amplifying the greenhouse effect still further. One current comparative model found that feedback of this sort from the biosphere would probably bring about an additional global CO_2 increase of 50–100ppm (in the most extreme model, up to 200ppm) by the year 2100 (Friedlingstein et al., 2006). Analysis of climate history data too suggests that feedback mechanisms of this sort in the carbon cycle can exacerbate the greenhouse problem (Scheffer et al., 2006).

5.3.5
Conclusions

Of all the impacts of climate change discussed in Chapter 5, the non-linear effects are among those that may have the most far-reaching negative consequences for human lives and societies. A radical change in the monsoon circulation or the collapse of the Amazon rainforest could fundamentally alter agricultural production in Asia and Latin America, incurring incalculable economic costs and triggering migration movements. A major weakening of the thermohaline circulation could have unforeseeable global consequences for habitats and markets. The tipping points outlined here are therefore associated with large-scale and complex risks of a social, economic and political nature that affect whole regions of the world. The level of uncertainty concerning these tipping points is also exceptionally high. Preventing the associated non-linear effects will first and foremost require curbing further climate warming. This is an urgent necessity as it is, in view of the climate-induced environmental changes that are slowly but progressively occurring. In this sense, consideration of the tipping points 'just' gives one more piece of ammunition in support of the argument that comprehensive and resolute climate protection is absolutely vital.

Conflict constellations 6

6.1
Methodology

6.1.1
Selection and definition

The conceivable forms of conflict driven by climate change can be grouped into typical constellations. Their characteristic configurations have the potential to emerge in various regions of the world. WBGU defines conflict constellations as causal linkages at the interface between the environment and society, which interact dynamically and are capable of inducing social destabilization or violence.

Various sources have been consulted in the construction of these conflict constellations: modelling of anticipated climatic changes; experiences of environmental conflict; findings from causal research into war, conflict and the dynamics of violent conflict escalation; and analytical studies on state fragility and multipolarity (Chapters 3 and 4). The aim is to identify key factors and mechanisms for the eruption of violence in the context of environmental change, with a view to deriving recommendations for action on conflict avoidance and resolution.

WBGU identifies three major areas in which climate change is expected to cause critical developments, namely the depletion of freshwater resources, the impairment of food production, and an increase in weather extremes. Also, as an indirect consequence of these developments, migration may increase in the affected regions. Informed by these considerations, WBGU derives the following conflict constellations, and proceeds to analyse each constellation as to whether it can trigger or amplify conflicts and social destabilization.

- *Conflict constellation 'Climate-induced degradation of freshwater resources':* Over 1.1 thousand million people lack secure access to sufficient drinking water. The situation may deteriorate further in some regions of the world as climate change will cause greater variability in precipitation. This will have impacts on the availability of water resources (Section 6.2).
- *Conflict constellation 'Climate-induced decline in food production':* Currently over 850 million people in the world are undernourished. This situation can become more acute as a result of climate change, especially in developing countries. Any decline in regional or supraregional food production may lead to food crises (Section 6.3).
- *Conflict constellation 'Climate-induced increase in storm and flood disasters':* Natural disasters and the damage caused by such incidents have increased dramatically in the last four decades. There is a particular likelihood of increased storm and flood disasters as a result of climate change (Section 6.4).
- *Conflict constellation 'Environmentally induced migration':* Changes in the environment as a consequence of climate change are expected to contribute to increased migration in future (Section 6.5).

6.1.2
Using narrative scenarios to identify security risks

On the basis of the conflict mechanisms identified, possible trends towards destabilization and violence are worked through in the form of geographically explicit, narrative scenarios. The ultimate aim is to derive preventive strategies (Box 6.1-1). For each constellation, both a 'fictitious confrontation scenario' and a 'fictitious cooperation scenario' are developed, and the strategic bifurcations which differentiate these scenarios are identified. A particular focus of analysis is the consequences of those developments, actions and decisions which produce lock-in effects. This method, i.e. the use of scenario technique to 'play through' potential future developments, is well established as a standard technique used by large corporations, e.g. insurance companies, and by military strategists. The goal of these scenarios is not to put forward the most accurate possible vision of the future (scenarios are not forecasts).

> **Box 6.1-1**
>
> **Scenarios and forecasts**
>
> Forecasts are scientifically derived statements on the probable course of future events within a given time span. The more complex the field of study and the longer the time span, the more uncertain forecasts become. In these circumstances, scenarios are the preferred method; on the one hand they point up the range of possible future developments and focus attention particularly on causal processes, influencing factors and strategic decision points. Scenarios, unlike forecasts, consist of a hypothetical sequence of events which is constructed in order to illustrate causal linkages. They are not backed up with probabilities. There are different types of scenario. The typical steps in creating scenarios are
>
> – defining the parameters (type of scenario, time span, etc.),
> – scenario field analysis, including identification of influences and analysis of key factors,
> – scenario projection, including assessment of how key factors will develop,
> – condensing the scenarios into consistent narratives,
> – and evaluation of the scenarios.
>
> In the report, narrative scenarios are developed which 'tell the stories' of possible future developments along alternative trajectories. As far as possible, they are based on forecasts about individual influencing factors (e.g. changes in the climate). Other influencing factors are deduced from empirical findings in relation to other processes. The aim is to envision how developments might potentially unfold and to identify pointers for strategic political choices at critical moments.

Instead, they are intended to contribute to identifying, and hence avoiding, the most risk-laden trends. The selection of regions for the scenarios is derived from the analyses in Chapter 5. The scenarios begin around the year 2020. Their time horizon extends to the year 2050, and in certain instances as far ahead as 2100. Some of the risks relating to the second half of the century have a low probability but high destructive potential (tipping points, Section 5.3).

CLIMATE CHANGE UP TO MID-CENTURY LARGELY PREDETERMINED

Until around the middle of the century (2020–2050) it is hypothesized that climate change will consistently advance, forming the 'backdrop' to possible socio-economic developments. Because of the inherent response times of these dynamics, there may already be drastic regional climate impacts, e.g. an increase in the frequency or severity of extreme events. The exact nature of the repercussions on the ground is determined by bifurcations in the socio-economic arena, e.g. whether or not a trend is identified and corresponding adaptations are made, or whether general political and social stability improves or deteriorates. During this period, the scale of climate change itself can only be influenced within narrow margins.

NON-LINEAR ENVIRONMENTAL CHANGE POSSIBLE FROM MID-CENTURY

In developing scenarios for the middle of the century onwards, it is assumed that climate change will begin to take on a new quality: in this period, its scale will depend critically on the decisions and actions taken in the first half of the century. In principle, different trajectories are conceivable for the development of environmental change. Moreover, under conditions of intense climate change, possible non-linear climatic dynamics such as the failure of the Indian monsoon, shifts in the Inter-Tropical Convergence Zone or the weakening of the North Atlantic Current may lead to further bifurcation of developmental trajectories. Effectively this would represent dramatic alteration of the biogeophysical 'backdrop' against which socio-economic developments are played out. It would pose major social challenges which could bring forth either cooperation or conflict. Section 5.3 presents some examples of non-linear environmental trends with a low probability of occurrence but extremely high destructive potential.

6.1.3
Deriving recommendations for action

For every conflict constellation and its associated scenarios, recommendations are derived for action which may help to avoid or alleviate the given set of problems. These include not only strategies for avoiding and adapting to climate change, but also aspects of good governance which make a difference to crisis prevention. Although some of the repercussions considered will only occur in the distant future, the aim is to develop recommendations for action for the present and the near future.

6.2
Conflict constellation: 'Climate-induced degradation of freshwater resources'

6.2.1
Background

6.2.1.1
Brief description of the conflict constellation

The conflict constellation 'Climate-induced degradation of freshwater resources' explores the causal linkages between environmentally induced changes in water availability and conflicts over water, destabilization of communities and migration. It deals primarily with changes in water availability which are caused either directly or indirectly by climate change (Section 5.2.1) and the scale or timing of which lead to the assumption that the necessary adaptation could overwhelm water management systems. The conflict constellation is based on the assumption that the adaptation process itself (that is the restructuring of water management), and in particular a failure of water management systems to adapt to the new situation (that is a massive degradation of the water supply), harbours potential for conflict. This could trigger or aggravate intrastate or international conflicts of interest, which under adverse conditions could erupt into violence. Conflicts of interest over freshwater resources have seldom led to violent interstate conflict ('water wars') in the past. In fact, on many occasions intergovernmental agreements have even been the inspiration for cooperative solutions (Section 3.1). A number of examples do exist, however, where local or regional conflicts over water have turned violent, particularly in cases where no formal rules or agreements on the use of the water resources had been agreed (Horlemann and Neubert, 2007). As such conflicts increase in frequency, so too does the risk of provoking or worsening social destabilization (Chapter 3; Carius et al., 2006).

This conflict constellation is expected to arise more frequently in future for two reasons. First, climate change will impact on the water balance in many regions to such an extent that its availability will deteriorate in terms of quantity or seasonal distribution. Second, the mounting aspirations of the world's growing population will substantially increase the demand for water. In some regions this steadily widening gap is already prompting considerable additional social conflict, which can contribute to destabilization. These causal linkages are analysed below.

6.2.1.2
Water crises today and tomorrow

Overall freshwater use has increased almost eightfold over the last century (Shiklomanov, 2000), and is continuing to grow by about 10 per cent each decade. Humans already utilize or control over 40 per cent of the renewable, accessible water resource (MA, 2006). A distinction is made between 'green' and 'blue' water. Green water is the moisture in the soil which is sustained by precipitation and is available for rainfed agriculture. Blue water is the more easily accessible surface water (lakes, rivers) and groundwater which can be tapped for irrigation, industrial purposes and domestic use (IWMI, 2007). Allowing for significant regional differences, agriculture accounts for approximately 70 per cent of the world's 'blue' water use, industry about 20 per cent and domestic users the remaining 10 per cent (Cosgrove and Rijsberman, 2000; Fig. 6.2-1). The significance of agriculture, especially irrigated farming in the developing countries, makes water a key element for food security and poverty alleviation (WBGU, 1998; Section 6.3). Moreover, hydroelectric power accounts for about 19 per cent of all the electricity generated worldwide, and in some countries the proportion is over 90 per cent (WBGU, 2004).

Increasingly, the problem is not simply overuse, but also water pollution. Agriculture (salination, input of nutrients and sediments), industry and private households (nutrients and pollutants) put lakes, rivers and coastal waters under stress, causing significant health and development problems – currently these are most conspicuous in China (Section 7.7). Water is also crucial to the functioning of natural ecosystems. Competition between water-use sectors is apparent here. The growing demand for water for human needs has a profound effect on aquatic ecosystems in particular. These are considered to be highly endangered, in turn causing the failure of vital ecosystem goods and

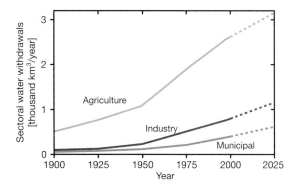

Figure 6.2-1
Water use by sector.
Source: IWMI, 2007

> **Box 6.2-1**
>
> **Integrated water resources management**
>
> Preventing supply bottlenecks and water crisis and ensuring an optimal use of water resources is a task requiring effective water policy and a sustainable water management system (WBGU, 1997; Dalhuisen et al., 1999). The concept of integrated water resources management (IWRM) has taken on significance in this respect. IWRM seeks to move beyond previously isolated sectoral and often inefficient usage systems. Water, land or soil and their associated resources should be developed in a sustainable fashion and administered in one cross-sectoral, participatory process. IWRM is a relatively new concept inspired and promoted mainly by international forums of research and water policy (especially the Global Water Partnership; GWP, 2000). It offers strategic points of reference for solving water crisis, despite persisting conceptual weaknesses and the circumstance that the political dimension is not yet adequately taken into consideration. The term 'water governance' is also used parallel to IWRM, and sometimes at a higher strategic level (UNESCO, 2006).
>
> Under IWRM the various tools and options for both supply and demand are clustered to form a holistic strategy which is customized to suit local circumstances (UNESCO, 2003, 2006). Technical options are targeted primarily at expanding availability or improving the quality of water. For instance, storage in natural and man-made reservoirs can balance out seasonal fluctuations, additional water resources can be made available by using groundwater reserves or desalinizing sea water, and wastewater treatment facilities can ensure adherence to pollutant limits. With large-scale engineering solutions especially, however, it can be difficult or even impossible to comply with aspirations of sustainability, e.g. when major dam projects are contingent upon large-scale human resettlement or substantial environmental damage (Box 6.2-2). The current focus of water policy and IWRM is shifting from supply to an improved use of 'green water' and integrated ecosystem approaches (IWMI, 2007). IWRM should ensure the efficient and environmentally sustainable use of water (the economic and ecological dimension), the equitable allocation of water resources and the maintenance of democratic opportunities for participation (the social and political dimension; UNESCO, 2006). IWRM is therefore embedded in the broad debate on both poverty reduction and equitable and sustainable development (WBGU, 2005; Neubert and Herrfahrdt, 2005).
>
> In spite of significant efficiency reserves and promising options, the water sector will not be able to solve the problems alone. Reforms in other sectors, such as agriculture, human settlement and industry are needed in order to ensure sustainable water resources management. A rational price policy, geared towards the cost of provision of the resource, or at least reflecting its scarcity, is vital (Dalhuisen et al., 1999). Minimum standards for individual access to drinking water should be taken into account (WBGU, 1998). Agricultural policy should focus on rain-fed agriculture (green water resources) and utilize the precious resources of blue water as sparingly as possible. This can be achieved by improving the coordination of water usage in terms of timing, raising the efficiency of water use and managing demand with a targeted pricing policy. Price signals and awareness-raising initiatives are also suitable tools for promoting water-saving technologies in private households, supporting process water recycling in industry, or reducing the rates of loss in the pipe infrastructure.

services, such as water purification, flood protection and fish stocks (MA, 2006). Groundwater tables are falling in many water catchment areas, many major rivers are overused, badly polluted and biologically depleted – some are so overused that they hardly reach the sea any more (e.g. Yellow River in China; Colorado in North America).

A third of the world population is affected by water scarcity. Approximately 1.1 thousand million people do not have access to safe drinking water. Water pollution is an equally critical problem: 2,600 million people do not have access to basic sanitation facilities, posing a serious threat to human health (UNDP, 2006). Up to 80 per cent of all diseases occurring in developing countries have to do with a lack of clean water (WI, 2004). Gastrointestinal infections are the most prevalent, claiming 2-3 million lives worldwide each year, mostly children (Gleick, 2003). Against this backdrop a Millennium Development Goal was agreed: to halve by 2015 the proportion of people without access to safe drinking water and sanitation (MDG 7, Target 10). With sufficient effort, this goal is globally achievable, although the situation is unlikely to improve in some regions (e.g. sub-Saharan Africa, the Southwest Pacific; WHO and UNICEF, 2004; WBGU, 2005).

The main cause of water crisis today is poor management of available resources (Cosgrove and Rijsberman, 2000; UNESCO, 2006). While connection levels to the water supply are almost 100 per cent in the industrialized countries, water management in developing and newly-industrializing countries is frequently overburdened, and hardly anywhere has an integrated water resources management system (Box 6.2-1) been implemented. There are major differences between urban and rural areas, where the supply is typically worse. There are also substantial regional differences. For example, the situation in sub-Saharan Africa is much worse than in parts of Asia and Latin America. A distinction should be made here between regions with 'physical water scarcity' and those with 'economic water scarcity' (Fig. 6.2-2; Box 6.2-1). Where there is physical water scarcity, even efficient technology is incapable of meeting current needs sustainably (mainly North Africa, the Middle East and Central Asia). This phenomenon is likely to spread to other regions in future as a result of climate change (Sections 5.2.1 and 6.2.2.1). In regions with economic water scarcity, the water resources

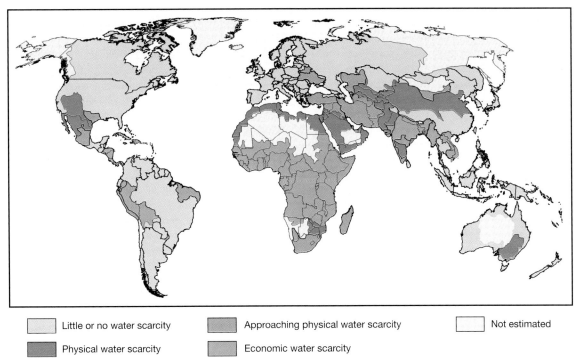

Figure 6.2-2
Areas of 'physical' and 'economic' water scarcity.
Source: IWMI, 2007

available would be sufficient to cover the per capita requirements as well as other needs, but the lack of institutional capacity or financial resources prevents its efficient distribution and use (IWMI, 2007).

Water use is expected to rise significantly around the world. The proportion used by each sector is unlikely to change dramatically between now and 2025. The greatest element of uncertainty here is the future expansion of irrigated farming. In view of the rising demand for water and the changes in regional water availability as a result of climate change (Section 5.2.1), water resources management – already overstretched in some areas – will also be faced with more formidable challenges in future (Cosgrove and Rijsberman, 2000).

6.2.2
Causal linkages

6.2.2.1
From climate change to changes in water availability

The principal impacts of climate change on water availability are shifts in precipitation patterns and rising temperatures. But indirect effects such as a potential change in plant evapotranspiration also lead to changes (Sections 5.1 and 5.2). The expected shifts in precipitation patterns vary considerably according to region, but the tendency will be for rainfall to decrease in the already arid sub-tropics and to increase at high latitudes. In many regions, however, the absolute changes are extremely difficult to predict (Fig. 5.1-4). Impacts can differ in seasonal terms, for instance an increase in precipitation in winter and a decrease in summer. Moreover, there is a general tendency towards heavier precipitation events, even in regions experiencing an average decrease in rainfall. In many regions a greater variability in precipitation is anticipated, which will increase the risk of both flooding and drought (IPCC, 2007a, b). At the same time rising temperatures almost everywhere increase evaporation, which tends to reduce the total amount of water available. The overall effect, the climatic water balance, varies enormously in regional terms (Fig. 5.1-3). In regions which experience snowfall in winter the temperature increase can lead to a seasonal shift of meltwater flows, i.e. the rivers will carry a larger proportion of their total annual flow of water in winter. About a sixth of the global population lives in regions which are likely to be affected (Barnett et al., 2005). Additional threats to water availability come from rising sea levels, which can lead to enhanced saltwater intrusion into coastal groundwater aquifers.

This chapter will focus on regions which rely on glacier meltwater for their water supply during the dry season. The glaciers act as a reservoir to accumulate seasonal precipitation, which then becomes available as meltwater in the dry season. How effectively they buffer stream discharge can be seen in the Río Santa area of Peru: those water catchment areas with larger fractions of glacier cover have a less variable seasonal runoff (Mark et al., 2003; Section 6.2.3.1). As the glacier melt intensifies, more meltwater is initially discharged than is replenished by precipitation, causing an increase in runoff. This additional runoff can be significant. It is estimated, for instance, that half the annual runoff from the Yanamarey Glacier in Peru's Cordillera Blanca stems from a net decrease in glacier mass caused by glacier shrinkage. This glacier is projected to disappear within the next 50 years (Mark et al., 2003). However, once the glaciers have melted away, not only will this additional source of water disappear, but also the storage of seasonal precipitation which can lead to a marked decrease in runoff during the dry season. In the Andes alone tens of millions of people are dependent on glacier meltwater during the long dry season, and it is anticipated that many small glaciers in Bolivia, Ecuador and Peru will disappear completely during the next few decades (IPCC, 2007b). The ice masses in the Himalayas and the Karakoram are also melting. Hundreds of millions of people in India and China alone are expected to be adversely affected by the resultant changes in water supply (IPCC, 2007b).

In some regions, therefore, the volume of available water will decrease in absolute terms. But even regions where the total quantity of water remains constant or even increases may also experience significant problems as a result of the shift in seasonal or interannual variability of precipitation or runoff. Climate change in many places is ushering in a situation where past experience can no longer be projected into the future (IPCC, 2007b). The lessons of the past are therefore losing their relevance as a base for planning water management. They could be replaced by model-based projections of regional water yield and demand which take climate change into consideration. Depending on the region, however, such projections are plagued by uncertainty. Accordingly, therefore, a deterioration of the planning basis for water management is to be expected in many places.

6.2.2.2
From changes in water availability to water crisis

There are many reasons for water management's inability to maintain an adequate supply, leading to a water crisis. The core problem is a lack of adaptive capacity to absorb changes in water supply and demand (Klaphake and Scheumann, 2001). Supply shortfalls of water resources are typically triggered by rapid, sweeping changes in water yield and demand which can overwhelm even relatively well-developed management systems. Various dimensions of influence are relevant here, each of which is affected by a range of key factors (Fig. 6.2-3).

One of the key factors for regional shortfalls in supply is a surge in demand. Rapid population growth and migration (Section 6.5) may cause regional *[1] Demand-side dynamics*. This not only directly increases the domestic demand for tap water, but at the same time indirectly increases the need for water for food production, e.g. regional (irrigated) agriculture (Section 6.3). Economic growth and its attendant rise in per capita consumption is a further factor that can drive rising demand for water by the agricultural sector. At the same time, as regional producers become more integrated on world markets and the overseas demand for food and other primary goods (including timber) increases, water-intensive production at home can also rise. The demand for water for irrigated farming (usually the largest regional user) can thus escalate within a short period of time (UNESCO, 2006). Increases in income and consumer spending, along with export growth, account for a growing demand for water for industrial production processes (e.g. energy, paper or the chemical industry) and for the service industries (tourism in particular).

Other key factors for supply crises are insufficiencies in the dimension of *[2] Water management's institutional capacity*. Poor management capacity plays a major role in creating water crises. The failure of integrated management in particular is attributed to funding shortfalls and a lack of qualified personnel – as well as poor coordination of management activities, fragmented institutional structures and a lack of suitable tools. Water management's room for manoeuvre can also be restricted by political-institutional constraints. These could include the specific requirements of other fields of policy (e.g. agricultural, environmental or settlement policy), which must be considered when decisions are made about increasing reservoir capacities as a potential adaptive measure of water policy. A limited public awareness of the water crisis and the lack of participation of the non-governmental sector are other contribut-

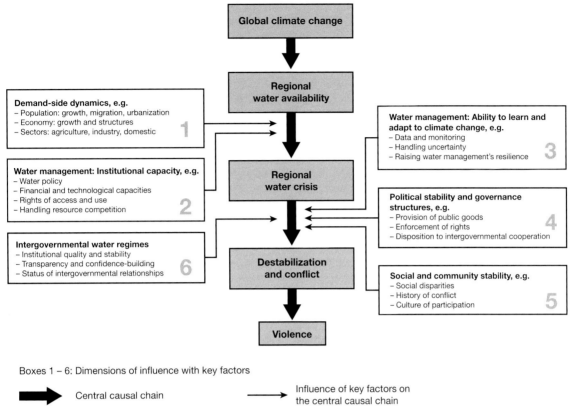

Figure 6.2-3
Conflict constellation 'Climate-induced degradation of freshwater resources': Key factors and interactions.
Source: WBGU

ing factors to the failure of integrated management approaches (Neubert, 2002; UNESCO, 2006).

Poorly defined and enforced rights of access to and use of diminishing water resources are one indicator of a failing water management system. Another indicator is a de facto allocation of these rights among competing water-users that is unsatisfactory in terms of efficiency and distributional equity.

Regardless of the material and coordination aspects of institutional capacity, the likelihood of management failure is greater when planning criteria partially or wholly ignore the impact of climate change on water availability. Key factors for a supply crisis can thus be found in the lack of *[3] Water management's ability to learn and adapt to climate change*. It is true that adaptive deficiencies and consequent water crises continue to be more the outcome of inefficient management practices than physical water scarcity (UNESCO, 2006). However, on the basis of the findings from Chapter 5 and Section 6.2.2.1 it cannot be ruled out that climate change is unleashing drastic and hitherto unseen changes to regional availability, which are putting water management systems under great pressure. The problems inherent in forecasting these changes are highlighted in Section 6.2.2.1. If past practice continues, and water management planning is still based on data of water availability and variability observed over the past 100 years, then unabated climate change can be expected to threaten water supplies and cause water crises in the long term (IPCC 2001, 2007b; UNDP, 2006). In view of this looming development WBGU is assuming that – unless the threat is countered by water policy measures – the scarcity and/or variability of water availability will acquire greater significance as a catalyst for crisis.

The different physical, economic and political-institutional causes of adaptive insufficiencies and water crisis are often found together, making the afflicted regions doubly vulnerable. It can be seen, therefore, that until now water crises have increasingly occurred in developing countries which are already burdened with the task of reducing poverty. These countries are often disadvantaged by both unfavourable physical conditions and inadequate water management systems. In recent times the deterioration of the water framework in certain industrialized countries has also been attracting attention, e.g. in southern Europe (WWF, 2006). Unlike the developing countries, however, the issue here is more a case of wrong structural decisions by management than a lack of facilities.

6.2.2.3
From water crisis to conflict and violence

A water crisis increases the probability of competition between water-use sectors and also – in the absence of systems regulating such competition – the likelihood of water conflict. The physical, political and social dimensions *[1–3]* described in Section 6.2.2.2, the key factors of which are relevant to water crisis, are equally valid for water conflict. Water scarcity does not inevitably lead to conflict and violence. As described in Section 3.3, the critical factors for conflict are found in the dimensions of *[4] Political stability and governance structures*, as well as *[5] Social and community stability*. In other words, the degradation of water resources and attendant water supply crisis can conspire with other socio-economic and political factors to provoke conflict.

At the same time a wide range of examples shows that interstate conflicts over the use of scarce water resources have also led to intergovernmental cooperation (Wolf et al., 2003). This can most clearly be seen on rivers, lakes and aquifers shared by several states, where a regime of transboundary water resources management is required to regulate water rights (Gleick, 1993; Sadoff and Grey, 2002; UNESCO, 2003).

When conflicts occur, the water shortage in itself is less important than the capacity of the water management system to adapt to new requirements. The likelihood of conflict increases when rapid system changes in water use outpace *[2] Institutional capacity* (Wolf, 2006).

In addition to the physical changes outlined (Section 6.2.2.1), political factors are the main contributors to rapid system changes in water use. Political measures which are implemented unilaterally, in other words without consultation with the other users, and which lead to a redistribution of regional water resources, increase the likelihood of conflict. What is crucial here is that the resulting redistribution takes place within a relatively short period of time, putting great adaptive pressure on those affected while changes to the supply infrastructure are slow to implement (Wolf, 2006).

Restructuring at an institutional level is one example of such political measures. For instance, water management reforms were triggered by the founding of new states in post-colonial Africa and South Asia, and in Central Asia following the collapse of the Soviet Union. Claims to water resource use, formerly under intrastate control, had to be reorganized by the newly independent states (Wolf, 2006). Infrastructure projects such as the construction of dams or canals, however, are an even more frequent driver of water conflict. Conflict resulting from forced migration following the building of dams is an additional factor (Box 6.2-2).

The task of defusing conflict caused by water scarcity or redistribution, or preventing violent conflict, places a variety of demands on *[2] Water management's institutional capacity*. These demands differ according to whether the water resources are used only within national borders or internationally.

INTERNATIONAL PERSPECTIVE: 'WATER WARS' OR COOPERATION?
A lack of *[6] Institutional quality and stability of intergovernmental and international water regimes* to control the use of shared water resources is a key factor behind the development of water conflicts. On an interstate level, disputes over shared water resources can aggravate already-tense relationships and give rise to violent conflicts. More often, however, cooperative interaction over water use improves the relationship between the states involved (Section 3.1; Sadoff and Grey, 2002; Wolf, 2006).

Historically the examples of cooperation between states over water resources have far outnumbered those of conflict (Wolf et al., 2003). Benefit-sharing has proved to be an invaluable concept for initiating cooperation (Klaphake and Voils, 2006). Coordinating and optimizing the use of water over state borders brings additional benefits to all the riparian states involved and, when the resources are also fairly allocated, leads to a win-win situation for all participants and to incentives in favour of cooperation. Other positive outcomes of the cooperation include an improved water management system, the conservation of freshwater ecosystems, improved efficiency in the agricultural and energy sectors or a political (peace) dividend (Sadoff and Grey, 2002; Philips et al., 2006).

It is apparent that a cooperation agreement is more likely to be concluded when certain benefit-sharing factors, such as economic advantages, are foreseeable. Until now this has been especially true for agreements over water infrastructure – dam building for shared power generation or projects to exploit additional water resources, for example. On the other hand, few agreements have so far been made to promote adaptive responses to increasing regional water scarcity, in other words agreements in the sense of burden-sharing instead of benefit-sharing (Klaphake and Voils, 2006). It therefore remains unclear to what extent benefit-sharing can function, and promote

> **Box 6.2-2**
>
> **Dams and conflict**
>
> Many of the major dam projects of the past have fulfilled their economic promise (e.g. electricity generation, water storage for irrigated farming, floodwater control or expansion of navigable waterways) and have therefore made a significant contribution to development. However, such projects often have ecological and social disadvantages which directly impact on people's lives. For instance, the flooding of fertile land or settlements has forced many millions of people from their homes, and the environmental side effects of dam-building have often been considerable. Past projects have frequently failed to ensure that the interests of both the beneficiaries and the communities affected were adequately coordinated, so that many dam projects became the catalyst for conflict – sometimes violent conflict.
>
> Carius et al. (2006) report that in at least 15 of 73 case studies presented on environmental conflict, dam-building contributed to conflict ranging from longstanding non-violent protest to threats of war between states (Section 3.2). Such interstate conflicts are mainly defined by the different interests of upstream and downstream states over the allocation of water from transboundary rivers. Until now these conflicts have not directly led to war, but they have led to diplomatic conflict which was usually resolved by negotiating and signing treaties. Within countries the situation appears to be quite different, however. When a lack of participation leads to an inequitable distribution of the economic advantages and negative ecological consequences of dam-building, there is a potential for substantial social conflict. A range of case studies shows that local resistance to the construction projects has met with government repression and violence (WCD, 2000).
>
> Since the 1970s local protest against major dam projects has been the subject of international publicity. Increased political pressure led to a change of thinking among national and international policy-makers. Multilateral funding institutions such as the World Bank accordingly attached much greater weight to the environmental and social acceptability of dam projects. International discussion on sustainability reached its peak with the analysis and recommendations of the World Commission on Dams (WCD). Despite a difficult political setting, the Commission marshalled representatives of disparate interests in an international forum and generated a set of guidelines against which major dam-building projects could be assessed (WCD, 2000).
>
> The Commission's report and its recommendations have experienced a positive echo in most countries (with the exception of e.g. China, India and Turkey). The recommendations on the participation of affected communities and the strengthening of local institutions (mediation processes and tribunals, environmental impact assessments before commitment to projects) are seen as appropriate measures to minimize the conflict potential of major dams, and to harness their sustainable potential in the long term. However, progress on implementing the WCD recommendations is very slow. The World Bank, for example, is again promoting major dam-building projects which NGOs view with scepticism, without applying the WCD recommendations. Controversial dams are also increasingly being privately funded without adequate sustainability assessment.
>
> Sources: WBGU, 1998, 2004; WCD, 2000; Carius et al., 2006

cooperation, under conditions of increasing climate-induced water scarcity. Nevertheless transparency and information sharing serve as a basis for cooperation. Information helps the parties involved to become more fully aware of the inefficiency of uncoordinated, unilateral activities, and to identify their mutual interests (Grossmann, 2006).

Limited knowledge of actual water usage, coupled with a lack of confidence in the opposite party, often provides a breeding ground for conflict (UNDP, 2006). Ultimately, gathering and sharing scientific knowledge, as well as confidence-building measures, are invaluable components of an effective international water management system. As far as potential interstate conflict is concerned, therefore, it is important that there is adequate institutional capacity, in other words effective joint management mechanisms and contractual rules, to enable the cooperative solution of conflicts over use. The status of intergovernmental understanding also plays a role here (dimension *[4] Political stability and governance structures*).

The extent to which interstate water wars could gain in importance in future is debatable. Only a state with high political and military capacity which is situated downstream of a watercourse would have an incentive to wage war (Wolf, 2006). Moreover, according to research on the causes of war (Section 3.3), at least one of the participating states must be autocratic in structure, because it is widely assumed that democracies do not solve conflict by resorting to violence. This overall combination of factors is to be found hardly anywhere around the world. Governments are apparently aware that it seldom makes strategic or economic sense to wage territorial wars aimed at gaining control of water resources outside their own borders. However, water is not often an isolated, stand-alone problem in foreign affairs, but is closely connected with other socio-economic issue areas (UNDP, 2006). It is therefore plausible to assume that when certain constellations of factors prevail, it is possible for disputes over water to contribute directly to interstate violence.

Intrastate perspective: Violent conflicts over water

The risk of violent conflict over water resources is estimated to be greater at an intrastate or local level than at an interstate level (Wolf et al., 2005). Supply problems contribute to the risk of conflict. But, as outlined in the key factors of dimensions *[4] Polit-*

ical stability and governance structures, and *[5] Social and community stability*, it is only in conjunction with adverse socio-economic constellations that they trigger violent conflict.

Intrastate conflict can be driven by two different categories of supply deficiency: either an inadequate quantity of water which provokes competition between water-use sectors, or a poor quality of available water resources. The potential for conflict often goes hand in hand with social discrimination in terms of access to good quality water.

Competition between users can arise between social groups, economic sectors or administrative units. In conflict between social groups, the main focus is on defending or challenging traditional rights of water use. This can most often be seen in arid and semi-arid regions, where the conflict is often between farmers and nomadic herders (Carius et al., 2006; Flintan and Tamrat, 2006). A typical cause of conflict between economic sectors is when water resources are redirected from agriculture to supply growing cities and the industries which are located there (Molle and Berkhoff, 2006; UNDP, 2006). In particular, *[1] Demand-side dynamics* can exacerbate existing competition between water-use sectors. Competition and an uncertain access to good quality water do not inevitably lead to conflict. The question is whether or not a local or regional water management system displays *[2] Institutional capacity* and is capable of dealing with looming intrastate water conflict, thus forestalling violent disputes and social destabilization.

Furthermore, restrictions in the supply of good quality drinking water can aggravate already-existing social conflict and contribute to the escalation of violence (Wolf, 2006). The history of conflict in Central Asia clearly shows that adverse health effects caused by degrading water quality have indirectly influenced social unrest (Carius et al., 2004; Giese and Sehring, 2006). Examples show that the catalyst for an intrastate escalation of violence is not often a poor water supply, but the sudden restriction of access to water, which usually impacts on poor population groups in particular. Such abrupt changes can be provoked by water management restructuring such as privatization, and the price increases which usually result. Existing social disparities, a history of conflict and a strong perception of disadvantage over the allocation of water resources are further drivers of violent conflict. The violent protest ignited by water privatization in the Bolivian city of Cochabamba is a well-known instance of this type of conflict (Lobina, 2000; Wolf et al., 2005). Also, when contained by draconian police state measures as was the case in Cochabamba (e.g. a local state of emergency was declared), local or regional water conflict can disrupt or compromise social and economic development in the affected areas over a prolonged period. It can have a destabilizing impact at a national level, and from there spread to the international level (Wolf et al., 2005; Conca, 2006). The avoidance of such conflict relies heavily on key factors as described under dimension *[5] Social and community stability*. Ultimately, the risk of violent intrastate disputes occurring over water is also determined by local conditions. The risk can be reduced by ensuring that people in disadvantaged areas have access to a safe and stable supply of clean water at local level, thus averting threats to life and health (Wolf, 2006).

INTERACTIONS WITH OTHER CONFLICT CONSTELLATIONS

Examples show that conflict related to the overuse of water resources often goes hand in hand with the overuse and degradation of soils (Carius et al., 2006). This applies particularly to the states of the Sahel zone and southern and eastern Africa (Bauer, 2006). For this reason, the interaction with the conflict constellation 'Climate-induced decline in food production' (Section 6.3) takes on greater significance (Section 8.1). Water crisis and water conflict are also the driving forces behind migratory movements of people. The examples cited in Carius et al. (2006) make it clear that migration can occur at different stages in the causal chain. First, drought in certain regions can elicit movements of people to less affected regions. The resulting rise in population and demand for water can overwhelm the water management system in the destination area. Second, migration can occur as the response to a collapsing water management system. Third, migration can be the reaction to (violent) conflict over access to scarce water resources. This includes forced resettlement as a consequence of dam projects, which in turn can be a driver of conflict escalation (Box 6.2-2).

6.2.3
Scenarios

Chapter 7 identifies regions where – in WBGU's opinion – unabated climate change combined with the socio-economic situation could lead to violent conflict. Two of the regions which appear to be very susceptible to the causal chain of this conflict constellation are examined below: the region surrounding Lima, the capital of Peru (Section 7.9), and Central Asia (Section 7.5). Fictitious scenarios of confrontation and cooperation are outlined for both regions (Section 6.1) in order to highlight the need for action and identify preventive and adaptive approaches.

6.2.3.1
Glacier retreat, water crisis and violent conflict in the greater Lima area

BACKGROUND

Peru and its capital city of Lima are heavily dependent on glaciers for their water supply. More than half of Peru's population lives in the arid coastal region to the west of the Andes, where Lima is also situated. The average rainfall in this region is only about 10mm a year (Chambers, 2005; Mitchell et al., 2002). Approx. 80 per cent of the water resources in the coastal region come from glacier meltwater (Coudrain et al., 2005). In the coming decades Lima's water supply will come under pressure from two directions: from both the increasing demand for water as a result of population growth, and also from the greater variability in water availability as a result of glacier melt. Lima, with a population of more than 7 million, projected to grow to about 12 million by 2030, is the second largest desert city in the world (UN, 2004). More than two-thirds of its water supply comes from the Río Rímac, which is fed by glacier meltwater. These glaciers lost approximately a third of their volume between 1970 and 1997 (Peru Cambio Climatico, 2001) as a consequence of rising temperatures (Coudrain et al., 2005). If the global warming trend continues unabated, the glaciers will disappear completely within the next few decades. Currently the Río Rímac carries 35m^3/s during the high water season and only 13m^3/s in the low water season. The local water authority, SEDAPAL, anticipates a rise in demand from 21.9m^3/s (in 2000) to 25.5–30.1m^3/s (in 2030; Yepes and Ringskog, 2002). Groundwater is already tapped to boost supplies in the dry season (6 m^3/s) and water is transported via tunnels from the far side of the Andes watershed (>10m^3/s). Local laws provide that private households take precedence over irrigated farming in times of scarcity (Molle and Berckhoff, 2006). Currently about 85 per cent of the population is connected to the water network, and 60 per cent in the informal settlements of the poor (Golda-Pongratz, 2004). About a million people depend on tanker trucks or public wells for their water supply (Yepes and Ringskog, 2002).

Social inequality, under-employment and poverty are pervasive in Lima, and in the past these characteristics were accompanied by rising crime rates, drug-related crime and police corruption (Riofrío, 1996). Currently the Peruvian economy is growing substantially, mainly due to the export of gold and copper (Rabobank, 2005). However, its dependence on global commodity markets makes the country vulnerable. About four-fifths of the electricity used in Peru comes from hydroelectric power stations (World Bank, 2006a), meaning that electricity – and thus the entire process of development – relies on water resources which are fluctuating as a result of glacier melt.

At a political level, authoritarian structures have become firmly established in Peru; only at certain times in the past has power been in the hands of democratically-elected governments. The government's monopoly on the use of force appears to be largely intact. However, the range of public services, including investment in water management, is limited. Public institutions are inefficient and typically corrupt (Rabobank, 2005; Bertelsmann Foundation, 2003).

PLAUSIBLE DEVELOPMENTS IN LIMA

The fictitious scenarios are based on the following assumptions about Lima's future development:

- *Lima's population grows appreciably:* Lima's population growth puts constant pressure on the local authorities, but the city's limited infrastructure is often unequal to the demands. The burden of the resulting restrictions falls over-proportionally on the poorer population groups.
- *Water supply is fairly stable at first:* The glacier melt in the Lima basin leads to a higher average annual runoff at first, so that in spite of increasing demand, both water supply and hydroelectric power generation function throughout the year. The visible glacier melt attracts increasing public attention and triggers discussion on how to safeguard the water supply and allocate the diminishing water resources.
- *Economic growth and social inequality:* The commodities exports ensure that Peru and Lima enjoy a relatively stable economic period during the next few decades. The poor, however, benefit little from the economic development.

FICTITIOUS CONFRONTATION SCENARIO:
INADEQUATE PREVENTIVE ACTION BY THE STATE,
WATER SCARCITY AND DESTABILIZATION

The following scenario is conceivable: The Sullcon glacier, and other glaciers in the water catchment area, are past their peak rate of deglaciation. In the wet season from October to April the water volume of the Río Rímac is unusually high, but in the dry season from May to September the flow rate dwindles increasingly. Resource conflict heightens between water and electricity utilities. For too long the relevant decision-makers ignore experts' warnings and take no precautions to secure Lima's future water supply. The construction of a new pipeline along with additional storage tanks in highland areas is years behind schedule due to corruption and mismanagement. The hydroelectric power plants are now capable of producing energy only during the wet season. The looming power crisis in the dry season leads to hefty price rises for electricity. Major enterprises and the upper classes use their political influence to avoid some of the price increases. Industrial employers respond to rising electricity costs with wage cuts and layoffs, despite workers' protests. Sections of the informal economy which depend on electricity can scarcely afford to pay the soaring prices, so their income decreases. The discrimination of socially weak groups exacerbates the disparities which already exist.

The water prices in the informal markets have risen sharply. This not only fosters dissatisfaction among consumers, but also allows criminal structures to take hold. People are forced to expend more and more resources in order to gain access to water, resources which are then no longer available for economic development. SEDAPAL's water pipes are illegally tapped with increasing frequency, and police and private security services clamp down ruthlessly on the perpetrators. An integrated water resources management system fails, partially due to the political opposition of the privileged population groups. Despite official water bans, gardens are watered and swimming pools are filled in some upper class residential districts, even in the dry season. Public opinion hardens against the state institutions which are seen to protect the interests of the privileged circles. Civil society groups such as churches, civil associations and other non-governmental organizations band together to protest against the prevailing conditions. Eventually, when it becomes apparent at the start of a dry season that the water supply will seriously deteriorate compared to the previous year, there are mass demonstrations, some of which turn violent. The situation worsens in the wake of the excessive reaction of the state's security forces, and the violence escalates. Several lives are lost and state infrastructure is destroyed. Day-to-day life is paralysed for several weeks. Public order is superficially restored after a period of time, but a rift occurs in the protest movement and militant underground organizations are formed. In the process of the conflict the state's monopoly on the use of force is undermined and the quality of public services declines further, which in turn aggravates the social disparities. The society is increasingly destabilized and drifts into a civil war with many casualties and negative implications for intergovernmental relationships.

FICTITIOUS COOPERATION SCENARIO: ADAPTATION THROUGH INVESTMENT AND WATER RESOURCES MANAGEMENT
The following scenario is conceivable: The positive economic situation at the beginning of the 21st century allows the government the opportunity it needs to plan long-term investment in infrastructure. Supply and demand are carefully analysed, taking account of the impacts of climate change. The model shows that within a few decades, in the case of business-as-usual, the glacier water will no longer suffice to supply the growing city during the dry season. Adaptive measures based on an integrated water management system are accordingly formulated and – with international support – implemented without delay, both on the demand side (water savings through appropriate incentive schemes and technology) and on the supply side (construction of reservoirs and tunnels). Technical improvements in automation substantially reduce the price of pipeline construction. Solar thermal seawater desalination is now available on an industrial scale and can contribute to the supply of the urban population. Advisory and mediation programmes help to raise the approval rating for an effective integrated water resources management system. At the same time a targeted subsidies policy improves economic structures in rural areas and slows down the rural exodus. These measures cannot entirely prevent water shortages and allocation disparities, but Lima does not experience any acute emergencies. State authorities make a greater effort to ensure that water is utilized efficiently, without inequities in supply. Suggestions from civil society are welcomed, protest is institutionalized in order to avoid violent disputes. The adaptive potential achieved by this policy is adequate to counter the consequences of a moderate level of global warming.

6.2.3.2
Glacier retreat, water crisis and violent confrontation in Central Asia

BACKGROUND
The Central Asia region consists of the former Soviet republics of Kazakhstan, Kyrgyzstan, Tajikistan, Turkmenistan and Uzbekistan as well as the Uiguric autonomous region of Xinjiang in the People's Republic of China. The region's water supply relies heavily on glacier meltwater. While the mountain ranges – depending on their aspect – receive sufficient precipitation, the plains experience extremely low rainfall, so that the natural environment is dominated by deserts and semi-deserts. The plains therefore receive virtually all their water supply from the rivers (Giese and Sehring 2006; Section 7.5), which in summer are fed to a large extent (up to 75 per cent) by glacier meltwater. The climate in Central Asia is warming more rapidly than the global average, which is speeding up the glacier melt. For example, the glaciers in Tajikistan lost a third of their area in the second half of the 20th century alone (UNDP, 2006). The major rivers in the region, such as the Syr Darya and

the Amu Darya, flow through the territory of several countries of Central Asia. The mountainous republics of Kyrgyzstan and Tajikistan are the upstream states which control the headwaters. Further downstream, Turkmenistan and Uzbekistan are the heaviest consumers of these resources (Giese et al., 2004; Karaev, 2005).

Up to 90 per cent of the water resources in these countries is used for irrigated farming, which accounts for 75–100 per cent of the area under cultivation in the four republics (Bucknall et al., 2003; Ahmad and Wasiq, 2004). Recent figures show that agriculture accounts for 20–40 per cent of the GDP here (World Bank, 2006b) and therefore plays a major economic role. Cotton – the leading export crop – is a very water-intensive crop and, when coupled with inappropriate irrigation technology, its production leads to soil salination and decreasing productivity. In Uzbekistan 50 per cent of the irrigated land areas are already affected, and in Turkmenistan the figure soars to a huge 96 per cent (Bucknall et al., 2003). The fate of the Aral Sea symbolises this combination of dehydration, salination and pesticide pollution, which has proved to be an ecological and social disaster for the region (Létolle and Mainguet, 1996; Giese, 1997). The electricity in Kyrgyzstan and Tajikistan is based almost exclusively on hydroelectric power which, with the aid of foreign investors, is being further developed for export.

The states of Central Asia, which have been independent since 1991, are defined by largely closed markets, weak state structures and corruption (Karaev, 2005; CIA, 2006). Their past was plagued by conflict, in which struggles over land and water resources played a major role. Extreme social disparities mark a culture of exclusion from economic life which is controlled by a small, rich class, mainly from the old Soviet nomenklatura. In Tajikistan declining water quality caused by pollution was an important contributing factor to violent intrastate conflicts (Giese and Sehring, 2006; Wolf, 2006). Ethnic disputes and campaigns by separatist or religious-fundamentalist groups are destabilizing the area even further. Interstate conflict is rooted in the partly arbitrary demarcation of borders, and stems mainly from the use of the unevenly distributed water resources in the region (Giese et al., 2004; Kreutzmann, 2004).

PLAUSIBLE DEVELOPMENTS IN CENTRAL ASIA
The following assumptions have been made about the future development of Central Asia:
- *Ongoing climate change and glacier melt:* Temperatures in Central Asia are expected to rise sharply in future (IPCC, 2007a; Section 7.5). By 2050 about 20 per cent of the glaciers in some of the mountains may have disappeared, along with approximately a third of the total glacier volume (Giese and Sehring, 2006), with major implications for water supply.
- *Hydroelectric power affected in the long run:* The glacier melt, accelerated by climate change, at first results in increased water availability in the summer months. To avoid putting the existing hydroelectric power infrastructure at risk, major runoffs and flooding are allowed in the agricultural areas surrounding the lower reaches of the rivers. As the inflows from the melting process slowly decline, the energy sector is compromised. The economic consequences are serious.
- *Agriculture in crisis:* The dependence of agriculture on constantly diminishing water resources leads to rapid falls in the production of cotton and essential foodstuffs. Soil degradation and desertification increase as a consequence of inappropriate agricultural production practices. Unemployment, poverty and migration escalate in the rural areas. Pressure grows to introduce a sustainable resource management system for the agricultural sector (Herrfahrdt, 2004; Opp 2004), but it proves impossible to implement efficient land-use and water management systems.
- *More frequent sand and dust storms*: Settlements are increasingly threatened by sand and dust storms as a consequence of desertification and the scarcity of water. In regions contaminated with salt and pesticides, these storms cause major human health problems (Giese and Sehring, 2006).
- *Social disparities and weak state institutions still predominate:* The poorer classes continue to suffer from inadequate access to land and water resources, and do not benefit from export earnings. The power of the existing elites is unbroken, state structures remain weak, the state's monopoly on the use of force is ineffective, and the provision of public goods does not improve.
- *Potential for interstate conflict persists*: Recurrent drought and a high demand for water in the cotton monocultures of the downstream states conflict with the needs of the upstream states (hydroelectric power) and provoke disputes over the allocation of water resources.

FICTITIOUS CONFRONTATION SCENARIO: RESOURCE CONFLICT AND INTERSTATE TENSION
The following scenario is conceivable: In the coming decades regional temperatures rise significantly, leading to greater evapotranspiration in the low-lying, cultivated areas. Glacier melt speeds up in the mountains. Consequently dams at first experience increased water inflows, but as the glaciers continue to shrink, the flows reduce substantially. Interstate resource conflicts increase as the poorer upstream

states, to avoid rising global market prices for fossil fuels, attempt to meet their winter energy needs with hydroelectric power. In summer, therefore, less water is left in the reservoirs for irrigated farming in the downstream states, although their need is increasing due to the higher temperatures. The lack of investment in efficient irrigation technology aggravates the water scarcity. The upstream and downstream states underestimate their interdependency in terms of water and fossil fuels and try to assert their national interests unilaterally. Subsequent reprisals worsen the conflict. The downstream states, for instance, block the access roads to the upstream states and restrict their gas supply. While the conflict does not escalate into interstate war, the social disparities increase. Low-income populations in the rural areas of both states are the ones to suffer. Increased water shortages – whether due to increased drought or ineffective water management – deprive them of their livelihood. Poverty and rural exodus increase. Recurrent drought, a gradual deterioration in living standards as well as a lack of opportunity to participate at a political level contribute to increasing unrest in the rural population. Various religious and separatist groups exploit this unrest for their own ends, and the tensions frequently erupt in ethnic conflict. The ruling elites respond with intensified repression. The Fergana Valley with its relatively dense population, ethnic diversity and economic importance is one of the hotspots. A sequence of natural disasters reveals the extent of government incompetence in the region, allowing destabilization and violent conflict to become entrenched. The weak states are only partly able to maintain public order. Organized crime and international extremist groups take advantage of the power vacuum in the region to set up a base of operations.

FICTITIOUS COOPERATION SCENARIO: SUSTAINABLE RESOURCE MANAGEMENT AND IMPROVED ADAPTIVE CAPACITY
The following scenario is conceivable: The governments of the downstream states anticipate the drastic long-term increase in water scarcity caused by the glacier melt soon enough to promote economically-efficient structural change. The export earnings from the commodities sector are invested relatively successfully in an adapted, diversified economic structure. This includes the construction of wind-powered generators to harness the huge potential of wind energy, in Kazakhstan and Turkmenistan for instance, to benefit all the countries in the region and reduce their dependence on hydroelectric power. Supported by international development cooperation, this transformation process takes place without marked aggravation of social disparities. With international participation, sustainable resource management of the agricultural sector is gradually introduced, along with an integrated water resources management system. The effective implementation of this policy steadily reduces the extent of land degradation and water scarcity, and so prevents a widening of social disparities. The development of the industrial and service sectors strengthens internal civil society structures. Assistance from overseas along with political pressure from international organizations ensures that the internal economic and social processes do not founder. Confidence-building measures mitigate boundary disputes. In consequence the adaptive capacities of societies across the region improve.

6.2.4
Recommendations for action

Water crisis today is more a crisis of water management in many areas than a problem of hydrological resources (Section 6.2.1). Depending on population development and the extent of future climate change, however, the volume of water available for human use will decrease in many regions, or a greater variability of water availability will complicate its management. To achieve the objectives of food security, poverty reduction, economic development and ecosystem conservation in the face of ongoing climate change, the sectoral approach should be replaced by broadly applied principles of integrated water management. Climate change is only one of several factors which will put pressure on future water management. Even where climate change does not lead directly to reduced water supply, it engenders a great degree of planning uncertainty. Its implications will be felt most strongly by systems which are already suffering from water shortages or greater variability (IPCC, 2001). It is therefore essential both to overcome existing water management deficiencies, and to align water management explicitly to the future challenges. Climate change impacts, in particular, need to be more fully incorporated in management strategies than has been the case in the past. This is the only way to contain conflicts over water.

Along the causal chain from climate change to changed water availability to violent conflict over water (Fig. 6.2-3), there are various options for political action to avert conflict. The features of dimensions *[4] Political stability and governance structures*, and *[5] Social and community stability* have a substantial bearing on whether a crisis deteriorates into violence or whether destabilization can be forestalled. This applies across the board, regardless of the cause of the crisis, and is equally relevant to all conflict constellations.

MANAGING DEMAND-SIDE DYNAMICS

WBGU assumes that a slowdown in the demand for water will relieve water management systems in general, especially in regions where the available water resources decrease or become less stable. The following recommendations are aimed at achieving such a slowdown in demand:

- *Improving water productivity:* Water productivity must be improved in all sectors; there is great potential for this in the developing countries in particular. But in the OECD countries too, there are still substantial efficiency reserves in regions experiencing water scarcity. The agricultural sector has special relevance, because it is the largest water user. Starting points are the improvement of infrastructure and techniques, such as employing efficient irrigation methods to increase production from the quantity of water used, as well as the further development of plant varieties (improved water utilization, tolerance to drought and salt). An improved efficiency of rain-fed agriculture, e.g. through technologies for 'water harvesting,' can also help to minimize pressure on irrigated farming. Both cultivation methods should be complementary. Increased efficiencies in water utilization, however, need to be incorporated within an integrated water management strategy, so that efficiency gains lead to an actual decrease in demand.
- *Abolishing blanket subsidies for water:* Too low prices which do not adequately reflect the economic costs of water provision encourage the overuse of resources. Such blanket subsidies can be reduced without undermining poverty reduction (see dimension 2).
- *Taking virtual water trade into account:* The demand for water in regions subject to water crisis can also be reduced by cutting exports of 'virtual water,' in other words thirsty crops which need large amounts of water to produce or cultivate (e.g. citrus fruits, cotton). At the same time water-intensive products can increasingly be imported from countries which experience copious rainfall and employ more efficient production methods. Whether these options are practicable for an individual developing country need to be analysed in each case.

SAFEGUARDING THE INSTITUTIONAL CAPACITY OF WATER MANAGEMENT

To ensure that the water management system is equal to future challenges, it is important to eliminate the deficiencies which are already apparent. Earlier studies by WBGU have made detailed recommendations on this topic (WBGU, 1998). Selected core areas are outlined below:

- *Strengthening water sector institutions*: To prevent water crisis, mainly in developing countries, improvements are needed to the institutional and financial infrastructure of the water sector, including wastewater management. Thus, a sound and secure supply of good quality water can be ensured and local water conflict prevented.
- *Adopting integrated water management systems:* Climate-induced changes in the water balance (Section 5.2.1) can intensify competition between water-use sectors. Therefore it is important both to slow down demand and to distribute supplies efficiently, while resolving conflicts of interest cooperatively and allocating rights of access and use equitably. Integrated water management provides the appropriate framework to do this. Integrated management of land and water resources can help to improve water productivity – especially in arid environments. It should not only guarantee the supply of drinking water, but also meet the requirements of agriculture and other branches of industry, as well as conserve the natural ecosystems. Participatory procedures that include all stakeholders and promote institutional learning are essential. Developing countries which lack the institutional capacity to implement an integrated water management system should receive appropriate assistance in the context of bilateral and multilateral development cooperation.
- *Ensuring access to water for all:* Regardless of whether the supply system is in public or private hands, the state must ensure that each and every citizen has access to clean drinking water. The Millennium Development Goals on drinking water and sanitation are only the first step. As the water sector displays many characteristic features of a public good, there should be a strong regulatory body to protect the public interest through price regulation and investment. Blanket subsidies should be avoided (dimension 1), but targeted assistance – to connect poor households to the network, for instance – may be required (UNDP, 2006).

STRENGTHENING WATER MANAGEMENT CAPACITY TO ADAPT TO CLIMATE CHANGE

Facilitating the adaptation of the water sector to climate change does not necessarily require new technological or institutional solutions, but does call for new procedures by which the possible courses of action in terms of water management can be evaluated and selected (IPCC, 2001). It is important that adaptation to climate change is established as a process in itself, not simply as the implementation of measures identified in a one-off manner. This includes making an expedient connection between measures to reduce

greenhouse gas emissions and adaptive measures, and taking interactions into account (BMU, 2007).
- *Using regional models and improving the data base:* To adapt water management to the implications of climate change, planning needs to be based on more than simply past experience in terms of average precipitation or variability of precipitation and runoff. Findings from regional models which take account of climate change should also be taken into consideration. As there are currently no reliable precipitation forecasts for many regions (Fig. 5.1-4) and such uncertainties are likely to persist in the future (IPCC, 2007b), the use of scenarios and ensemble modelling should be intensified. In addition, improved regional data gathering (e.g. measurements of climatic parameters such as precipitation and temperature) can be deployed to improve and adapt regional models. Against this background, high priority should be given to the regular analysis and provision of scientific knowledge about the regional implications of climate change for water availability. International cooperation is of vital importance here, particularly for the developing countries. It needs to be considered whether the international community could maintain a database containing interpreted data, and make it widely accessible as a basis for water management.
- *Aligning water management to action under increased uncertainty:* In many cases, targeted action need not wait for the development of suitable models. Measures that improve adaptation to existing climate variability often also improve adaptability to future climate impacts. This applies in particular to improvements to local water storage capacity (e.g. regeneration of aquifers, conservation of freshwater ecosystems, decentralized water reservoirs) and systems to distribute the stored water. However, major storage facilities – such as dams – can themselves be a catalyst for violent conflict (Box 6.2-2), and can also have major undesirable social and environmental side effects (WCD, 2000; WBGU, 2004).
- *Taking account of climate impacts on water management in development cooperation:* Climate change aggravates already strained regional conditions and presents a major extra challenge for the water management systems in many developing countries. This problem has been acknowledged by the German government (BMZ, 2006a, b). It should be systematically taken into account in bilateral and multilateral cooperation in the water sector. The new focus of the German Federal Ministry of Education and Research (BMBF) on providing assistance to activities in the field of integrated water resources management is a promising move in this regard. The developing countries in particular, often forced to take short-term, reactive action due to their limited economic and institutional resources, frequently lack the understanding and capacities needed to cope with the impacts of gradual climate change. It is therefore vital that appropriate publicity be given to the options available and the need to adapt to climate change, in order to raise awareness among decision-makers and society in general (GTZ, 2003).

IMPROVING THE QUALITY AND STABILITY OF INSTITUTIONAL REGIMES OF INTERGOVERNMENTAL WATER COOPERATION

Even though fears of increasing conflict over transboundary waterways ('water wars') have so far proved to be unfounded, it is important to promote measures that will continue to avert violent interstate conflict.
- *Taking pro-active measures to promote intergovernmental cooperation:* Initiating intergovernmental cooperation is often a lengthy process. With physical shortages already foreseeable in many places, there is only a limited window of time to set this process in motion. Establishing a data pool about jointly-used water resources is an appropriate starting point for a cooperative process of transboundary water management (Grossmann, 2006). Weak institutional capacity and underfunding is, however, a major obstacle to cooperation (UNDP, 2006). International assistance should help to improve the conditions for cooperation.
- *Promoting benefit-sharing:* A strategically applied benefit-sharing approach can help to prevent resource conflicts and foster cooperative solutions. The opportunities for benefit-sharing in the light of regionally diminishing water resources warrant investigation. In general, however, the approach offers a good alternative to national sovereignty claims and/or existing bilateral agreements based on absolute quantities of water.
- *Mainstreaming cooperation over transboundary waters within international organizations:* Many international lakes and rivers lack both an official framework defining cooperation and key elements of an institutional structure. In order to foster cooperation over transboundary waters, specific strategies and resolute measures are required from international organizations (incl. World Bank, UNDP, UNESCO). This core task therefore needs to be established or strengthened within the international organizations themselves.

6.3 Conflict constellation: 'Climate-induced decline in food production'

6.3.1 Background

Global environmental change – especially climate change – is capable of changing environmental conditions to the extent that particular countries or global regions face the threat of a significant reduction in agricultural production (IPCC, 2007b). In certain circumstances this causes local or regional food production to decline, resulting in food crises with the potential to amplify or even trigger destabilization and violent conflicts (Homer-Dixon et al., 1993; de Soysa et al., 1999). The most vulnerable to such crises, even today, are the many developing countries in which large sections of the population live directly from agriculture. But in future, some major newly industrializing countries are also potentially threatened: in India, for example, 60 per cent of the population is employed in agriculture, and in China, 49 per cent (CIA, 2006). At the same time, India and China are home to around half of the world's current total of 850 million undernourished people. Europe and North America may only feel minimal effects initially; later, however, their security situation could be directly affected by structural changes on global agricultural markets or by migration in response to hunger or violence.

6.3.1.1 Global food production: Future trends in supply and demand

Global food production has more than doubled between 1961 and 2003 (Fig. 6.3-1). Cereal production accounts for a considerable proportion of this growth. Thanks to the 'Green Revolution' it has reached two-and-a-half times its original level (MA, 2005a): global cereal production has grown faster than the world population (Nussbaumer, 2003). This rise in global food production over the last 40 years was attributable principally (approx. 80 per cent) to increases in land productivity. In developing countries, clearance of new land accounted for 29 per cent of growth (1961–1999), still a comparatively high share. Particularly in Africa, the growing demand for food is still mainly met by clearing new land for crop production: two-thirds of production increases are attributable to the reclamation of new land (MA, 2005a). Changes in demand patterns which have been observable for some years are another key factor in global agricultural production: meat consumption will increase in developing countries, while it is expected to stagnate at a high level in the OECD countries (OECD-FAO, 2006). The lower the overall consumption of meat, the more effectively agricultural land and cereal yields can be utilized for the benefit of human nutrition. Originally, humans began to keep livestock for its ability to transform waste products, or plants such as grass which humans find inedible, into milk or meat. But today, the animals that are fattened for meat production are largely fed on cereals. The bulk of world cereal production is used for the feeding of livestock. Only around 10 per cent of the cereal fed to livestock is converted into meat mass.

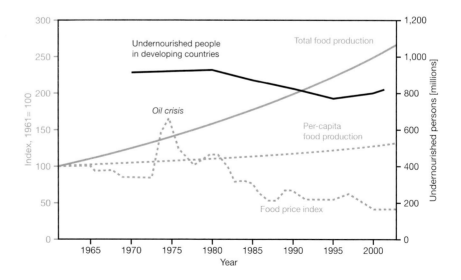

Figure 6.3-1
Global food production 1960–2003.
Source: Le Monde diplomatique, 2007

The scenarios from the Millennium Ecosystem Assessment assume a global rise in demand for food of 60–85 per cent by 2050, based on predicted population growth and assumptions about trends in consumption patterns (MA, 2005a). At the same time, the MA scenarios assume that global food production and per-capita food yield will rise until the year 2050. This is subject to regional variation, per-capita food yields (staple foods) are likely to stagnate or decline in North Africa, the Middle East and in sub-Saharan Africa. It is expected that these regional shortfalls in food production can be offset by means of imports (MA, 2005a).

The FAO arrived at a similar assessment: it forecasts that the annual growth rate for world cereal production will rise from 1 per cent today to 1.4 per cent by 2015, thereafter falling to 1.2 per cent (1970s: 2.5 per cent per annum; 1980s: 1.9 per cent per annum; 1990s: 1 per cent per annum). In the estimation of the FAO, many developing countries will become increasingly dependent on imports. Whereas between 1997–1999 developing countries imported around 9 per cent of their cereal requirements, this is expected to reach a figure of 14 per cent in the year 2030 (FAO, 2002; OECD-FAO, 2005, 2006). According to the FAO, some 80 per cent of the expected future production increases in developing countries will be achieved by means of intensification, i.e. higher yields per hectare, more mixed cropping, and shorter fallow periods (FAO, 2002).

At the same time, the area of agricultural land – according to the Millennium Ecosystem Assessment assumptions – will continue to expand in developing countries but is more likely to contract in industrialized countries (MA, 2005a). Viewed on a global scale, in principle there is sufficient agricultural land – about twice as much as is currently being farmed, in fact (MA, 2005a). For forest and ecosystem conservation reasons, however, only part of the total amount could or should be used. The FAO study *World Agriculture – towards 2015/2030* sees vast areas of potential land in Africa and South America, whereas in South Asia the supply of agricultural land has largely been exhausted. However, these assessments do not take account of the effects of land degradation through soil erosion and deforestation or through climate change (FAO, 2002, 2003; OECD-FAO, 2005, 2006).

The FAO also assumes that irrigated farming will take on greater importance in future. In the period 1997–99, 20 per cent of agricultural land in developing countries was already being irrigated. This land produced some 60 per cent of total yields. The irrigated area in developing countries is expected to rise from 202 million hectares (1997/99) to 242 million hectares by the year 2030. The land suitable for irrigation in developing countries covers 402 million hectares, according to FAO data. In Latin America and Africa, in particular, only half of the potential area is being used (FAO, 2002). It remains unclear whether this assessment took account of the risk of salinization and whether it refers only to land that can be used sustainably rather than, say, primary forests.

Viewed on a global scale, according to the FAO – and without taking the effects of climate change into account – overall food security will continue to improve in future. Globally, food supply and demand will rise until 2015. Growth in both consumption and production of most food and feedstuffs will outstrip the average, global and annual rates of population growth (OECD-FAO, 2005, 2006). According to the forecast, the proportion of undernourished people will fall from 17 per cent today to 11 per cent in the year 2015, and to six per cent in 2030. As a result of population growth, however, the numbers of people affected will decline only slowly in absolute terms (FAO, 2002).

6.3.1.2
Changing framework conditions for global food production

GLOBAL WARMING AND AGRICULTURAL PRODUCTION
The repercussions of climate change on food production will vary enormously from region to region. According to the SRES scenarios (IPCC, 2000), warming of around 1.5–2.5 °C compared with pre-industrial levels can be expected by the middle of the century in the event that climate policy measures do not work. By the end of the century – depending on human actions – warming of over 6 °C (compared with 1990) is possible (IPCC, 2007a), adding another 0.5 °C to express the figure in comparison with pre-industrial temperatures. Generally it is in lower latitudes that food production is most severely threatened by global warming, particularly through loss of cereal harvests and insufficient adaptive capacities (IPCC, 2007b). In lower latitudes, even warming of 2 °C (compared with the pre-industrial level) can be expected to increase food insecurity (Hare, 2006). Under these conditions, there is likely to be a significant increase in the numbers of people threatened by famines in some developing regions of the world (Pilardeaux, 2004). In contrast, many but not all regions in middle and higher latitudes can initially expect an increase in agricultural production to result from warming of between 1 and 3 °C (henceforth always in relation to 1990). Once the mean global temperature warms by 2–4 °C, agricultural pro-

ductivity is likely to decline worldwide. Finally, a temperature rise of 4 °C or more can be expected to have major negative impacts on global agriculture (IPCC, 2007b).

According to a study by IIASA (quoted in FAO, 2005c), climate change in developing countries will result in an increase in drylands and areas under water stress by 2080. In Africa, 1.1 thousand million hectares of arable land under cultivation only afford crops a relatively short growth phase of less than 120 days per year. As a result of climate change, this area could expand by 5–8 per cent which would equate to another 50–90 million hectares of arable land in Africa. The IIASA study assumes a simultaneous 11 per cent decline in the area of rain-fed farming regions, with a consequent drop in cereal production. The study concludes that as a result of climate change, 65 developing countries which accounted for more than half of the world population in 1995 would lose a cereal production potential of some 280 million tonnes. In India, for example, the IIASA model suggests that 125 million tonnes of cereal production potential would be wiped out. This is equivalent to 18 per cent of India's current rain-fed cereal production. China, in contrast, would have the potential to increase rain-fed cereal production by 15 per cent (24 million tonnes) compared with its current level (FAO, 2005c). The majority of sources agree that other 'winner' regions will be North America, northern Europe, the Russian Federation and East Asia (FAO, 2005c) but not all authors share this optimistic appraisal (Long et al., 2006; Stern, 2006).

Under the assumption of a moderate change in the global climate (up to about 2 °C) it is widely believed that production gains in industrialized countries will be sufficient to make up for losses of production in developing countries (FAO, 2005c; OECD-FAO, 2005, 2006). Therefore the worldwide supply will remain roughly level. This effect is merely expected to trigger moderate price increases on world cereal markets (FAO, 2005c). Nevertheless, even minimal price increases could have noticeable repercussions in net food-importing developing countries, which will often inhibit development. Moreover, environmental factors in many of these countries are causing a growing shortfall in domestic supply. Therefore import volumes are increasing and there is growing risk of overstretching these countries' economic capacity. Beyond this, a parallel rise in world market prices could occur for other reasons. For instance, if there is greater demand from the EU for staple foods on the world market, e.g. if the European agricultural market is liberalized, or production-incentivizing agricultural subsidies are abolished, or if production conditions in Europe deteriorate as climate change progresses, regardless of the initial phase of production growth.

Degradation and depletion of soil and water resources

In the next 30 years, not only climate change but various forms of degradation of the natural environment, e.g. desertification, salinization or freshwater depletion, will have grave consequences for agriculture. In South Asia and northern Africa, for example, where population growth is high, land reserves for agriculture have already largely been exhausted (FAO, 2002). Soils are being degraded by erosion and salinization, whilst overuse of freshwater resources is lowering the groundwater table, severely limiting agricultural production. The construction of reservoirs and dams has substantially improved access to freshwater resources in many regions of the world, yet per-capita water availability is declining.

The degradation and depletion of freshwater resources currently affects 1,000–2,000 million people (MA, 2005a). UN estimates put the number of people without access to safe drinking water at 1.1 thousand million (Section 6.2.1). Water abstraction in many OECD countries has been dropping since the end of the 20th century, a trend that is expected to persist in the 21st century. This can be explained by saturation of per-capita demand, efficiency improvements, and stabilization or reduction of population size. On the other hand, in regions outside the OECD, water abstraction is expected to increase steeply. Economic growth and population growth are the critical factors influencing this trend. The scenarios for the Millennium Ecosystem Assessment assume an increase in global water abstraction of 20–85 per cent by 2050. Already, 5–25 per cent of global freshwater use is deemed unsustainable because it exceeds the natural replenishment rate. Worldwide, irrigation agriculture is particularly threatened, and 15–35 per cent is considered to be unsustainable.

Nevertheless, the assumption is made of a 5–7 per cent increase in global water availability (Latin America 2 per cent, former Soviet Union 16–22 per cent) because global warming accelerates the global water cycle (MA, 2005a). This, however, does not necessarily imply any improvement in access to clean and safe drinking water.

Structural changes to world agricultural markets

The environmentally induced change in agricultural production conditions is expected to be accompanied by structural upheavals on world agricultural markets. A key factor will be the transition of China and India, both highly populated countries, from self-sufficiency to significant sources of net demand on world agricultural markets (Heilig, 1999; Fischer, 2005; Worldwatch Institute, 2006). Forty per cent of the world population lives in these two countries. Since 1985, China and India have imported no more than six per cent and three per cent of their cereal requirements respectively. The per-capita cereal-growing area in China (600m^2) and India (650m^2) is already relatively low by international standards (by way of comparison: USA 1900m^2). Given that the land suitable for agriculture has largely been utilized, the populations of both countries will continue to grow in the next two decades and urbanization is increasing, a decline in their per-capita cereal-growing area is a certainty. The trend towards rising net food imports will be accentuated by chronic overuse of water resources in India and in northern China. Added to that, soil degradation is reducing land productivity in both countries, but particularly in China. A further problem is the increase in land sealing due to construction activity.

LAND-USE COMPETITION BETWEEN ENERGY CROPS AND FOOD PRODUCTION

It is unclear whether land-use competition between energy crops and food production will be detrimental to food security. Competition may be heightened by factors relating both to fuel scarcity and to climate change mitigation. Demand for bioenergy may rise in order to cushion the impact of declining supplies and rising prices of fossil fuels, and due to a switch to biofuels. This would stimulate more widespread production of energy crops. As yet it is impossible to predict whether this would initially utilize abandoned and marginal farmland, or whether in the long term it would appropriate land currently used for food production. The outcome depends in no small measure on the strategic choices made in energy policy.

6.3.2
Causal linkages

The following discussion will focus on factors with a critical bearing on the conflict impacts of declining food production and reductions in food supply. A number of dimensions of influence will be defined, under which relevant critical factors can be grouped. Neither the definition of these dimensions nor the selection of the relevant critical factors makes any claim to be exhaustive.

6.3.2.1
From environmental change to declining food production

Starting with dimension *[1] (Regional) environment* (Fig. 6.3-2), the first question to address is the extent to which adverse changes in environmental conditions could have consequences for food production. The critical factors under this dimension include the regional climate (temperature and precipitation trends; scale, frequency and geographical distribution of weather extremes) and the availability and accessibility of natural resources, such as soils (e.g. soil depth and fertility, slope and usable area of land) and water (e.g. volume, temporal variability). The impact of environmental changes on food production depends critically on the configuration of factors under dimension *[2] Regional production*. The quality of these agricultural production conditions is primarily a function of local agro-ecological conditions and land use. Critical factors are the area of land that can be used sustainably, the level of soil fertility, the degree of damage already caused by soil degradation, the availability of freshwater, the heat and drought tolerance of plants and animals, the resilience of the agricultural system to pest outbreaks and the type of management (e.g. monoculture or mixed cropping, irrigated or rain-fed farming). The viability of agricultural production is also determined by the adaptive capacity of agricultural systems to changing climatic conditions, by virtue of crop selection, new breeds and varieties or other yield-stabilizing techniques.

A similarly crucial role is played by dimension *[3] Competing regional demands and land-use needs*, particularly the competition between food production and other land-use forms (e.g. energy crops, timber production). The more profitable the alternative forms of use appear, the greater the danger that food production will be crowded out. This applies particularly to marginal agricultural regions with rising population density. Utilization conflicts between different producer groups harbour the potential for outright conflict and violence. This is evident, for example, in the land-use competition between producers of different foods in Africa. In conditions of growing land and water scarcity, the zones of contact between nomadic livestock farmers and settled arable farmers are increasingly the scenes of violent conflict (Oxfam, 2006). There are more forms of competing demand than the conflict between food and non-food agricultural production. The same factor comes into play between meat consumption and the consumption of plant-based foods (Section 6.3.1.1). For example, if the observable trend for rising meat consumption continues, not least in newly industrializ-

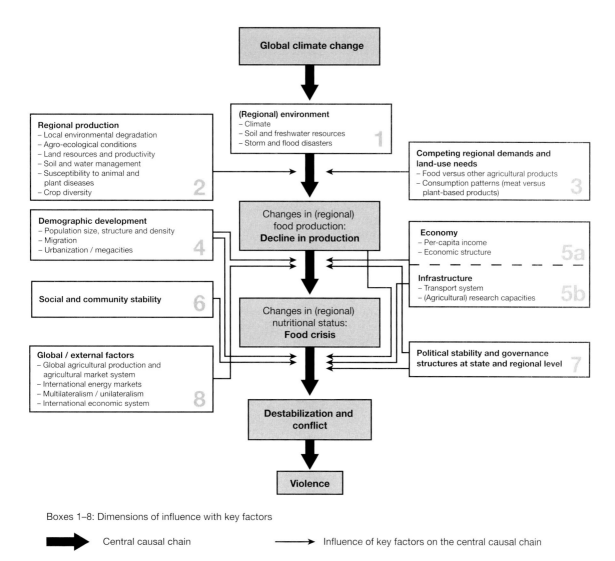

Figure 6.3-2
Conflict constellation: 'Climate-induced decline in food production': Key factors and interactions.
Source: WBGU

ing countries such as China (Hubacek and Sun, 1999; FAO, 2002; Hubacek and Vazquez, 2002), there will be a growing probability of food shortages. This factor is not adequately taken into account in FAO forecasts (Section 6.3.1.1).

6.3.2.2
From declining food production to food crisis

Whether the decline in food production can lead to a food crisis is also critically determined by dimension *[4] Demographic development*. Influence dimension *[5a] Economy* is crucial to the capability of a country or a region to cope with declines in agricultural and food production without succumbing to crisis or even violent conflict. The higher a country's economic output and per-capita income, the more readily it can resort to agricultural imports to compensate for drops in production. High economic output also makes it possible to render adequate transfers to affected groups and thus mitigate the potential for conflict. The structure of the (regional) economy, including its export structure, is at least equally significant (Bender et al., 2003). If the agricultural sector's contribution to gross domestic product (GDP) is low and GDP is high, economically affected groups have a good chance of tapping into alternative income sources. The greater the part played by agriculture in employment and value creation in the economy as a whole, the more resounding the impact of a decline in agricultural production on GDP, especially in terms of the numbers of people negatively affected by falling yields (ADB, 2006b). This is particularly serious in economically weaker countries in which the export of agricultural products is a major source of foreign currency income. In the event of a noticeable decline in GDP, these countries have a dwindling ability to avert such crises or withstand them without violence. Equally, economic stability is a key factor: high levels of underemployment and high inflation increase the probability of violent conflicts. The level of state and private saving and the level of wealth or debt also play a role: if debt is very high, the possibility of obtaining credit to finance increased expenditure on food or imports is appreciably limited (BIS, 2006).

Another prominent dimension of influence is *[5b] Infrastructure*. The quality of agricultural infrastructure (e.g. supply networks for fertilizers, seed, loans), transport infrastructure and institutions responsible for disaster mitigation is vital, to prevent destabilization and conflict risks in the event of a food crisis (KfW, 2002). Moreover, the extent and quality of research capacity and agricultural research infrastructure are decisive to whether an imminent decline in agricultural production in the wake of environmental change can be averted through adaptation of the agricultural system or the use of new breeds and varieties (Senat, 2005; Long et al., 2006).

Dimensions of influence *[6] Social and community stability* and *[7] Political stability and governance structures* refer primarily to the quality of statehood and governance and to the stability of the political system. These critically determine whether food crises culminate in destabilization and violence (Section 3.3). The more able and willing a state is to engage in international cooperation and to be involved in supranational forums and international decision-making bodies, the higher the probability that it will be able to cushion the impacts of agricultural production failures and food shortages by obtaining external supplies.

Dimension of influence *[8] Global/external factors* comprises the extent as well as the regional and sectoral structure of global agricultural production (FAO, 2006a). If climate change gives rise to production failures on a large scale, prices on world agricultural markets can be expected to rise, leaving many poor countries financially overstretched. If worldwide food production falls to a level that jeopardizes the nourishment of the world population, declines in regional production or shortfalls in supply are likely to become increasingly difficult to make up with imports. Factors such as a growing world population, changing patterns of consumption and increasing competing global demand for land use (Sections 6.3.1.1 and 6.3.1.2) will only serve to amplify such a global shortfall in food supply. Conditions in other world markets, e.g. in agrochemicals, fuels or machinery, may also influence the course of a food crisis, due to the interdependency of markets and corresponding price responses. This makes access to other countries' markets another key factor if they are procurement or sales markets with a direct bearing on the region's economy (dimension *[5]*) or the technical aspects of crisis management (dimension *[2]*). International energy markets and prices can have a key influence, particularly if high energy prices set incentives to expand energy-crop production, making foodstuffs scarcer and thus also more expensive (IEA, 2006a, b).

6.3.2.3
From food crisis to destabilization and violence

Influence dimension *[4] Demographic development* is relevant at this stage of escalation also: in cities more frequently than in rural areas, food shortages and food price hikes lead to violent conflicts ('bread riots') (Abel, 1974; Lummel, 2002; Hecht, 2004). Migration is another key factor (Miyan, 2003). On the one hand, migration away from a region can defuse land-use rivalries, e.g. between market-oriented agricultural producers and subsistence farmers or between nomadic livestock farmers and arable farmers, can reduce utilization pressure on local resources and hence lower the risk of violent conflicts over resource use. Yet migration can also be the cause of violent confrontations when the presence of migrants in transit and destination areas adds to the scarcity of resources (Section 6.5).

Other major influences on a society's conflict potential are dimensions *[5a] Economy* (Section 6.3.2.2) and *[6] Social and community stability*. Of particular importance here are the distribution of wealth and income, and the opportunity to participate in societal decision-making (Imbusch and Zoll, 2005). For example, this is manifested in landless people's movements where protests can culminate in violence. Taken to the extreme, if the majority of the population is hungry while a small minority remains almost unaffected by food shortages, outbreaks of violence are more likely than if the entire population is negatively affected (Kaplan, 1985). Statistics on food crisis victims permit few conclusions about the potential for violence: in the major famines that claimed the most victims, the most typical feature was mass migration or escape from the famine, whereas violence played a lesser role (Box 6.3-1).

The robustness of civil society structures is likewise a key factor. The history of the conflict can influence the further course of the conflict constellation: depending on how they have been dealt with, past

> **Box 6.3-1**
>
> **Examples of destabilization and violence resulting from crop failures and food crises**
>
> POTATO BLIGHT, FAMINE, EMIGRATION AND LIBERATION STRUGGLE IN IRELAND
> From the 17th century, the potato became an increasingly important food in Ireland. Potatoes gave higher yields per unit of land than cereals and their preparation was less labour-intensive. At the beginning of the 19th century, the potato was Ireland's primary staple food and was predominantly grown as a monoculture. In around 1840, a fungus (*Phytophthora infestans*) introduced accidentally from the USA caused the potatoes to rot in the fields. The spread of the pathogen was favoured by short crop rotations in the mid-19th century coupled with a mild, damp oceanic climate with infrequent ground frosts. The potato blight that broke out in the years 1845/46 and lasted until 1849/50 was climate-supported, since it thrived in unusually hot summers. It caused a major famine with over one million fatalities. Subsequently over a million people emigrated to the United States. During this period Ireland was a British protectorate. English landlords in Ireland exploited the crisis and the associated wave of death and emigration to boost their land holdings. Moreover they were responsible for exporting the usable remainder of the potato harvest. Part of the grain harvest was also exported because the Irish population could not have afforded to buy the flour. Although there were isolated uprisings against local authorities and landowners (Abel, 1974), it was the 'transgressions' of the English during the potato famine – i.e. refusing to send food aid or subsidized food and expropriating the minimal potato harvest for export – that motivated the Irish liberation struggle against English rule.
>
> ETHNIC CONFLICTS, RESOURCE SCARCITY AND CHRONIC FOOD CRISIS: GENOCIDE IN RWANDA, 1994
> Several studies show that the scarcity of natural resources was a key driver in the genesis of the genocide in Rwanda in 1994 (Homer-Dixon, 1995; Gasana, 2002; Percival and Bigagaza et al., 2002; Verwimp, 2002). Gasana (2002) illustrates how, as a result of soil degradation, continuing population growth and unequal land distribution, the environmental crisis in Rwanda in the 1980s developed into a nationwide crisis: starving people were fleeing to other countries such as Tanzania, and discontentment was rife among the population. This gave radical forces an opportunity to escalate ethnic rivalries into a political power struggle. Bigagaza et al. (2002) come to very similar conclusions in their analysis. In their view, the underlying cause of the ethnic conflict was a conflict over the control of scarce land.

experiences of conflict can either help to avert violent conflicts or play a part in conflict escalation. The same applies to the experience of coping with food crises (Verwimp, 2002). Other key factors are current social problems (e.g. underemployment, precarious livelihoods), the crime rate and the extent of overt or latent conflict (e.g. between ethnic groups). These increase the risk of increasing rivalries over land use or food crises escalating into violent conflicts or the looting of relief supplies (Section 6.4). A review of historical cases also showed that where societies are already destabilized, there is a strong risk that environmentally induced conflicts could 'boil over' or that relatively localized violence might expand on an uncontrollable scale. Furthermore, the pre-existence of (violent) conflicts increases the probability that environmental changes will result in declining food production and, subsequently, food crises (e.g. Lefebvre, 1932; Bigagaza et al., 2002; Diamond, 2005; Box 6.3-1).

A significant aspect governing the potential for violence arising from food shortages is whether the region is predisposed towards conspiracy or scapegoat theories; whether, for instance, demagogues succeed in convincing larger groups of people that certain 'enemies within' (e.g. ethnic, religious or political minorities) or enemies elsewhere are responsible for the deterioration of the situation, such that violence breaks out against those deemed to blame (Sommer and Fuchs, 2004). The perpetrators of violence are not necessarily the poorest people or those most severely affected by a food crisis, but may also be members of the middle class who are threatened with social downgrading as a consequence of price rises (Box 6.3-1). Social stability also manifests itself in effective disaster mitigation and crisis management structures, as well as the degree of involvement of people affected by the crisis. Social participation is thus a critical factor.

Dimension *[7] Political stability and governance structures* exerts a further major influence on societal conflict potential. The risk of deteriorating security due to an environmentally induced decline in food production can be influenced by institutional factors. There is less likelihood of this constellation taking a negative course if the state performs its core functions well. The state should conduct both its internal and external affairs with sovereignty, and ensure at least adequate provision of elementary public goods (e.g. infrastructure and public safety). Furthermore, the state should guarantee the protection of the citizens' rights to resource utilization, property ownership and participation. The extent of corruption is a further critical factor. As regards the danger of cross-border violence, the critical question is how the state has conducted relations with its neighbours currently and historically: has the relationship been dominated by longstanding hostility or even military conflict, or has the situation long been characterized by successful cooperation or peaceful coexistence? This is crucial to the course of the conflict.

Finally, dimension *[8] Global/external factors* is of importance because international economic relations and their multilateral frameworks (the International Monetary Fund, World Trade Organization, etc.) can exert a substantial influence over the conflict resolution or escalation process.

6.3.3
Scenario: Agricultural production crisis, food crisis and violence in southern Africa

BACKGROUND
In the following subsections, fictitious, narrative scenarios of confrontation and cooperation are developed, in order to highlight the need for action and indicate possible prevention and adaptation measures (Section 6.1.2). Agriculture is crucial to African economies: two-thirds of the population work in agriculture, the structure of which is mainly smallholder-based with an emphasis on subsistence farming (Spencer, 2001). Around one-third of the population in sub-Saharan Africa is malnourished or undernourished (FAO, 2005c), a high proportion which is unlikely to diminish for the foreseeable future. Against the global trend, per-capita food production in Africa has been in decline for over 20 years. A breakthrough along the lines of the 'Green Revolution' in Asia did not occur in Africa. Although absolute production has been increased, these gains have not kept pace with population growth. The per-capita area of agricultural land fell between 1965 and 1990 from 0.5ha to 0.3ha (IPCC, 2001; UNEP, 2002b). In addition to resource depletion, key causes of the weaknesses in the agricultural sector include deficiencies in agricultural policy, lack of investment in agriculture, and neglect of rural development (urban bias). In southern Africa in particular, extremely inequitable land distribution is a problem. Irrigated farming is a comparatively minor factor in southern Africa: only 10 per cent of agricultural production (excluding livestock farming) comes from irrigated agriculture. The FAO estimates that a total of less than 30 per cent of suitable land in Africa is currently irrigated (FAO, 1997). Of Africa's current area of land under irrigation, half (around 12 million ha) is already in need of rehabilitation (FAO, 1997). The geographical distribution of surface and groundwater in Africa is very uneven. Moreover, there is huge variability in precipitation levels. In 1990 at least 13 countries suffered from poor water supply or water scarcity (for drinking water and irrigation). According to forecasts, this number will double by 2015 (EU Commission, 2005). Africa possesses 17 per cent of global forested areas. The deforestation of these areas by timber cutting and agricultural clearance is advancing without restraint.

The main thrust of agricultural development over the last thirty years has involved bringing marginal areas into cultivation, but not without a certain amount of habitat destruction such as clearance of forests and drainage of wetlands. Conversion measures of this kind are the principal cause of soil degradation (EU Commission, 2005). Drylands are especially prone to soil degradation: two-thirds of Africa's entire land area is in arid or semi-arid zones, of which up to 20 per cent are already affected by desertification (MA, 2005a). In arid regions of Africa, desertification and drought are a major risk factor for food production. Around 34 per cent of the African population lives in arid regions (Europe: 2 per cent). Altogether, Africa will be particularly hard hit by climate change (FAO, 2005c; Hare, 2006). The main factors are the expansion of arid regions and regions under water stress.

PLAUSIBLE DEVELOPMENTS BY 2020
Based on the arguments in Chapter 5 and Sections 6.3.1.1 and 6.3.1.2, the following assumptions are made about development in southern Africa up to the year 2020:
- Warming of the climate and the rising incidence of extreme weather events exposes agriculture to steadily declining yields. Population growth causes an additional decrease in per-capita food production.
- A majority of the working population remains in agriculture. No progress is made with developing alternative sources of income in the secondary and tertiary sectors.
- The plight of the rural poor, in particular, is greatly exacerbated by the agricultural crisis induced by climate change. People flock to the cities in growing numbers in search of viable alternative livelihoods.

FICTITIOUS CONFRONTATION SCENARIO: ANARCHY AND ARMED REBEL GROUPS DOMINATE THE SCENE
The following scenario is conceivable: from around 2020, food production in southern Africa is increasingly unable to keep pace with population growth. Food crises therefore become more frequent. The poor purchasing power of the predominantly rural population and the economic weakness of states make it untenable to import food, other than in the form of aid consignments. Declining yields from the rain-fed farming regions cause particular problems. New agricultural land continues to be cleared at the expense of intact ecosystems. The degradation of soils and utilization of water reserves impairs the livelihoods of subsistence farmers in many regions. The losses of yield due to rising mean annual temperatures and weather extremes (above all, droughts) cause subsistence farmers major problems. They can

no longer adapt quickly enough to changing environmental conditions. Traditional coping strategies are no longer effective.

After 2020 the investment necessary for the large-scale development of agriculture is not forthcoming; a 'Green Revolution' tailored to the needs of Africa fails to take off. Only the few countries with additional income from exporting mineral resources (e.g. South Africa, Botswana) manage to compensate for the lack of income-earning opportunities in certain locations with comprehensive 'food for work' programmes. In the other countries, the majority of the population remains in the subsistence economy. Both in urban and in rural regions of Africa, unemployment and underemployment grow. The informal sector which dominates the economies of southern Africa does not provide sufficient employment opportunities. On the one hand, this boosts internal migration on the African continent, whereby millions of people move from region to region in search of a livelihood producing 'corridors of destabilization'. Others attempt to get money together, at least for individual family members, to emigrate to the EU with the help of traffickers, in the hope that their remittances will provide better prospects of survival. Meanwhile, the EU steps up surveillance of its external borders, with the result that few migrants reach their destination. There is growing pressure on the transit countries of North Africa, where migrants are increasingly exposed to repression and violent assaults.

Slums in African cities grow rampantly and become breeding grounds for crime and violence. Already severe security problems in African metropolitan centres deteriorate so badly that the elites and the wealthy few can only live in compounds with paramilitary protection and self-contained utilities infrastructure. In the face of hopelessness, chronic food crises and the provocative wealth of the privileged few, by the middle of the century riots in the cities become commonplace, directed primarily against the fortified compounds of the rich. Soon the spark is ignited in rural regions where the nutritional status of smallholders has been deteriorating steadily. Growing numbers of marginalized people join groups of armed rebels and roam from place to place, looting as they go. Lawlessness prevails in growing areas of southern Africa. Destabilization culminates in outright civil war in several parts of the region, which is compounded by a complex constellation of anti-government political liberation fighting, ethnic conflict and drug-related crime by local warlords. Security is tipped beyond the point of no return; most of the region's metropolitan districts and broad swathes of countryside descend into anarchy.

FICTITIOUS COOPERATION SCENARIO: CONCERTED EFFORTS BY THE INTERNATIONAL COMMUNITY DEFUSE TENSION

The following scenario is conceivable: from 2020, the agricultural crisis caused by increasing environmental degradation and climate change in southern Africa brings the international community under growing pressure to act. Concerted international efforts succeed in initiating a 'Green Revolution' tailored to Africa's needs. At the same time, agriculture is successfully adapted to meet the challenges of climate change. Appropriate crop varieties are planted, vast irrigation potential is utilized, and land suitable for farming is brought into cultivation. A better water infrastructure supports the sustainable management of groundwater reserves. At the same time deforestation is successfully halted, because the prospering agricultural sector creates sufficient employment, rendering illegal logging and slash-and-burn forest clearance unrewarding by comparison. Per hectare yields in favourable areas rise while agriculturally marginal regions become depopulated. Granaries are developed on the model of those in South Asia. Thanks principally to cooperation with China, which clearly views Africa as a strategic partner in matters concerning the supply of resources and other goods, transport infrastructure is enhanced within Africa. This goes hand in hand with the development of granaries and rising production of agricultural products suitable for export. Repetition of the mistakes of the old 'Green Revolution' is successfully avoided: the transformation process has broad socio-economic impact and drastically lowers the number of undernourished people – despite continuing population growth up to the middle of the century. Domestic demand for foods can be met, whilst growth is registered in exports of sugar, coffee and other cash crops. World market prices remain comparatively stable. Export revenues rise due to the expansion in trade. The NEPAD process (New Partnership for Africa's Development) moves forward successfully, considerably improving inter-African cooperation. Prosperity rises in the cities of southern Africa by mid-century as small industries become established and produce for the world market. The cities provide numerous income-earning alternatives to agriculture, though initially these are mainly in the informal sector. The unrestrained growth of slums is first brought under control and then halted, thanks to decentralized approaches and participatory urban planning processes. In many of Africa's metropolitan centres, the homicide rate begins to fall after spiralling upward for many years. Migrant workers assist with the establishment of integrated coastal protection projects in the large coastal cities of Mombasa, Dar es Salaam, Maputo, Cape Town, Port Elizabeth and

> Durban. Democracy and good governance become increasingly well established.
>
> There is a sharp drop in the number of interstate and domestic violent conflicts. For the first time, a sizeable portion of the revenue from abundant mineral resources is spent for the public benefit. Per-capita income rises throughout the region. Energy crop production for modern biomass use gains importance in many parts of southern Africa. Another significant factor are the remittances sent home by African migrants who are increasingly finding employment in Europe, where workers are much in demand. They contribute up to 10 per cent of national income to their African countries of origin. In the course of the economic upturn in southern Africa, more and more migrants decide to return to their home countries. By the middle of the century, deforestation and land degradation have largely been halted; adaptation to climate change is successful.

6.3.4
Recommendations for action

Food crises and famines are not inevitable events and nor do they occur simply by chance (Nussbaumer, 2003). Certain levers offer particularly good prospects of resolving or preventing conflicts that may arise out of production or food crises. The dimensions of influence *[5b] Infrastructure, [6] Social and community stability* and *[7] Political stability and governance structures* are centrally important to measures for crisis prevention. Since even robust climate change mitigation policy can no longer prevent environmental change, agriculture needs to be adapted to changing production conditions (*[2] Regional food production*). No less importantly, effective disaster prevention systems need to be established. Generally there will be higher potential for societal conflict wherever there are inadequate opportunities to participate in civil society and the distribution of economic resources is marked by extreme inequalities. Action should therefore be rooted in policies aimed at development, social equity and participation in society. The quality of governance is ultimately critical to a country's ability to cope with food crises. Hence 'good governance', which comprises more than the provision of public goods and the protection of the population, is an essential element for any crisis prevention measure. The summary of recommendations for action below addresses factors where action is especially urgent.

- *Giving greater attention to climate change and environmental degradation in scenario development:* Climate change and environmental degradation are themes that have only been taken up by the FAO in recent years. In their global scenarios on the development of agriculture, and particularly agricultural potential, these factors have not been given sufficient consideration, a problem which could result in misleading forecasts, especially over the long term. Moreover, particularly under changing and dramatically fluctuating climatic conditions, yield stability is of the utmost importance in the development of an agriculture strategy. Statements on the future development of global food production and food security should therefore give due consideration to aspects of environmental change as significant factors.
- *Adapting agricultural development strategies to the challenges of climate change and uncontrolled environmental degradation:* The expected burden on the agricultural sector, especially in many developing countries, should be acknowledged by substantially upgrading the 'rural development' field of policy within German and European development cooperation. For example, it is astonishing that agriculture is not a significant theme of the EU Africa Strategy (EU Commission, 2005), despite the fact that a majority of Africa's people are economically and socially dependent on agricultural production, that yields will fall as a direct result of climate change, and that there is no realistic hope of market mechanisms alone (e.g. reforms to economic structures, but also market-driven insurance against harvest losses) delivering effective responses to the problems. Furthermore, in view of the potential tensions between food-crop and energy-crop production, it would be wise to proceed with caution in setting targets for energy-crop expansion in Africa, let alone steering development and energy policy towards the idea of promoting regional specialization in energy crops as the 'cash crops of the future'. Nevertheless, this appeal for the promotion of agriculture should not be interpreted as detracting from the development of other sectors. What is needed, instead, is not only a well-developed and adaptable agricultural sector, so that any climate-change-induced decline in agricultural production is kept to an absolute minimum, but also economic diversification, to enable management of the socio-economic consequences of unavoidable declines in agriculture.
- *Reforming global agricultural markets:* The reform of international (and national) agricultural markets should be tackled with renewed vigour, to bring about fundamental improvements in market access for developing countries and to enable the development of market-based production incentives in developing countries. Particularly in view of the suspended 2006 Doha Round of the

World Trade Organization (WTO), the centrepiece of which is the removal of barriers to agricultural markets in industrialized countries and the removal of development-inhibiting subsidies, this message is more imperative than ever. However, the price rises that can be expected in the wake of liberalization are most harmful, in the short to medium term, to net food-importing developing countries. This has been a known problem for some time. Back in the 1990s it was dealt with by the contracting parties to the General Agreement on Tariffs and Trade (now the WTO) when they approved compensation measures for least-developed countries suffering extreme negative effects due to agricultural price rises (pursuant to Article 16 of the WTO Agreement on Agriculture). Subsidies and cheap loans for food imports to these low-income countries, including effective financing instruments (e.g. funds) are just as appropriate as IMF and World Bank facilities and programmes. It was in 2001 that the affected countries suggested a 'revolving fund'; prolonged consultations are still ongoing (WTO, 2005a, b, 2006). In emergencies, the countries could borrow from such a fund in order to finance more expensive food imports. The monies later repaid – with or without interest – must be returned to this special-purpose fund. Economically strong countries could and should ensure that the fund is always adequately resourced if debtors are unable to service their debts within the agreed term.

- *Taking account of the growing dependency of affected developing countries on food imports:* The liberalization of agricultural markets and short-term compensation measures will not suffice to solve the long-term supply and demand problems of many developing countries. To better respond to the growing dependency of some developing countries on imported food, existing mechanisms such as the WTO should be used and further transfer mechanisms and finance instruments should be introduced as necessary to help to compensate for the additional financial burdens. Beyond this, such issues should be taken into account in international climate change mitigation policy. For example, it would be worth considering whether the countries that are essentially responsible for global climate change should compensate developing and newly industrializing countries for its negative consequences, such as rising world market prices and climate-change-induced declines in agricultural production.
- *Recognizing the risk potential of land distribution:* In the course of climate change, increasingly frequent droughts and desertification must be expected, particularly in drylands. Furthermore, per-capita availability of land in these regions will often diminish due to high population growth. According to the findings of environmental conflict research, the scarcity of resources such as water or soil frequently results in violent conflicts, with increasing probability where the distribution of land-use rights is very inequitable. The German federal government's development policy strategy should therefore incorporate a greater emphasis on careful processes of land reform. The status of rural development within development cooperation should be raised accordingly.
- *Giving constant attention to disaster mitigation:* The infrastructure for crisis relief should be improved. In the event of a food crisis, it is essential to guarantee the transport and distribution of foods and other elementary supplies, e.g. by the World Food Programme or other relief programmes.

6.4
Conflict constellation: 'Climate-induced increase in storm and flood disasters'

6.4.1
Background

Disasters in the wake of storms and floods are sudden onset disasters with drastic impacts on human lives and livelihoods. Therefore they harbour the potential to provoke far-reaching social and political changes. Even today, storms and floods cause fatalities in similar numbers to the victims claimed by earthquakes, tsunamis and volcanic eruptions combined (calculations based on data from CRED, 2006). As severe storms and floods increase in frequency, there is reason to fear the eruption of more frequent and more acute conflicts in future.

In order to analyse the potential conflict impacts that could be unleashed by natural disasters, a study was undertaken of the social and political repercussions of past storm and flood disasters. It emerged that of the major storm and flood disasters since 1950 with at least 1,000 fatalities, at least thirteen resulted in the escalation of existing conflicts. Brief descriptions of these cases are presented in Section 3.2.2. Natural disasters were found to have defused conflicts in three cases, none of which were storm or flood-related; instead they involved earthquakes and tsunamis which are not influenced by climate change (Section 6.4.2.3).

6.4.2
Causal linkages

6.4.2.1
From environmental change to increase in storm and flood disasters

The likely environmental changes, particularly the expected intensification of tropical cyclones, the increase in extreme weather events, and rising sea levels, have been discussed in detail in Chapter 5. But climatic-meteorological mechanisms and rising sea levels are not the only factors heightening the risks of disaster: deforestation in the upper reaches of rivers and land subsidence in major urban centres also contribute to this development. Further considerations are the growing concentration of cities and settlements in coastal regions, the high proportion of economically weak population groups in developing and newly industrializing countries, and the particular vulnerability of the urban and industrial infrastructure.

DEFORESTATION OF RIVER CATCHMENTS
No form of land cover does more to balance the discharge behaviour of streams and rivers than forests. They intercept and store the bulk of the precipitation, and so attenuate the direct run-off from heavy rainfall (FAO, 2005a). Localized studies show that direct run-off in forests sometimes amounts to less than one-thousandth of that on grassland (IPCC, 2001). Although forests cannot prevent floods completely, the peak run-off in forested river catchments is normally considerably lower than in unforested areas. Overall this effect is less marked in large river systems than in small ones. Nevertheless, in the case of the Yangtze, one of the longest rivers in East Asia, deforestation of the upper reaches was identified as a major cause of recurrent catastrophic floods (UNEP, 2002a; MA, 2005a). Similarly, in other regions the deforestation of catchment areas is acknowledged as a significant flood risk factor. This applies to Haiti, Honduras, Nicaragua and Bangladesh, to name but a few countries which have repeatedly been stricken by severe storm and flood disasters.

In many of these regions, deforestation continues unabated. Whereas the forested areas of Europe and North America have increased slightly on average in recent years, the same does not hold true for Africa, Latin America (including the Caribbean) and the Asia-Pacific region, all of which have registered decreases, sometimes on a major scale. Forested area is shrinking by 0.7 per cent per year in Africa and 0.5 per cent per year in Latin America and the Caribbean (UNEP, 2002b). In certain countries the figures are higher. In Honduras, almost 3 per cent of the remaining forests are disappearing every year (calculations based on data from FAO, 2005b).

LAND SUBSIDENCE
Many coastal cities have grown on the estuaries of large rivers. These are traditionally locations of great economic and political importance as transshipment sites between marine and inland waterway transport routes. Furthermore, river estuaries normally offer a variety of benefits for agriculture (e.g. fertile soils and irrigation potential). A specific problem of such sites is that the geological substrate is composed of recent, relatively uncompacted sediments. If an excessively large volume of groundwater is withdrawn from these sediment bodies, the substrate becomes compacted, resulting in subsidence of the land surface above (Nicholls, 1995; Galloway et al., 1999). Heavy loading can also lead to settlement of the sediments and hence to land subsidence. Both causes – the abstraction of groundwater and heavy loading – are consequences of a typically urban pattern of production and consumption: due to the poor quality of surface water, the bulk of urban water requirements are often met by means of groundwater. Moreover, urban building activity creates substantial land loading. According to Nicholls (1995) and Klein et al. (2002), at least eight to ten of the twenty-one largest coastal cities have experienced marked subsidence in the 20th century, including Tianjin, Shanghai, Osaka, Tokyo, Bangkok, Manila, Jakarta and Los Angeles. Subsidence has occurred at rates of up to 1m per decade (Nicholls, 1995), with top rates of 30cm per year recorded in Shanghai (Hu et al., 2004). But this phenomenon is not confined to megacities alone: according to Barends et al. (1995) more than 150 regions worldwide are at risk. In China alone, Hu et al. (2004) estimate the number of affected cities and districts at 45.

In many cities, land subsidence is perceived as a problem primarily because it results in damage to buildings and infrastructure (Galloway et al., 1999). But problems also arise where cities are built only a few metres above sea level and any subsidence implies a precarious rise in the relative sea level (Coplin, 1999). The potential repercussions of such a rise in relative sea level were seen when New Orleans was hit by Hurricane Katrina. New Orleans has subsided significantly in recent decades leaving some districts at up to 3m below sea level (Burkett et al., 2005). Many districts could only be kept permanently dry by constructing dams and operating large pumping stations. When Hurricane Katrina crossed the Gulf of Mexico in August 2005, the dykes of neighbouring Lake Pontchartrain were breached and the city was flooded (Section 3.2.2).

Although the special vulnerability of New Orleans does not automatically apply to every other coastal city, there are numerous cities including Naga (the Philippines), Bangkok (Thailand) and Semarang (Indonesia) in which major areas lie below sea level (Douglas, 2005; Phienwej and Nutalaya, 2005; Effendi et al., 2005). But even in areas above sea level, land subsidence increases the risk of severe floods. On the one hand, the gentle gradient of the slope slows the run-off of rainwater, on the other hand, it also makes the affected areas more vulnerable to storm surges and rising tide waves from the sea. The disadvantage of obvious countermeasures such as dyke construction is that the land behind such defences can only be drained with pumping stations, barrage and sluice systems.

Alongside these reactive measures, it is therefore essential to address the principal cause of land subsidence, namely the overuse of local groundwater resources. Although Shanghai, Tokyo and Osaka have already achieved notable successes in this area, land is still subsiding relentlessly in Tianjin, Bangkok, Manila and Jakarta (Nicholls, 1995; Klein et al., 2002). Furthermore, progressive urbanization and increased demand for water in cities can be expected to cause similar problems in other cities. In this connection, Klein et al. (2002) name the cities of Rangoon (Myanmar) and Hanoi (Vietnam) as especially vulnerable owing to the high cyclone risks in south and southeast Asian coastal regions (Fig. 6.4-1).

6.4.2.2
From more frequent storm and flood disasters to crisis

The International Strategy for Disaster Reduction of the United Nations defines a disaster as 'a serious disruption of the functioning of a community or a society causing widespread human, material, economic or environmental losses which exceed the ability of the affected community or society to cope using its own resources' (UNISDR, 2006). From this definition it follows that natural disasters are usually associated with a temporary local collapse of state functions. The devastation of infrastructure blocks external consignments of relief, the water and energy supply is disrupted and hospitals are overstretched. In situations like this, the scope of the government to act is often so heavily impeded that it is no exaggeration to talk about a total collapse of state functions. This loss of state power to intervene in disaster situ-

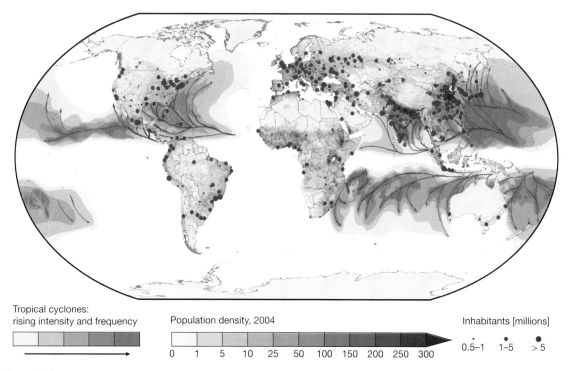

Figure 6.4-1
Tropical cyclone threat to urban agglomerations.
Cartography: Cassel-Gintz, 2006.
Source: WBGU

ations can result in problems for political stability in a number of respects.

On the one hand, at no other time is a population more reliant on external support than during and after a disaster. The call for public assistance with the response to the direct consequences is part and parcel of almost all major natural disasters. If the obstacles to intervention mean that the needs of disaster victims can only be met slowly and inadequately, however, in most cases there will be mounting frustration and dissatisfaction with the government in office. On the other hand, in many disasters it is obvious that damage and casualties are not attributable purely to the physical forces of nature: poorly implemented building codes, housing development on vulnerable sites, inadequate mitigation measures and unclear emergency instructions are often identified in the aftermath of disasters as obvious failures of government, and discussed as such in public (Drury und Olson, 1998). These or similar issues contributed to political crises in Haiti (1954), East Pakistan (1970), Bangladesh (1974 and 1988), Bihar (1987), Orissa and West Bengal (1999), West Bengal (2000) and New Orleans (2005) (Section 3.2.2).

The collapse of state functions always poses a huge challenge for a government. This is all the more acute when natural disasters strike, because the political decision-makers are expected not only to implement relief measures but also to ensure security in the disaster areas. Indeed the breakdown of state functions may well be exploited by non-state actors in the pursuit of their own interests. A relatively common phenomenon in this context is looting. Most often, this involves people acting out of necessity to obtain essential goods for survival; in some cases, however, thefts of valuables and acts of violence are also reported. These or similar incidents have been documented for disasters in Bihar (1987), Nicaragua (1998), China (1998), Orissa and West Bengal (1999), Venezuela (1999), West Bengal (2000), Haiti (2004) and New Orleans (2005) (Section 3.2.2).

6.4.2.3
From crisis to destabilization and violence

MECHANISM 1: ESCALATION OF EXISTING INTRASTATE CONFLICTS

Drury and Olson (1998) establish a clear relationship between disasters and political unrest. Although disasters can occasionally consolidate the political leadership in its position, as a rule disasters tend to heighten dissatisfaction with the ruling government. In disaster situations governments often lose their capacity to act. Any mismanagement and incompetence on the part of the government and the administration are likely to be ruthlessly exposed during and after disasters (Drury and Olson, 1998). Poorly implemented building codes and emergency planning, delays in the arrival of relief, and misuse of aid funds are typical examples of state failure in these situations. Pain over the loss of family members and property combined with the blatant failure of the responsible authorities can have such an effect on public opinion that political stability is seriously jeopardized in the medium term. These kinds of incidents marked the course of conflicts in Haiti (1954), East Pakistan (1970) and Bangladesh (1974 and 1988) (Section 3.2.2). It is striking that in none of these cases did new conflicts arise. Rather, events acted as yet more damning evidence to justify the criticism of the ruling government by its political opponents.

As a generalized assumption, then, the risk of conflict is especially high when a disaster coincides with existing or growing intrastate political tensions. Particularly where there is poor governance and a well-organized opposition, a disaster is likely to act as a catalyst.

MECHANISM 2: CONFLICTS GENERATED BY SOCIAL TENSIONS IN POWER VACUUMS

The temporary breakdown of state functions can be abused by a wide variety of groups for their own ends (e.g. looting by gangs). In past disasters, however, reports of violence and looting have often been blown up out of proportion by the media and have not reflected the real situation in the disaster area (Auf der Heide, 2004). In such situations, the risk of armed uprisings by rebel groups also tends to be low, because they are equally exposed to the organizational and logistical constraints imposed by the natural disaster. There is no known case in which foreign armies or rebel groups have exploited a natural disaster to mount an invasion.

The descriptions of lootings referred mostly to the theft of essential goods for survival. Nevertheless, in some disaster areas it went as far as systematic thefts of consumer goods and the use of violence. These lootings took place in conditions of temporary anarchy, closely linked with the timing of the transitory collapse of state functions. Looting generally begins about 48 hours after the actual disaster (Ebert, 2006) and ends when state functions are restored. It is striking, however, that a disaster-related breakdown of state functions need not necessarily bring looting and higher crime in its wake. There tend to be more reports of a wave of mutual goodwill within affected societies and a fall in the crime rate (Fuentes, 2003; O'Leary, 2004). Lootings of consumer goods are mostly carried out by perpetrators from outside the disaster zone (O'Leary, 2004).

O'Leary believes that a raised crime rate is an expression of social tensions and that, by the same token, societies with social tensions are more prone to anarchy-like conditions following disasters. This is difficult to verify with reference to historical examples (Section 3.2.2) because little reliable information is available on the nature, scale and targets of reported lootings. Nevertheless it can be assumed that at least a proportion of the lootings in Venezuela (1999) and New Orleans (2005) involved the theft of consumer goods rather than the meeting of basic needs. Social tensions were apparent in both regions even before the onset of the disasters.

Although there is no definitive historical evidence of anarchy-like conditions persisting in the longer term after natural disasters, various authors cite Haiti as one example (Diamond, 2005). The country was struck by a series of very severe natural disasters (1954, 1963, 1994, and twice in 2004) and its central government ceased to command any notable monopoly on power some years ago. Considering the incidents in 1954 and 2004 (Section 3.2.2), moreover, it can be assumed that natural disasters were important factors in this development. If regions with weak state structures and major social inequality are hit by severe natural disasters, a permanent loss of state functions is certainly a possibility.

A frequent problem with steps undertaken to restore state functions is incomplete information from the disaster area. Inaccurate or distorted reports of looting and violence can result in the disproportionate use of security forces and harsh treatment of suspected perpetrators. A further difficulty hampering the response is that looters are often virtually indistinguishable from other disaster victims, and their motives for looting – survival versus criminal intent – cannot readily be verified (Auf der Heide, 2004). A historical example that can be cited is the deployment of the army in Venezuela (1999), but security forces were similarly deployed in Bihar (1987), Anhui (1998) and New Orleans (2005) (Section 3.2.2).

MECHANISM 3: DE-ESCALATION THROUGH RELIEF AND NEGOTIATIONS

Sudden-onset, severe natural disasters tend to generate huge, sometimes global, media interest. In most cases this leads to the mobilization of international assistance. As the following examples show, often assistance of this kind is forthcoming even across conflict fractures:
- After a severe earthquake in 1999, Turkey accepted extensive offers of assistance from Greece. Relations between Turkey and Greece are permanently strained due to the unresolved issue of Cyprus.
- After the Bam earthquake in 2003, the Iranian government accepted assistance from the United States. Since the fall of the pro-western Shah regime and the occupation of the American embassy in Tehran in 1979, relations between the two states have been extremely tense.
- After the tsunami in December 2004, the government of Sri Lanka consulted with the Tamil rebel organization LTTE on the distribution of relief consignments in the disaster area. LTTE separatists have been fighting an armed struggle against the Sri Lankan central government since 1976.
- In the wake of the tsunami disaster of 2004, a peace agreement was reached between the Free Aceh Movement (GAM) and the Indonesian central government. The movement and the Indonesian army have been fighting since 1976 over the status of the province of Aceh.
- After a severe earthquake in the Pakistani area of Kashmir in October 2005, the Indian and Pakistani governments agreed on relief efforts for the disaster area. These two nuclear powers had previously fought two wars over the unresolved issue of Kashmir.

Indisputably, in the cases of USA – Iran und Sri Lanka – LTTE, the offers of assistance have not led to any major defusion of tensions. However, the success of negotiations between the Indonesian government and the Free Aceh Movement and between the Indian and Pakistani governments prompted the Worldwatch Institute (2006) to suggest treating disasters as a peacemaking opportunity.

On closer consideration of the three cases evaluated positively, the following common features become clear: all three conflicts were in phases of reduced tension at the time of the disaster and were playing out between two clearly identifiable actors:
- The Cyprus conflict had been stuck in an unresolved state since 1974. Particularly on the Turkish side, renewed escalation was unthinkable due to its efforts to gain accession to the EU.
- Following a two-year onslaught from the Indonesian army, the rebels in Aceh were so severely weakened that carrying on their armed struggle would have been doomed to failure (Aspinall, 2005). After the disaster, the Indonesian government also came under huge international pressure to end the crisis.
- In Kashmir the disaster struck during a phase of ongoing negotiations between India and Pakistan, which a few months earlier had already produced the outcome of a direct bus connection between the two parts of Kashmir.

Against this backdrop it can be assumed that where conflicts are being fought between two clearly defined parties and have already crossed their zenith, natural

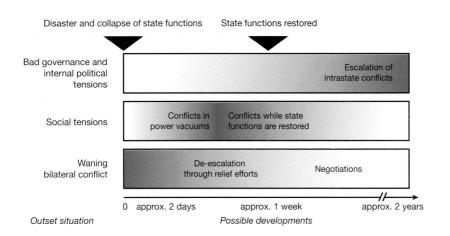

Figure 6.4-2
Characteristic time sequence of disaster-induced conflict mechanisms. Intensity of colour indicates intensity of conflict.
Source: WBGU

disasters certainly represent opportunities to overcome entrenched political-ideological differences. In the case of conflicts involving several parties or an increasing level of tensions, there appears to be relatively little prospect of de-escalation.

6.4.2.4
The time sequence of disaster-induced conflict mechanisms

Figure 6.4-2 shows a simplified classification of the mechanisms presented and their progression over time. The collapse of state functions is both an inherent component of large-scale natural disasters and a precondition for the conflict mechanisms described. The latter are linked to specific outset situations, but can also occur in combined forms in certain circumstances. Whereas de-escalation may begin from the first offer of assistance, generally looting will only start to happen two days after the disaster event. Intrastate conflicts will not normally escalate until after the main direct consequences of the disaster have been responded to.

Overall the preconditions under which conflicts will be triggered or amplified can be summarized in terms of key factors, which are briefly outlined below. The typical mechanisms of action are presented schematically in Figure 6.4-3.

The dimension of influence *[1] Physical threat* to a location or a region depends substantially on the intensity and frequency of extreme events. The level of threat is not purely a function of individual factors like precipitation levels and wind speed, but also depends on geographical location (e.g. low-lying coast or river valley). Other key factors that come into play are, in many cases, partly influenced by human activities. Examples are the deforestation of river basins, which increases peak run-off (the highest water level on a water gauge during a flood), and the subsidence of cities due to increased groundwater abstraction.

Influence dimension *[2] Vulnerability to extreme events* groups together the key factors which determine whether an event will actually unleash catastrophic effects or whether people and infrastructure will escape largely unscathed. A significant key factor here is population density and the concentration of property and assets in areas at risk. The flooding of an uninhabited river valley is not, in itself, a natural disaster. The devastation of critical infrastructure facilities such as hospitals, transport and communication nodes by extreme events, on the other hand, is a crucial feature of disasters. However, vulnerability is also determined by economic, social and organizational characteristics. Besides the nature and extent of disaster preparedness, factors such as poverty, economic structure and educational level are also critical. Past experience has shown that disaster exposes poor population groups to a disproportionately higher risk than wealthy population groups (UNISDR, 2004).

Whether the temporary breakdown of state order in disaster situations is abused by non-state actors for criminal ends depends essentially on dimension of influence *[3] Social stability.* In many societies, disasters lead to higher levels of internal cohesion and widespread helpfulness (Fuentes, 2003; Auf der Heide, 2004); in different milieus, they lead to the emergence of criminal gangs and looters. It is assumed that a raised crime rate is most likely in places where social tensions and crime have been known problems beforehand. This assumption is supported by observations in the USA (Auf der Heide, 2004). Social tensions are often the expression of large income disparities, a high unemployment rate, and social and ethnic segregation. Social inequality is also cited as a key factor with negative repercussions for the stability of post-disaster states (Drury and Olson, 1998). The explanation offered is that in societies with severe social inequality, poorer sections of the population

Figure 6.4-3
Conflict constellation:
'Climate-induced increase in storm and flood disasters':
Key factors and interactions.
Source: WBGU

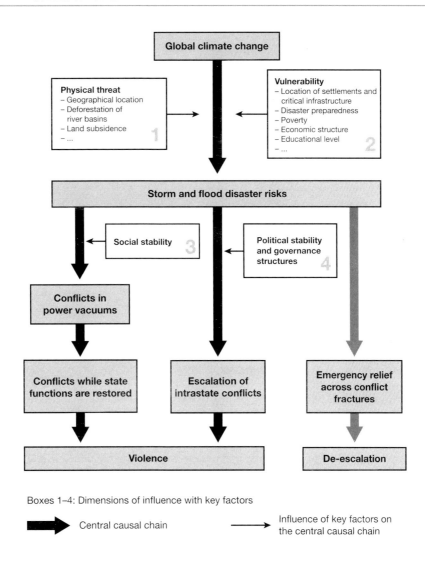

have little reason to believe that relief will be distributed fairly, and are therefore more motivated to take action in the streets to claim their share of financial assistance. This linkage does not apply to sections of the population in extreme poverty, however. During and after natural disasters, they are too occupied with tasks of daily survival to be able to devote time and energy to political activities.

Influence dimension *[4] Political stability and governance structures* refers to the quality of the government concerned. If, in the normal course of events, it performs all its essential functions largely to the population's satisfaction, then as a rule a disaster comes as a relatively minor shock to the political system. Especially rich nations also have the advantage of being able to respond to crises relatively quickly by mobilizing substantial resources, which in turn keeps public dissatisfaction at bay. If dissatisfaction with the government is rife even before the disaster, and is then heightened by mistakes in disaster mitigation and in the response and reconstruction phases, then a natural disaster may trigger a crisis of government. If, on top of this, disaster strikes during conditions of political instability and civil conflict, natural disasters can act as a catalyst and determine the trajectory of the conflict. This is especially the case when the government is already weakened through attacks on its public image, and opposition groups succeed in demonstrating and effectively publicizing government failings relating to disaster mitigation and relief. According to Drury and Olson (1998) strong, repressive regimes are less threatened in this constellation of events because they can largely block any strengthening of opposition groups.

Natural disasters can also have positive repercussions for the trajectory of conflicts. This is conceivable when arrangements are made to deliver relief across existing front-lines of conflict, and when the conflict has already entered a phase in which none of the parties can gain any particular advantage from carrying it on. A review of historical cases shows that de-escalation after a natural disaster has only occurred in conflicts which were unequivocally bilateral.

6.4.3
Scenarios

In the following subsections, fictitious narrative scenarios of confrontation and cooperation are developed for China and the Gulf of Mexico, in order to point up the need for action and indicate possible prevention and adaptation measures.

6.4.3.1
Storm and flood disasters in China

BACKGROUND
Throughout history China has been exposed to major disaster risks. Since 1950 it has endured 23 storm and flood disasters with more than 1,000 casualties. During this period, over two million people have lost their lives in storms and floods, more than in any other country (CRED, 2006). According to information from the Ministry of Civil Affairs, 200-400 million people every year are directly affected by natural disasters (earthquakes, floods, tropical storms, droughts or land slips). The government is making efforts to counteract flood risks, in particular, with structural measures (dams, bank reinforcements). Higher volumes of direct run-off, more heavy rainfall events, rising sea levels and tropical cyclones of growing intensity are calling the effectiveness of these measures into question, however (Section 7.7). The problem is further exacerbated by localized land subsidence. Many of China's economic centres are built on unstable sediments in close proximity to coasts. The building load and the uncontrolled abstraction of groundwater is causing subsidence of several centimetres per year in a large number of cities and districts. This further increases the potential for catastrophic floods (Section 6.4.2.1).

In the recent past, China has experienced disasters which caused at least temporary and regional loss of confidence in the political leadership. Examples are the Yangtze flood disaster of 1998 (Section 3.2.2) and the chemical accident on the Songhua River in northern China in 2005.

The real growth rate of the Chinese economy has averaged nine per cent over the past ten years, bringing China out of developing country status and into the ranks of the world's leading economies. Between 1981 and 2001 there was a threefold increase in average per-capita income. This rapid growth has led to a clear reduction in poverty: over the same period, the proportion of the population living in extreme poverty fell from 64 per cent to 17 per cent (Jenkins, 2005). Nevertheless, nowhere near the entire population is currently benefiting from these advances. Economic development is taking place almost exclusively in the urban centres in the coastal region. Many regions of the interior continue to exhibit typical traits of developing countries. According to the China Human Development Report (UNDP, 2005a), at the end of 2005 China's urban-rural income disparities were the greatest in the world. Recently the income gap has widened dramatically on a very short timescale, the like of which has only been observed in the central Asian republics and in Russia. In rural regions, labour migration to the coastal industrial zones is one of the few alternatives to employment in agriculture. According to various estimates, some 200 million people in China are employed as migrant workers (People's Daily Online, 2007). Often they earn an income only minimally above the absolute poverty line.

In environmental respects, too, China's economic development is having considerable repercussions. There is extreme urban air pollution, with most cities recording two to six times higher than acceptable levels according to the WHO. China leads the rest of the world in sulphur dioxide emissions. State emissions reduction programmes conflict with the rising demand for energy, which is why even outdated coal-fired power stations remain in commission. One consequence of the severe air pollution is acid rain, which affects 30 per cent of land area, principally in the south. In some 60 per cent of Chinese rivers, water quality is so poor that people should avoid coming into contact with it. In many of the country's cities, drinking water quality poses a health hazard due to the presence of toxins, and three-quarters of lakes have been contaminated by the extremely high use of fertilizers and pesticides in Chinese agriculture. Around 30 per cent of industrial effluent and 60 per cent of urban wastewater is channelled into rivers without purification. Much of China's domestic, industrial and special waste is dumped on insecure landfill sites. Furthermore, China is a destination country for waste exports, particularly plastic and electronic waste, yet the recycling sector remains largely unregulated and lacks expertise in good recycling practices. Chemical plants are often located in direct proximity to the water reservoirs for densely populated areas, representing a huge potential hazard to the water supply. Frequent accidents in industrial facilities also contribute to environmental pollution. The China Human Development Report (UNDP, 2002) concluded that if broad-scale environmental degradation persisted, it could potentially undo all the positive economic development achieved.

A marked increase in social unrest has been recorded in China for the last few years. In 2005, according to official data, there were 87,000 social conflicts and protests (Willmann, 2006). The head of China's State Environment Protection Administration (SEPA) speaks of as many as 510,000 incidents over the same period. The primary causes of these protests are said to be the expropriation of agricultural land for industrial use, the pollution of agricultural resources, and conflicts between company managers and migrant workers (Cody, 2005a, b). The central government is certainly conscious of the social and ecological challenges it faces, and is attempting to bring economic growth into harmony with sustainable development by means of appropriate legislation. So far the government has been successful in presenting itself as competent to solve the problems. Nevertheless, it is questionable whether the relevant policies will be implemented effectively in future, because economic development still tends to be given absolute priority, especially at local authority level.

PLAUSIBLE DEVELOPMENTS BY 2020
Based on this background, the following represents a conceivable set of developments up to 2020:
- The east coast is increasingly hit by severe typhoons. This is also the region where the country's economic potential is concentrated.
- Economic disparities between the east coast and other parts of the country continue to grow. It is not possible to create sufficiently attractive jobs in regions of the interior. Internal migration intensifies further. Political efforts to equalize these disparities are hampered by local bureaucracy.
- At the same time, there is growing dissatisfaction among the farming population, which feels disadvantaged by the industry-friendly policy of local functionaries and sees the basis of its livelihood jeopardized by pollution and degradation of the local environment.
- In the next two decades, assets and functions continue to be concentrated in the metropolitan centres and special economic zones on China's east coast. This prompts the Chinese government to launch various initiatives to promote the less-developed provinces in the west of the country. These largely fail, however, due to the inefficient bureaucracy and the overwhelming economic attractiveness of the existing industrial zones. Per-capita income continues to rise in coastal metropolitan areas such as Hong Kong, Shenzhen, Shanghai and Tianjin. The principal beneficiaries of this trend are established entrepreneurs and highly skilled workers.
- Disadvantaged groups in the population – primarily the rural population – become more prone to conflict, reaching a point where confrontations with the regime are no longer spontaneous but the result of more advance preparation by various groupings. Discontentment also rises among the unemployed, underemployed, migrant workers and ordinary workers. Several hundred thousand low-qualified workers are unable to find work at all, or only for very low pay. Older workers have particular difficulty in surviving in this system and are often completely reliant on the meagre support family members can offer. Meanwhile the middle class that has now emerged achieves a distinct improvement in its standard of living. The years of economic upturn at the turn of the millennium seem to have come to an end.
- The system is stabilized primarily by private enterprises which maintain close and profitable relations with the various levels of the public administration. The People's Liberation Army has become increasingly politicized since 1989, gaining in strength and benefiting at the same time from economic liberalization by setting up a private financial empire under its auspices. Economic growth continues at a rate of several percentage points per year. Numerous trading partners protect themselves against the flood of goods from the People's Republic by means of covert import quotas, in the attempt to control their balance of trade deficit with China.

FICTITIOUS CONFRONTATION SCENARIO:
MAINTENANCE OF POWER AND NATIONALISM
The following scenario can be imagined: the Chinese state cannot fully cope with the challenges due to structural weaknesses, and is increasingly perceived by the population as powerless to act. Consequently the government concentrates the bulk of its energy on remaining in power and attempts to sideline civil society actors in the political arena using repressive methods. For the most part, social and environmental problems remain unresolved. Water, air and soil pollution, crop failures due to droughts in the north of the country, bottlenecks in the energy supply and the exploitation of workers dominate international reporting. The east coast is hit by increasingly strong typhoons and floods. In many cases, this results in damage to industrial facilities and to extreme pollution of the environment in localized chemical spill incidents. Although the state attempts to improve its disaster mitigation, the growing divide between government and civil society militates against any comprehensive and far-sighted approach to disaster mitigation.

In the year 2025 the Chinese east coast is swept by an unusual series of severe typhoons. The Pearl River Delta region is particularly hard hit: storm tides of unprecedented volume surge up the network of rivers in the delta, penetrating far inland, and torrential rains swell rivers to record levels. The first typhoon is enough to flood large areas of the region's settlements and industrial plants. Before the situation in the disaster area has normalized, a second typhoon devastates the Fuzhou section of the coast.

The authorities are completely overwhelmed due to the scale of the disasters. In a few locations, which are actually less at risk, compulsory evacuations are carried out. Meanwhile the residents of other settlements are abandoned to the mercy of the floods. The response is painfully slow in many places, and relief consignments take more than a week to reach the most seriously affected regions. Desperate victims of the disaster loot shops and warehouses. Riots break out at relief distribution centres. When the government orders the People's Liberation Army to restore public order in the disaster areas, there are repeated clashes between the incensed population and the security forces. Eye-witnesses report of shots fired into crowds and executions of suspected looters. At the end of the hurricane season the situation in the disaster areas settles down, but the population's confidence in the political leadership remains badly shaken. Furthermore the incidents have repercussions for the country's economy. Although the government prioritizes reconstruction of the disaster regions as quickly as possible, foreign investors take an increasingly sceptical view of the Chinese coast as a suitable production location. Aside from the hazards of typhoons and floods, it is above all the tense internal political situation that convinces many economic decision-makers that their planned investments will be better deployed in other countries. Consequently the country goes into an economic and political recession. With rising unemployment and social deprivation, public revenue declines and is no longer sufficient to finance core social institutions. The legitimacy of the government is questioned more and more openly, and various provincial governments publicly distance themselves from the central government. The government invokes a policy of national unity and blames the growing instability on various ethnic minorities and neighbouring Russia. Observers interpret this step towards nationalism as a deliberate distraction from the unresolved domestic policy challenges. Subsequently, military operations are launched in the western border provinces.

FICTITIOUS COOPERATION SCENARIO: REFORMS AND DEVELOPMENT

The following scenario can be imagined: the Chinese leadership responds vigorously to the challenges of political and economic transformation. Comprehensive reforms of the administration create increasing transparency about administrative decisions. Legal security is strengthened, and there are moves to upgrade democratic policy elements, especially at local level. Civil society actors are deliberately encouraged to help resolve social and environmental problems. Increasingly they are seen within political decision-making processes as important contributors of ideas. Through high levels of investment in the environmental sector, in many places China achieves significant reductions in the scale of environmental pollution, improving the quality of life in the urban centres. By boosting energy efficiency and massively expanding the use of renewables, progress is also made with reducing emissions of greenhouse gases. Nevertheless, the east coast is hit by increasingly strong typhoons and floods. In many cases, this results in damage to industrial facilities and extreme environmental pollution resulting from chemical accidents. Learning from this experience, disaster preparedness is comprehensively reorganized. For the first time, priority is given to mitigation, and civil society groups are involved.

In the year 2025, the Chinese east coast is swept by a series of severe typhoons. The Pearl River Delta region is particularly hard hit. Storm tides of unprecedented volume surge up the network of rivers in the delta, penetrating far inland, and rivers swell following torrential rains. The first typhoon floods parts of the region's settlements and industrial plants. Two weeks later, a second typhoon devastates the section of coast near Fuzhou.

Despite certain weaknesses, the official response to events is confirmed as competent. In cooperation with the population there is a controlled evacuation of residential areas at risk, while the army constructs dams to protect infrastructure and settlements and ensures that supplies of food, water and medicine are maintained. At the end of the hurricane season, the government convenes a national dialogue process in order to discuss the future management of such disasters. One key outcome is the insight that a response to the growing risks of disaster is only possible within a multilateral framework. Consequently China steps up its involvement in United Nations bodies. The Chinese economy shows continuing growth, not least because of the high degree of legal security and general political stability.

6.4.3.2
Hurricane risks in the Gulf of Mexico and the Caribbean

BACKGROUND

Owing to the relatively warm water temperatures at the sea surface, hurricanes have regularly struck the Gulf of Mexico and the Caribbean. Areas at risk include the Caribbean islands as well as the neighbouring coasts of Central America and the USA. In particular, the Gulf Coast of the USA and Mexico is predominantly low-lying. Protection from storm surges is afforded only by a few wetland areas, offshore sandbanks and islands. In many locations, however, these are subject to rapid erosion (USGS, 2005). In the USA especially, a growing concentration of material assets along the coastline is evident. This is explained by the prestigious nature of the location for leisure pursuits and the presence of several economic centres (Houston-Galveston, New Orleans, Miami). Settlements on the American Gulf Coast are also undergoing a process of subsidence in many locations. Parts of New Orleans are already 3m below sea level, and some districts of Houston-Galveston are also at or below sea level.

The Gulf of Mexico is the site of substantial oil and gas resources, which are extracted with the help of drilling platforms and a wide-ranging system of pipelines. Oil refineries are concentrated on the coast. They not only process the locally drilled oil but also convert the bulk of oil imports into high grade products for the US market (Jones, 2005). The entire oil and gas infrastructure in the Gulf of Mexico is considered highly vulnerable to strong hurricanes. When Hurricane Katrina tore through the Gulf of Mexico, 90 per cent of refinery capacity had to be shut down. Oil rigs are regularly evacuated as a result of storm warnings, at which point oil and gas extraction is largely suspended. Depending on the extent of the damage, it can take several weeks for full drilling capacity to be restored. In 2005, the USA was more than usually dependent on imports of refined oil products (Jones, 2005). There were also major delays in bringing the majority of refineries back into operation due to the extent of storm and water damage. In the summer of 2005, two severe hurricanes swept the Gulf of Mexico one month apart (Hurricane Katrina and Hurricane Rita), seriously impairing drilling capacity for months afterwards (Energy and Environmental Analysis, 2005). Hurricane Ivan (2004) triggered numerous marine mudslides which caused considerable damage to coastal pipelines (Münchener Rück, 2006).

The 2005 hurricane season was characterized by an abnormal number of very strong hurricanes. At the same time, higher than average surface temperatures were measured in the Gulf of Mexico. Many scientists assume that hurricanes will increase in intensity as climate change progresses (Section 5.1.3).

PLAUSIBLE DEVELOPMENTS BY 2020

In the light of this background, the following set of developments up to 2020 is conceivable:

- Surface temperatures in the Gulf of Mexico continue to rise. As a consequence, there is a rise in the statistical average frequency of Category Four and Five hurricanes, the two highest severity levels. The hurricane season is seen to begin earlier and earlier and to end later in the year.
- The concentration of property and assets in threatened areas remains high. Insurance companies withdraw from the storm and flood insurance market for private customers, because the expected claims potential far exceeds the premium levels that the market will bear. Insurance cover can only be kept in force for major industrial installations with the help of state guarantees. Once private assets in areas at particular risk are no longer insurable, citizens shift the focus of new private investment to better protected regions further inland. Initially, however, continued expansion of the industrial installations takes place. Although the state draws up and implements sophisticated disaster preparedness plans for important infrastructure facilities, the privately operated oil and gas infrastructure remains largely beyond the scope of such measures. The industry is far more concerned with adapting its refinery capacity for crude oil from new oilfields, and with further expansion. No thought is given to shifting the industry's geographical focus. Property owners in risk zones on the south and east coast of the USA face a difficult plight, since their property is practically unsaleable. Many residents are convinced that upgrading the physical infrastructure would be an adequate response to the heightened storm and flood risks. Underlying this attitude is the vain hope that threatened and now uninsurable property could at least still be saved.
- Overall, the US government has taken the risks of climate change more seriously since the experience of Hurricane Katrina. Nevertheless, the authorities are still responding too slowly. Warnings issued by the National Oceanic and Atmospheric Administration, the US Geological Survey and the Federal Emergency Management Agency are, in isolated instances, prompting the relocation of certain functions from hazardous areas. Overall, however, a substantial proportion of expensive infrastructure and material assets remains in the coastal zones.

- Around the year 2020, hurricane experts discuss adding a sixth category to the Saffir-Simpson hurricane rating scale. The reason for this proposal is the observation that hurricanes with wind speeds of over 300km/h have occurred repeatedly.

FICTITIOUS CONFRONTATION SCENARIO: STORM, REFUGEE AND OIL CRISES

The following scenario can be imagined: in each year between 2020 and 2030 there are almost two dozen maximum severity hurricanes in the Gulf of Mexico and the Caribbean. In addition, up to four serious hurricanes occur every year in the Atlantic. Major parts of the Atlantic coast of Florida, Georgia, North and South Carolina and the American and Mexican Gulf Coast are repeatedly devastated by storms. Many coastal sandbanks, islands and wetlands suffer severe damage from hurricane storm surges, and are left undefended at the mercy of erosion. As a consequence, populated coastal regions are increasingly exposed to the force of storm floods. Miami, New Orleans and numerous smaller locations are so badly devastated by various disasters that an overwhelming proportion of the population moves away for good. Those who remain behind are predominantly sections of the population on low incomes. The wealthier population retreats into specially designed protected settlements. These are still near to the coast but are constructed on artificially banked up plateaus with all essential infrastructure facilities and capable of functioning independently. Similarly, the Caribbean islands and the countries of Central America experience a rising incidence of storm and flood disasters. Their governments find themselves increasingly unable to cope with the humanitarian consequences of these events. In Honduras, Jamaica and the Dominican Republic there are recurrent violent protests by disaster victims against ruling governments. Independent observers witness the three countries sliding towards 'Haitian conditions'. At the same time, ever-increasing flows of migrants head for the USA. Especially after the end of each hurricane season, many people attempt to reach the US coast on boats and rafts.

A Category Six hurricane hits Houston-Galveston, destroying a large amount of its oil and gas infrastructure. The American government tries to subdue oil price increases by selling strategic oil reserves, but given the long-term loss of such a major proportion of refinery capacity the strategy only succeeds for a period of a few weeks. In the generally strained global political climate, various states engage in rivalry with the USA and exploit the weakness of its economy and its government by delaying previously promised consignments of oil. In broad sections of the political sphere in the USA, these steps are taken as clear provocation. There are mounting calls for military intervention to safeguard US interests.

FICTITIOUS COOPERATION SCENARIO: TRANSNATIONAL COOPERATION

The following scenario can be imagined: major parts of the Atlantic coast of Florida, Georgia, North and South Carolina and the American and Mexican Gulf Coast are repeatedly devastated by storms. Many coastal sandbanks, islands and wetlands are severely damaged by hurricane storm surges. In many locations, nevertheless, far-sighted coastal and biotope protection programmes are able to prevent complete erosion. In view of the scale of damage, US Congress passes a three-stage emergency programme. As the first priority, it specifies that all public and private facilities in areas subject to storm and flood hazards must be fitted with certain physical protection measures within six months. Next, disaster mitigation must be made the overriding regulatory principle for zoning plans within two years. Finally, local and national environmental protection and emissions control programmes must be implemented within eight years. To implement the short-term measures first and foremost, state funding is provided for low-income sections of the population while non-implementation is sanctioned with fines. Despite criticism from some quarters, the programme mobilizes major investment in physical and organizational disaster preparedness projects. After just a few years, a decline in the average extent of damage is registered in relation to the statistical mean, which is remarkable since no fall in the frequency or intensity of hurricanes has yet been recorded. Just six years after the programme is adopted, economists and disaster-preparedness experts agree that the value of the damage already prevented far exceeds the volume of investment committed to the first and second phases of the programme.

There is a similar rise in the incidence of storm and flood disasters on the Caribbean islands and in the countries of Central America. Their governments find themselves increasingly unable to cope with the humanitarian consequences of these events. After every hurricane season, many people attempt to reach the US coast using boats and rafts. In response to this ongoing problem, and encouraged by the positive results in its own country, the American government initiates a broad-scale disaster preparedness and mitigation programme for this region. Bilateral technical support and advice is offered on the implementation of individual projects, and cooperative links are strengthened on the political and civil society levels. Consequently, political relations with all the United States' southern neighbours improve.

> A Category Six hurricane hits Houston-Galveston, destroying part of its oil and gas infrastructure. Although large amounts of American refinery capacity are put out of action, oil price increases can largely be prevented, not least thanks to short-notice supplies promised by Venezuela.

6.4.4
Recommendations for action

In many cases, preventive disaster preparedness measures are limited to the construction of physical infrastructure. This is far from utilizing the full potential for comprehensive mitigation and there is often considerable scope for improvement, particularly in terms of organizational preparedness and simulation exercises, school-based education, land-use planning and advance clarification of decision-making structures. In bilateral and multilateral cooperation with disaster-prone states, relevant approaches should be supported and promoted.

- *Minimizing urban land subsidence:* In coastal cities and other rapidly growing urban areas, land subsidence is becoming an increasing problem. In conjunction with sea-level rise and the growing frequency of extreme weather events, this is significantly exacerbating disaster risks. In order to keep subsidence within bounds, in many locations it is necessary to develop alternatives to the current water supply. To this end, experience from countries with similar problems should be pooled and made available to the rapidly growing coastal cities in newly industrializing countries. Appropriate contact partners and coordinators would be international alliances of cities and local authorities such as the 'Local Governments for Sustainability' network (ICLEI).
- *Fostering sustainable consumption patterns to conserve forests:* Deforestation in river catchment areas and the destruction of mangrove forests almost always lead to a higher risk of flooding. The conservation of forests and coastal ecosystems is therefore an issue of the utmost importance. Particularly in poorer regions, it is essential to reconcile the interests of different actors (e.g. forest industries, agriculture, upstream and downstream riparian communities). It is also necessary to determine how new product groups and changes in consumer behaviour can be harnessed in support of forest conservation. For instance, heightened demand for non-timber forest products (e.g. for the cosmetics and pharmaceutical industry) may upgrade the economic value of intact forests. Rapidly growing demand for sustainably produced foods (organic and fair trade certification) is another trend that can be harnessed to exercise more influence over the conversion of forests into agricultural land. Last but not least, the certification of timber products offers an opportunity to boost the economic viability of sustainable forestry. In addition to the necessary scientific analysis of the issues, a conducive policy setting needs to be established to promote sustainable consumption patterns.
- *Setting up 'people centred early warning systems':* Early warning systems are pivotal components of disaster prevention. Effective reduction of the number of victims and level of damage relies upon warning systems which are technically functional and ensure that information reaches the potentially affected population in good time. In addition to the technological challenges, administrative information and decision-making channels must be defined and well-rehearsed. Many examples show that translating relevant warnings into practical instructions that the public can easily understand is often a critical bottleneck. The idea of involving potentially affected population groups is often overlooked, but this is the only way of tailoring warnings to the underlying cultural and political realities. Furthermore, long-term education and training programmes are required in risk-prone areas, to be sure that actors in the affected areas are clear about the allocation of responsibilities in the event of a disaster warning, and about what the people affected are expected to do in such a situation. Relevant approaches should be driven forward on all levels and integrated with one another (Sections 9.1.3 and 10.3.4.1).
- *Using disaster mitigation as a lever for good governance:* Effective disaster mitigation presupposes long-term, far-sighted cooperation between numerous societal and political actors. Although in most cases government bodies bear the principal financial and organizational responsibility, comprehensive strategies also rely on close cooperation with civil society groups and the private sector. Within the political system, a clear allocation of responsibilities and decision-making authority is an absolute prerequisite for effective mitigation and response measures. This applies both to cooperation between ministries, government bodies and public authorities and to the allocation of powers between national and local level.
Particularly in developing and newly industrializing countries, these aspects of governance are often less than adequately developed. This acts as a major constraint upon disaster mitigation and indeed upon political and economic development of all kinds. Furthermore, disaster mitigation is an especially useful vehicle for improving gov-

ernance structures. Ideally, if disaster-prone states undertake the necessary reform efforts, support will be channelled to many political and societal groups, enabling the emergence of a climate in which entrenched institutional dynamics can be overcome with comparative ease. With regard to reform efforts in Germany's partner countries, it is therefore recommended that greater emphasis should be placed on disaster mitigation within the framework of consultancy strategies.

- *Promoting targeted conflict resolution after natural disasters:* Where conflict-prone regions are hit by natural disasters, there are major risks that conflicts will escalate further. The role of the national government and the international community must be to provide rapid and effective humanitarian assistance and to counteract any growing resentment in the population. Besides alleviating human suffering and avoiding the escalation of conflicts, in these situations it is entirely possible to motivate the conflicting parties to enter into negotiations. Particularly in cases where the response to a humanitarian crisis requires cooperation across conflict fractures, the international community should press for cooperation to be consolidated and extended.
- *Adapting to unavoidable climate change in industrialized countries too:* Although developing countries are the most commonly affected, industrialized countries must not cling to a false sense of security. They, too, have a vulnerability which can pave the way for security problems or magnify them. Productivity in modern industrial societies is dependent to a great extent on transport, energy and information networks. These infrastructure networks are hugely important in a globalized and highly networked world, but also highly sensitive. This is demonstrated time after time by power outages, transport hold-ups caused by weather events, or overloading of telecommunications lines.

WBGU recommends a review of disaster mitigation instruments in industrialized countries in relation to the challenge of advancing climate change. In the planning of highly sensitive infrastructure, the expected consequences of climate change must be taken into account. For example, important infrastructure and new residential development should not be located in flood-prone areas. A review of existing installations should be undertaken to see if relocation would be advisable. For coastal cities, in view of rising sea levels and the potentially increased threat posed by storms, coastal defence plans need to be reviewed and adapted to the requirements of climate change.

6.5
Conflict constellation: 'Environmentally induced migration'

6.5.1
Background

The flight of hundreds of thousand of people from Hurricane Katrina was described by the Earth Policy Institute in Washington in September 2006 as 'the first documented mass movement of climate refugees' (Brown, 2006). Although a direct link between anthropogenically induced climate change and the occurrence of an individual hurricane cannot be scientifically proven, the case nevertheless provides potent evidence of the scale that environmentally induced migration has already reached.

Migration is one of the oldest coping strategies for dealing with changeable environmental conditions (Box 6.5-1); it may, for example, be a response to periods of drought. The perceived increase in the intensity and geographical scale of environmental change has nevertheless led several authors to speak of environmentally induced migration as a new type of phenomenon (Bächler and Schiemann-Rittri, 1994; Suliman, 1994; Nuscheler, 2004).

The opportunities and risks associated with environmentally induced migration are hotly debated. The effects of migration in general are multifaceted. Migrating workers often make an important contribution to the economic and social development of their host country. The money they send home also has a significant positive influence on the economic and social situation in their country of origin. But despite the positive effects of migration, public discussion is dominated by a discourse of risk that is weighted with emotionally loaded metaphors. Especially since the terrorist attacks of 2001 migrants are often viewed by their host countries as a threat to internal security (Nuscheler, 2004; Faist, 2005; Tränhard, 2005; Leighton, 2006). It needs to be stressed, though, that media treatment and consequent public perception of this issue is often not based on the results of scientific analysis. For example, the West African boat people who land on the coast of Spain frequently attract a great deal of media attention in the EU, although numerically they represent only a fraction of global refugee and migration movements.

6.5.1.1
Structure of the conflict constellation

At the centre of this conflict constellation, migration represents a typical transfer mechanism linking grad-

> **Box 6.5-1**
>
> **Migration – definitions and trends**
>
> CONCEPTS AND DEFINITIONS OF MIGRATION
> The term migration covers in general all forms of displacement and flight that take place voluntarily or involuntarily and across or within national borders (Jahn, 1997; GCIM, 2005; UNHCR, 2006a, b). If migration takes place within a country it is referred to as internal migration. Where internal migrants are forced to leave their places of origin for reasons that would qualify them to be recognized as refugees under the Geneva Refugee Convention, they are termed Internally Displaced Persons (IDPs) (cf. with regard to the definition Geissler, 1999).
>
> Migration movements can also be categorized according to the precipitating factor (war, political persecution, natural disasters), the distance of move (migration within a country, south-north migration), the duration of displacement (temporary departure from the region of origin, permanent emigration), the goal of the migration (rural-urban migration, south-south migration, south-north migration) and the outcomes in the countries of origin and arrival (successful integration in the destination country, return to country of origin, conflict) (Black, 2001; Kröhnert, 2003; Paul, 2005; Clark, 2006).
>
> INTERNATIONAL MIGRATION – FIGURES AND TRENDS
> Migration is now a major structural element of a world characterized by ever more complex economic, political and cultural linkages. Nevertheless, changes in the pattern of international migration are apparent – triggered by modern communication networks, demographic trends and increasing workforce mobility. Whereas in the past the geographical range of migration was relatively small and migration was often a unique event in an individual's life, migration today often represents a repeated and reversible process of collective action. Groups of individuals, such as households or organizations, make use of complex migration networks or are drawn in to institutional migration systems that have developed in the course of history (Hillmann, 1996; Massey et al., 1998; Gogolin and Pries, 2003).
>
> The number of international migrants has risen constantly in recent decades; there are now estimated to be 191 million such people (GCIM, 2005; IOM, 2006). Between 30 and 40 million of these are regarded as 'irregular migrants', who thus constitute 15-20 per cent of the total number of international migrants. This means that overall some 3 per cent of the global population does not live in its countries of origin.
>
> Refugees are included in the migration statistics. The UNHCR estimated that in 2006 there were 8.4 million registered refugees worldwide. Refugee numbers fluctuate sharply over time because even within the space of a few months the size of refugee flows can change considerably depending on local socio-economic and political conditions. The UNHCR (2006a) currently puts the number of Internally Displaced Persons at 23.7 million.
>
> It is anticipated that migration will gain in importance in the future (IOM, 2006; UNHCR, 2006b), with qualified workers benefiting from the advantages of migration while the barriers for the impoverished and poorly educated become higher (UNFPA, 2006). As yet, the much-quoted storming of 'Fortress Europe' by a rush of migrants from Asia and Africa has not taken place (UNDP, 2002; Nuscheler, 2004). Despite the relatively small number of Asian and African refugees who manage to reach Europe, the issue has attracted increasing attention in recent years. It is clear that even these small numbers are politically sensitive.

ual environmental degradation or weather extremes on the one hand and conflict on the other. The conflict constellation describes a relationship between humans and the environment that is divided into two phases. In the first phase people leave their homeland because of environmental degradation or extremes of weather. This environmentally induced migration may be triggered directly by environmental changes. This is the impact mechanism on which the conflict constellation focuses. Alternatively, migration may be indirectly triggered by environmental changes; this occurs if environmental changes lead to the outbreak or escalation of conflict and this in turn leads to migration (Sections 6.2, 6.3 and 6.4). In the second phase of the conflict constellation environmentally induced migration can trigger conflicts, or escalate existing ones, in the regions that migrants come from, pass through or travel to, or in their region of origin when they return. A spiral of environmental degradation (or weather extremes), migration and conflict that spreads to other regions can be set in motion if these conflicts are themselves linked to the destruction of the natural environment and unleash migration movements in their turn.

6.5.1.2
Environmentally induced migration as a core element of the conflict constellation

TERMINOLOGY

The term 'environmental refugees' is often used in connection with the issue of environmentally induced migration. El-Hinnawi (1985) describes environmental refugees as 'those people who have been forced to leave their traditional habitat, temporarily or permanently, because of a marked environmental disruption (…) that jeopardizes their existence and/or seriously affected the quality of their life.'

The appropriateness of the term *environmental refugee* is, however, much disputed in the literature. There is criticism of the assumption that environmental changes are the sole cause of the migration (Keane, 2004). There is as yet no empirical evidence that this is the case. Instead, there is a complex interaction in which environmental changes play a part alongside political and socio-economic factors (Lonergan, 1998; Biermann, 2001; Black, 2001; Castles, 2002). The peasant who flees because his field can no longer be farmed leaves his home because of poverty and because of a lack of other opportunities for earning a living – ultimately, therefore, he can be described as an economic refugee. The person who flees because ecological problems have triggered social disorder, as in Rwanda, is at the end of the day fleeing from war and violence. These people are therefore refugees of war. The term 'environmental *refugee*' is also criticized because it is misleading from a legal point of view. According to Art. 1A no. 2 of the Convention relating to the Status of Refugees of 1951 (Geneva Refugee Convention, supplemented by the Protocol of 1967) 'refugees' are people who had to leave their country of residence because of 'well-founded fear of being persecuted for reasons of race, religion, nationality, membership of a particular social group or political opinion'. These criteria do not apply to migration of which the primary causes are environmental; those affected by environmentally induced migration are therefore not protected by the guarantees accorded to refugees in international law.

Both the United Nations Refugee Agency (UNHCR) and the International Organization for Migration (IOM) have therefore spoken out against the use of the term 'environmental refugees'. The preferred phrase is 'environmentally displaced persons', defined as 'persons who are displaced within their own country of habitual residence or who have crossed an international border and for whom environmental degradation, deterioration or destruction is a major cause of their displacement, although not necessarily the sole one' (IOM, 1996; Keane, 2004). Taking account of the above-mentioned criticism and drawing on the definition of UNHCR and IOM, the German Advisory Council on Global Change (WBGU) uses the term 'environmental migrant', because the term 'migrant' has a much broader meaning than the legal term 'refugee' (Marugg, 1990; Jahn, 2000). The term 'environmental migrant' will be used here to describe anyone who migrates because environmental changes either (1) have such an unfavourable effect on living conditions that previously achieved income levels and standards of living can not be maintained or (2) destroy structures that are necessary for the maintenance of these levels and standards (Lonergan, 1998; Wenzel, 2002; Salehyan, 2005). Depending on the type of environmental change and the courses of action open to those affected, a distinction can also be made between planned environmentally induced migration in the face of gradual environmental degradation and sudden environmentally induced migration as a response to extremes of weather. The distinction is, however, not a rigid one (Hugo, 1996; Clark, 2006).

CURRENT ESTIMATES OF ENVIRONMENTALLY INDUCED MIGRATION AND FORECASTS OF THE FUTURE SITUATION

Figures for the number of environmental migrants worldwide vary depending on the definition and source data used. For example, Myers (1993, 2002) arrived at a figure for the middle of the 1990s of at least 25 million environmental migrants; he expects there to be 50 million by 2010 and up to 150 million by 2050. These figures have been frequently quoted – by, among others, the International Red Cross (IFRC, 1999), the IPCC (2001) and the United Nations University Institute for Environment and Human Security (UNU-EHS, 2005). In the literature, however, they have been harshly criticized for being inconsistent and impossible to check and, above all, for failing to take account of opportunities for adapting to the effects of climate change. Protecting coasts by dykes, for example, reduces the number of coastal dwellers who are forced to migrate because of floods caused by the rise in sea level and thus reduces the number of environmental migrants (McGregor, 1993; Kibreab, 1994; Black, 2001). Nuscheler concludes that it cannot be known how high the number of environmental migrants currently is (Nuscheler, 2004). Stern, however, stresses that within the context of current climate models Myers' estimates can be regarded as plausible (Stern, 2006).

The structure of environmentally induced migration can be described with greater certainty. It is thought likely that most such migration currently takes place within national borders and that this will continue to be the case in future. Environmental migrants are therefore more likely to be internally displaced persons rather than migrants who cross national borders. Most cross-border environmentally induced migration will probably take the form of south-south migration; no trend towards large south-north migrations has been identified (UNPD, 2002; Nuscheler, 2004; Clark, 2006; UNFPA, 2006).

It is also thought likely that gradual environmental degradation will cause significantly more people to migrate than weather extremes. The link between gradual soil degradation and migration is already recognized in the Desertification Convention (UNCCD) (Prologue and esp. Art. 17 Section 1 (e) UNCCD; UNCCD, 1994).

Despite many uncertainties about the delimitation of the definition of environmentally induced migration and its measurement, one fact remains certain in the face of the expected effects of anthropogenically induced climate change: the part played by environmental degradation and weather extremes as causes of migration will increase (Lonergan, 1998; Biermann, 2001; Nuscheler, 2004; Salehyan, 2005; Stern, 2006; IPCC, 2007a).

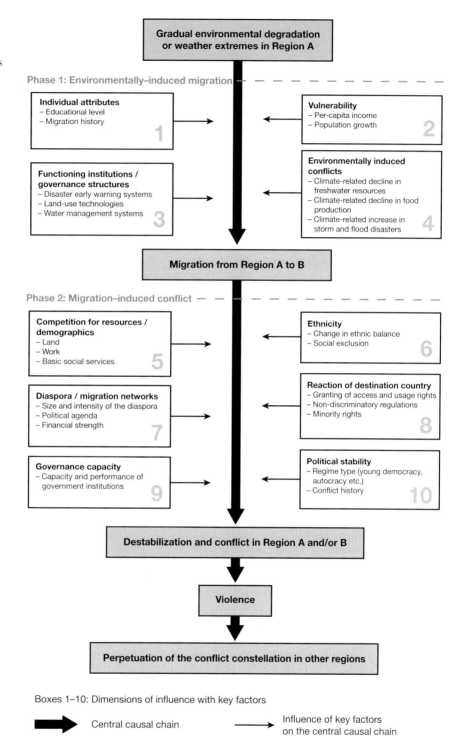

Figure 6.5-1
Conflict constellation:
'Environmentally induced migration':
Key factors and interactions
Source: WBGU

6.5.2
Causal linkages

In research into environmental conflict, the environmental change–migration–conflict linkage is one of the most frequently mentioned scenarios and topics for case studies (Nordås and Gleditsch, 2005).

6.5.2.1
From environmental change to migration

The first phase of the conflict constellation is triggered by environmental changes (Sections 6.2, 6.3 and 6.4). These are so marked that they undermine the basis of people's livelihoods, jeopardize their future income or have detrimental impacts on their health. As a result, those affected are forced to emigrate (El-Hinnawi, 1985; Hugo, 1996; IOM, 1996; Keane, 2004). The extent to which environmental changes actually give rise to environmentally induced migration depends both on personal characteristics of the affected individuals and on various external conditions (Phase 1 of Fig. 6.5-1):

Influence dimension *[1] Individual attributes*: Environmentally induced migration is usually chosen by an individual as a coping strategy. The decision to migrate is therefore significantly determined by individual attributes such as age, level of education (e.g. knowledge of foreign languages), the degree of traumatization after natural disasters (Grote et al., 2006) and the subjective perception of general structural conditions (Lee, 1972; Kröhnert, 2003). In addition, case studies reveal the importance of a history of migration; people are more likely to leave their home region if there is already awareness within the extended family or the community of the use of migration as a coping strategy (Epstein and Gang, 2004).

Influence dimension *[2] Vulnerability:* The extent of environmentally induced migration is also influenced to a significant extent by the vulnerability of those affected – that is, by the extent to which the effects of environmental changes and a range of political and socio-economic factors have an impact on them. These factors may overlay or reinforce each other. For example, families' economic vulnerability may be increased by unfavourable aspects of the regional economic structure or the level of regional economic activity, such as a low or markedly fluctuating per capita income, unequal rights of ownership, restrictions on access to markets (labour, credit or sales markets) or the absence of social security arrangements (Kalter, 2000; Clark, 2006).

Influence dimension *[3] Functioning institutions and governance structures:* Consideration of the way in which natural disasters are dealt with serves to show that the extent of migration is significantly determined by the functionality of the relevant local and national institutions. In countries that lack early warning systems or evacuation plans, extreme weather events cause relatively greater damage and compel more people to flee than is the case in countries that are institutionally well prepared for emergencies (Section 6.4). The same applies to the problem of gradual environmental degradation. For example, it has been shown in Section 6.3 that continuing soil degradation can be avoided through efficient land-use technologies and land-use systems.

Influence dimension *[4] Environmentally induced conflicts:* Migration can also be triggered by environmentally induced conflicts. In this case the link between environmental change and migration is indirect, because the initial consequence of environmental change is conflict (Bächler, 1998). It is only in a second phase that those affected are forced to flee regions that have become violent and politically unstable. At first glance these refugees appear to fall into the category of refugees of war. In fact, however, they belong to the category of environmental migrants, because environmental changes are the cause of their migration. From this point of view conflict can be interpreted as a transfer mechanism (Homer-Dixon, 1999; Salehyan, 2005). Reference is frequently made in this connection to the example of Rwanda (Box 6.3-1). Some scientists take the view that the genocide of 1994 was only in part a consequence of the politicization of the issue of ethnicity among the groups involved; they see an additional cause in the impoverishment of large sectors of the population as a result of soil erosion and high population density (Percival and Homer-Dixon, 1995; Gasana, 2002; Diamond, 2005).

6.5.2.2
From migration to conflict

If environmental changes do indeed lead to environmentally induced migration, one still needs to identify the mechanisms by which such migration can trigger or exacerbate conflict. The recent past yields little evidence that large migration movements have been the cause of conflicts. There is nevertheless some empirical evidence that migration can – sometimes significantly – increase the likelihood of conflict (Goldstone, 2002). Whatever the situation, it is evident that different types of migration involve different security risks (Lohrmann, 2000). Thus a sudden mass exodus after extreme weather events and planned environmentally induced migration in the face of gradual environmental degradation present differing challenges to the societies that are affected (Phase 2 of Fig. 6.5-1).

Influence dimension [5] Competition for resources / demographics: The likelihood of migration-induced conflict increases if environmental migrants have to compete with the resident population for scarce resources such as land, accommodation, water, employment and basic social services, or if the immigrants are perceived as competitors. A competitive situation of this sort is particularly likely to arise in regions in which population growth is strong (Hugo, 1996; ODI, 2005; Urdal, 2005; Clark, 2006; Sections 3.3.2.2 and 3.3.3.1). Competition for resources can also trigger conflict when migrants return to their regions of origin. It is possible, for example, that returnees may lay claim to their land or property, only to find that ownership or rights of use have in the mean time been granted to others (Hensell, 2002; Black and Gent, 2006; Hansen, 2006).

Influence dimension [6] Ethnicity: The likelihood of ethnic conflict may increase if the arrival of environmental migrants upsets the 'ethnic balance' in the region. This change of ethnic balance may lead to xenophobia and violent confrontation, or further destabilize an already labile ethnic situation (Homer-Dixon, 1999). Goldstone (2002) also refers to 'clashes of national identity' that may be caused by immigration if the dominance of one resident ethnic group is threatened. Ethnicity may in addition exacerbate conflict if ethnic differences are exploited for political purposes within the context of migration (Lohrmann, 2000) or if membership of an ethnic group leads to social exclusion of the migrants in their country of arrival (Bade, 1996; Rydgren, 2004; Section 3.3.3.2).

Influence dimension [7] Diaspora / migration networks: In recent years scientists and politicians have highlighted the presence of a diaspora as a possible cause of conflict (Lohrmann, 2000). A diaspora describes a religious or ethnic group that has left its traditional homeland and is dispersed over large areas of the globe. Collier and Hoeffler (2004) find there to be a strong positive correlation between both susceptibility to repeated conflict and conflict intensity and the size of the diaspora abroad. Diaspora migrants can play a significant part in the financing of conflict and hence contribute directly to the initiation, escalation or prolongation of conflict within a country. If they are members of rebel groups, they can also introduce weapons or violent ideologies in their transit or arrival countries. There is a risk that refugee camps or emergency settlements will serve as logistic platforms for rebel groups and be misused for strategic purposes (Lohrmann, 2000; UNHCR, 2006b). Existing conflicts are exported in this way to other regions.

Influence dimension [8] Reaction of the government of the arrival country / arrival region: The impact of the conflict factors described above depends to a significant extent on the response of the government in the arrival country. Even now governments are often unable to cope with the task of managing migration. In particular, affected governments face enormous challenges if their institutions are to be able to handle sudden and unexpected migration. This has been demonstrated by Spain's periodic problems in dealing with West African boat people and by the difficulties encountered by the US government in catering for 1.3 million internal refugees after Hurricane Katrina in August 2005. The risk of (ethnic) conflict may be reduced by non-discriminatory regulations, wide-ranging rights for minorities and opportunities for including minorities in political decision-making processes (Cramer, 2003; Addison and Murshed, 2003; Korf, 2005; Reynal-Querol, 2005; Nordås and Gleditsch, 2005; Regan and Norton, 2005). There is inherent potential for conflict if environmental migrants are treated as 'temporary guests' and in consequence experience restriction of their rights with regard to freedom of movement, freedom of residence, acquisition of property and access to the labour market (Jacobsen, 2002; Crips, 2003). On the other hand, the granting of widespread rights of access and use may, as has already been mentioned, itself provoke conflict – for example, if environmental migrants compete with locals in a difficult labour market (Martin, 2005). Where environmentally induced migration takes place suddenly after extreme weather events, disaster management plays an important part in determining the further unfolding of the crisis and eventual conflict (Jacobsen, 2002).

Influence *dimension [9] Governance capacity:* Governance capacity is particularly relevant in the context of environmentally induced migration because it is likely that the majority of environmental migrants will seek refuge in developing countries with weak governmental structures. The receiving countries usually have little hope of adapting to the changing situation without outside help, since their institutions are poorly developed and there is a general lack of resources and know-how. Situations in which migrants overstretch the capacity of the authority in the reception region are therefore a potential cause of conflict. In countries that are also prone to economic crises, the influx of environmental migrants further increases the likelihood of conflict (Reuveny, 2005).

Influence dimension [10] Political stability: Conflict research has also established that the likelihood of conflict within a country depends on the degree of political stability there (Section 3.3.6). Conflict becomes more likely if there has already been conflict in the recent past – that is, if there is a history of conflict. The likelihood of conflict also increases if there has recently been a change of government or if the country in question has only recently gained independence (Sambanis, 2001; Gates, 2002).

6.5.3
Scenarios

Fictitious, narrative scenarios of confrontation and cooperation will now be described for Bangladesh as well as for North Africa and the neighbouring Mediterranean countries. The aim is to illustrate the need for action and indicate how prevention and adaptation might be addressed (Section 6.1). These scenarios are based on facts relating to the current situation in these regions and probable future developments, taking account of the expected effects of climate change over the next few decades. The primary concern here is not to describe the scenarios that are thought to be most likely.

6.5.3.1
Environmentally induced migration and conflict in Bangladesh

BACKGROUND
Bangladesh is already affected every year by natural disasters. Its geographical position on the Gulf of Bengal and in the delta of the three great rivers, the Ganges, Bramaputra-Jamuna and Meghna, combined with precarious socio-economic conditions, makes the country vulnerable to floods. With an annual per capita income of US$440 Bangladesh is one of the least developed countries in the world. 144 million people live on 144,000km² of land – that is, on an area just 40 per cent the size of Germany. This makes Bangladesh the most densely populated country in the world. Storm floods, flooding during the summer monsoon and droughts repeatedly wreak destruction and force millions of people to migrate temporarily or sometimes even permanently. It is estimated that the erosion of river banks alone forces a million people a year to leave their homes (Zaman, 1991; Schmeidl, 1997). In addition, tropical cyclones and tornadoes cause damage to property worth several thousand million US$ annually (Ali, 1999; Section 6.6).

PLAUSIBLE DEVELOPMENTS BY 2020
Based on this background, the following represents a conceivable set of developments up to 2020:

- It is assumed that in the 2020s the economic situation in Bangladesh is extremely unstable and the country is embroiled in a major political crisis. As a result, initial positive developmental trends, such as the halving of poverty and broader integration into the world market, have not fulfilled their early promise. Political elites thwart the spread of democracy; urgently needed reforms are put on the back burner. Widespread corruption is a major hindrance to development. A broad middle class, which could institute economic and political reforms, is extremely slow to develop (Section 7.6). In addition, the annual costs of the damage caused by natural disasters now amount to almost 15 per cent of GDP. In 2020, Bangladesh could be one of the countries in which the effects of anthropogenically-induced climate change, floods, drought and coastal erosion are already very clearly visible.
- The sea level around Bangladesh has risen considerably – by about 12cm since 2000. This represents the greatest threat of all. As a result, the coastal regions around the Gulf of Bengal are constantly losing valuable residential land, fertile farmland and coastal ecosystems such as the Sundarbans – unique mangrove swamps which provide a livelihood for more than two million people in Bangladesh. Three million people have had to leave their homes; most of them have moved to the urban slums, significantly accelerating the process of urbanization. The most catastrophic damage is caused by storm floods; as a result of the rise in sea level these floods have greatly increased in recent decades in both frequency and destructiveness. Larger and larger flood waves batter the coast, carrying brackish water far into the country's interior. With the regular floods during the summer monsoon, this contributes to the flooding of large areas of the country. Bangladesh is sinking under water. The small number of defence works that were built at the beginning of the century as part of an integrated coastal zone management system and financed by international aid provide insufficient protection for the residential districts and business structures that are concentrated along the coast. Existing early warning systems are useless because the population is not trained to respond appropriately to warnings and the roads that would make rapid evacuation possible do not exist. There is neither the financial and technical capacity nor the political will to build modern structures. The World Bank has rejected a request to co-fund a major coastal protection project. Withdrawal to the interior of the country as an adaptive strategy is not an option; there is simply not enough space.

FICTITIOUS CONFRONTATION SCENARIO: DISASTER MANAGEMENT FAILS

The following scenario can be imagined: stronger and stronger cyclones affect the coast of Bangladesh. The Ganges delta, which has already been badly damaged by storm floods, offers little resistance to the destructive power of the winds. Wind speeds of up to 180km/h on land and flood waves up to 12m high cause devastation. Hundreds of thousands of people are killed. Millions of coastal dwellers are made homeless and flee the disaster areas.

After the disaster it becomes clear that the government of Bangladesh is helpless; effective emergency aid plans have never been drawn up. Moreover, lack of clarity in the allocation of responsibilities and in the exchange of information between the government, local contacts and international organizations regularly results in chaos. Because there are no roads in the affected regions or existing roads are blocked, the urgently needed aid is slow to reach the disaster victims. It also becomes clear that corrupt elites are misappropriating goods supplied as aid. The hurriedly erected refugee camps lack food, medicines, sanitation and doctors.

The international aid organizations take months to stabilize the situation locally. Even then the fate of the environmental migrants remains uncertain. Those who are physically and emotionally able to do so leave the refugee camp as quickly as they can. However, many of the landless farmers do not return to the coastal regions because fields have been destroyed by salinization, making farming – and hence the chance of earning a livelihood – impossible. Instead these people flee further into the interior of the country. But the conditions they encounter there are little better. The local population is itself suffering from hunger and a lack of medical supplies. Landlessness, underemployment and poverty are common. The new arrivals are perceived as competing for scarce resources. In some villages Hindu families are dispossessed by Muslim environmental migrants. The military looks on but does not intervene. The situation re-ignites ethnic conflicts between Hindus and Muslims that had supposedly been resolved through the granting of minority rights, causing some two million Hindus to migrate to neighbouring India.

It is not only the Hindu minority of Bangladesh who take refuge in India. Because of its geographical proximity and its generally better living conditions it is increasingly becoming a destination for environmental migrants. Many of these migrants cross the border illegally, often with the aid of people smugglers who know how to get through the border fence erected by India. Many people die in the attempt.

India itself has for many years been concerned about the stream of immigrants from Bangladesh. It is estimated that 40 million people are now living in the country illegally. Because of the historically strained relationship between the ethnic groups, the majority Hindu population rejects the Muslim migrants. The immigrants are held responsible for the 20 per cent rise in the crime rate and are suspected of supporting militant Islamic groups. They are not offered the prospect of long-term integration. Very few are granted a work permit. The situation is particularly tense in the Indian states of Bihar, West Bengal and Assam, which adjoin the border and which attract the most migrants. Immigration in these states is so strong that in some places the local population is in the minority and sees its livelihood threatened. As a result there is frequent local unrest.

The cross-border migration also gives rise to diplomatic tensions between the two countries. India claims that Bangladesh's failure to take appropriate steps to reduce environmentally induced migration is an attempt to solve the problem of rapid population growth and the associated social and economic problems at India's expense. Because the two sides are not prepared to engage in political dialogue, the environmental migrants become a security problem. The political conflict between the two states escalates; India threatens Bangladesh with 'humanitarian intervention' on the pretext that the environmental migrants represent a terrorist threat.

FICTITIOUS COOPERATION SCENARIO: CRISIS MANAGEMENT SUCCEEDS

The following scenario can be imagined: at the beginning of the 2020s the population becomes increasingly dissatisfied with the government, which appears incapable of dealing with the acute threat of climate change. The people demand the introduction of long-delayed political reforms and measures designed to protect them against natural disasters. Their anger is vented in demonstrations. At international level, too, political pressure grows. International donors press for more of the funds they have provided to be invested in disaster prevention instead of, as so often, being used only for emergency aid programmes and reconstruction measures. In addition, more stringent conditions must be met to access the funds, including the requirement for the receiving country to tackle corruption and promote democracy.

In Bangladesh these measures are successful. The government bows to political pressure and introduces the urgently needed reforms. As a first step the government approves an integrated coastal zone management system, to be set up in cooperation with international experts; it will take account of the concerns of the local population and anticipate environmental changes induced by climate change. Because Bangladesh has neither the financial means nor the technical know-how to implement the project, it turns to international organizations. At the world environmental conference in 2025 it successfully seeks international support for the erection of modern defence works. Subsequent negotiations among donor countries result in the initiation of a unique coastal protection project costing US$40,000 million; the funding is provided from the climate compensation fund set up in 2018. The industrialized countries, as the main emitters of global greenhouse gases, make payments into this fund; the money is used to finance adaptation measures in developing countries. This process takes particular account of the Sundarbans, which provide the west coast of Bangladesh with an irreplaceable natural protective wall against tropical cyclones. Reafforestation measures, laws against illegal logging and a sustainable land utilization plan contribute to the conservation of the ecosystem and thus to coastal protection.

In the summer of 2038 Bangladesh is affected by an unusually strong cyclone season. Within the space of a few weeks a number of powerful cyclones strike the coast of the Gulf of Bengal, causing widespread devastation. However, due to effective early warning systems there are only a small number of deaths. Despite this, the damage inflicted is severe, particularly on the coast; thousands of environmental migrants cannot return immediately to their villages and must be accommodated in refugee camps in the interior of the country. The government of Bangladesh acts quickly; immediately after the disaster it approaches the international community for support in providing for the environmental migrants. Although transport routes are blocked and the number of people in need is high, efficient disaster management plans and good cooperation between the government, international organizations and local contacts pay off – aid is quickly provided and in the months that follow reconstruction gets quickly under way. Nevertheless, an estimated one million people, disheartened by the destruction and the continuing threat of storm floods and cyclones in future, attempt to cross the border into neighbouring, economically prosperous India.

India has for years been concerned about the enormous influx of illegal immigrants from Bangladesh, and erects a high-technology fence almost 4000km in length along the border. This attitude on India's part and its reluctance to enter into political dialogue fuel a political conflict between the two neighbouring countries that has been simmering for years. The international community and the Association of South-East Asian Nations, of which both countries are associate members, is deeply concerned about the security situation in the region.

Because India's borders remain closed to Bengali environmental migrants even in the face of natural disasters, the international community increases the political pressure. After years of simmering conflict this at last results in negotiations, and a strategy paper on dealing with illegal migrants is published. This includes agreements under which Bangladesh undertakes to draw up plans for restricting emigration. In return, India undertakes to support Bangladesh with disaster prevention measures. The parties to the negotiations also reach agreement on plans for a transnationally coordinated coastal management system and the drawing up of a joint emergency aid strategy.

Under the eye of the United Nations implementation of the agreements proceeds promptly in the years that follow, which has the additional effect of improving diplomatic relationships between the two countries.

6.5.3.2
Environmentally induced migration and conflicts in North Africa and neighbouring Mediterranean countries

BACKGROUND
The region comprises the countries of North Africa which border on the Mediterranean (Morocco, Algeria, Tunisia, Libya and Egypt) and the Sahel zone (i.e. primarily the countries of Mauritania, Mali, Niger, Chad and Sudan, plus parts of Burkina Faso, Senegal, Nigeria, and Ethiopia, and Eritrea and Djibouti). The entire region is already beset by severe problems including freshwater shortages, soil degradation and desertification, whereby the situation is deteriorating steadily as a result of climate change as manifested especially by appreciable declines in precipitation (Sections 5.2 and 7.2). Because of the Sahel's location, desertification and the effects associated with it are particularly virulent there (Section 7.3).

A central factor in both the current and the future problems of the region is the demographic trend: all the countries named above currently have high rates of population growth and urbanization. Migration movements are already observable and take two forms: there is, firstly, internal migration from rural regions to the towns and cities, and secondly – and increasingly – cross-border, northwards migration, primarily of young people. North African countries play a special role in migration movements on the continent: they are themselves migration destinations (for both internal and cross-border migrants), but at the same time they are a transit area for people from sub-Saharan Africa and Asia attempting to reach Europe (transit migration). In certain locations, this has already given rise to problems including social unrest and attacks on migrants, human trafficking and an increasing number of fatalities from attempted sea-crossings. With regard to political stability it is noticeable that the entire region is affected – albeit to differing degrees – by weak governance structures. Radical fundamentalist religious movements pose a further major problem in all the countries studied.

PLAUSIBLE DEVELOPMENTS BY AROUND 2020
In view of this background, the following sequence of developments is conceivable for the period up to 2020.
- The already observable effects of global climate change (heat, reduced precipitation) exacerbate the problems of resources. Desertification and water shortages increase in the whole of North Africa and all the countries of the Sahel. The need for drinking water and water for irrigation increases, while precipitation (in the Atlas Mountains and elsewhere) continually decreases. Because of overuse of the soil, agricultural productivity drops sharply – including along the Mediterranean coast and in the Nile valley, the most fertile parts of the region. The density of population along the Mediterranean coast increases the environmental stress in this region in an unprecedented manner.
- In the Sahel the amount of agriculturally viable land declines significantly by 2025. In the rural regions of the Sahel and in the North African countries along the Mediterranean traditional homelands shrink rapidly as a result of desertification. The broad mass of the rural population is unable to take steps to compensate for the negative consequences of climate change. Strategies that have for centuries enabled local people to survive the prevailing climatic conditions are now inadequate. It is not only climate that is causing desertification: widespread poverty is a contributory factor, and on account of population growth an increasingly significant one. By 2020 the population of the Sahel countries will have quadrupled since 1960. Furthermore, it is estimated that between 2025 and 2050 the population of North Africa will increase by around another 50 million (UN DESA, 2005). Even if population growth is stabilized in the coming decades, the rural exodus and the concentration of population in the cities will continue to increase. The food security situation will become increasingly precarious as a result of environmental degradation, forcing entire villages to abandon their land and start a new life in the urban agglomerations.

FICTITIOUS CONFRONTATION SCENARIO:
DESTABILIZATION IN NORTH AFRICA IMPACTS ON EUROPE
The following scenario can be imagined: alongside the worrying demographic trend and the negative features that accompany it, the situation with regard to natural resources continues to deteriorate after 2020. The climate factor dramatically worsens the resource crisis, and by the middle of the 21st century the rise in sea level becomes an acute problem – particularly in Egypt, where the viability of coastal cities such as Alexandria and other cities of the Nile Delta is threatened. In the Sahel countries climate changes cause such decline in the productivity of former agricultural and cattle-farming regions that by the middle of the 21st century only a 'rump' of the population remains. Famines are an almost annual occurrence. Disputes over the allocation of the few remaining areas that can still be farmed give rise to regular outbreaks of violence between local ethnic groups, especially between those who were formerly nomadic and traditionally settled population groups.

With no prospects in any of the countries of the region, many young people see migration to Europe as their only opportunity. For migrants from the Sahel region the Maghreb countries are usually the first stopping-off point on the journey they hope will take them across the Mediterranean to southern Europe. Towards the middle of the 21st century the northwards migration of predominantly young men from the countries of the Sahel takes on the proportions of a 'Völkerwanderung', i.e. mass migration such as was seen in Europe during the Dark Ages. Every year hundreds of thousands of people from the Sahel and the tropical areas of West and Central Africa arrive in the North African coastal regions. As a result, enormous slum settlements housing stranded emigrants arise in the urban agglomerations of the Maghreb.

The year 2020 sees the start of serious social and political destabilization in all the countries affected by this migration. The rural exodus caused by desertification and drought leads to further urbanization; as a result the urban slum areas expand further and their population density is very high. The situation of economic hopelessness generates enormous potential for political destabilization among young people who have no prospects; the urban slums threaten to become lawless areas. This creates a breeding ground for the further radicalization and spread of extremist religious movements.

The governments of the countries most affected manage to stay in power by subjecting religiously oriented political movements to increasing repression. They do this with the support of various countries with which they have signed treaties on the extraction of oil, gas and other resources (uranium in Niger). The number of countries pursuing their own interest in resources in the region has steadily increased, with Russia and China joining the ranks of the traditional stakeholders such as France, Great Britain and the USA. The population of the region has become increasingly dissatisfied with what is seen as a 'sell-out' of national resources; as a result, attacks in the infrastructure of resource extraction are becoming more and more common. In the first decades of the 21st century China's political and economic power becomes increasingly significant. After the USA China is now the second largest economy in the world; in the pursuit of its resource interests it is beginning to do deals with those in power in the region, a move that is increasingly seen by western states as a threat to their own interests.

The drastically increased severity of the water shortage throughout the region leads to an armed international confrontation in the Nile catchment area: Egypt, which is 95 per cent dependent on water from the Nile, is concerned about Ethiopia's plan to safeguard its own water requirements by building a dam on the Blue Nile. The international community intervenes in an effort to negotiate a treaty that would regulate the use of Nile water on conditions that would be fair to all the countries involved, but the attempt founders. The Egyptian government then despatches troops to occupy the province of Amhara in northwest Ethiopia.

Developments in North Africa have a significant impact on Europe. The European countries that are the primary destination of migrants have a need for workers, but the need is far exceeded by the number of illegal immigrants. Because illegal migrants are not integrated into society, there is increased ghettoization of North African immigrants. Xenophobia increases, and the immigrants react to their difficult circumstances by turning in large numbers to radical religious groups (a choice favoured by the huge popularity of these movements in the migrants' home countries). European countries fear that immigration from North Africa will allow the infiltration of more and more members of extremist groups into Europe, and so they take steps to strengthen 'Fortress Europe'. Under the Common Foreign and Security Policy of the EU and NATO, increasingly high priority is given to measures to halt the influx of illegal migrants from the Mediterranean area. All the North African countries with the exception of Egypt have signed an agreement with the EU that commits them – in return for compensation payments – to stemming the tide of migration in their own country. With the support of European states, internment camps are set up on the edge of the Sahara; there are reports of revolts within the camps and of force being used in ways that infringe human rights.

FICTITIOUS COOPERATION SCENARIO: ADAPTATION SUCCEEDS WITH INTERNATIONAL HELP
The following scenario can be imagined: the positive scenario arises from similar changes in the natural environment. It is based, however, on the assumption that both at domestic level and internationally, the steps that need to be taken are promptly identified and effectively implemented.

With considerable support from the international community in sustainable methods of resource management and soil cultivation, desertification is slowed in many parts of the region and in some places even stopped. As part of these measures agriculture is extensified rather than intensified; job losses are prevented, however, because mechanization of agriculture is avoided and the market for sustainably produced products guarantees increased sales opportunities and higher prices on the world market. In large parts of the Sahel this now provides local communities with adequate income to ensure their livelihood; as a result what was previously a driving force of migration has significantly diminished. Conflicts between regional ethnic groups and central states – such as the hostility between the Tuareg and the governments of various Sahel countries that has been simmering for years – have been checked by the consistent promotion of participatory and federal political structures.

The rising revenue from the extraction of mineral resources provides North African states with the funds needed to introduce economic and political reforms, which lead to a reduction in social problems. There is a noticeable opening up of previously autocratic regimes. A similar process takes place in some of the Sahel countries, where money brought in by the resource extraction sector is used to finance improvements to social infrastructure. International cooperation also plays a significant part in the positive trend. Under the Barcelona Process of the European-Mediterranean Partnership, the Mediterranean area is established as a region, a move that paves the way for cross-border cooperation transcending religious and cultural divides and gives concrete form to this cooperation. An example of such collaboration is the migration policy model, which envisages that European countries will accept a certain number of labour migrants each year. It anticipates quotas for citizens from North African states and from other African countries. This defuses the situation. There is rapprochement between the affected countries themselves. As members of the Maghreb Union the North African states have agreed on the creation of a regional economic area and are working towards making this a reality. In the Nile catchment area the expansion of the 'Nile Basin Initiative' has resulted in a successful international cooperation programme that is making a significant contribution both to tackling desertification and to political stability in the region.

6.5.4
Recommendations for action

Environmental stress only develops its potential to trigger conflict when it is compounded by adverse social, economic and political circumstances (Clark, 2006). As shown, there are many key factors that exert an influence on the different phases of the conflict constellation. Thus possible political action strategies for the avoidance of conflict are correspondingly diverse in nature. Bearing in mind the analysis of causal linkages and the outcomes of the scenarios described above, possible strategies will now be described in more detail.

6.5.4.1
Avoiding environmentally induced migration

- *Avoiding dangerous climate change:* The selection of efficient and effective political strategies should focus first on the avoidance of environmentally induced migration. The most important strategy here is the avoidance of dangerous climate change (Section 10.3.2).
- *Reducing vulnerability:* Even large-scale avoidance measures will now not completely avert the effects of climate change. There is therefore a need for the development of regional and national strategies to reduce the vulnerability of the local population and structures to these effects. Such strategies are a way of reducing environmentally induced migration. Appropriate measures include not only efficient coastal and irrigation management systems but also disaster prevention instruments, early warning systems, disaster relief plans and plans for the coordination of reconstruction measures in the wake of natural disasters (Sections 6.2, 6.3, 6.4; WBGU, 2006). Emigration can also be reduced by exerting an influence on social, economic and political factors; for instance, by promoting economic growth and more equitable distribution of its benefits. Specific factors that reduce individuals' vulnerability are effective social and micro-insurance systems, adequate health provision and strong government institutions (WBGU, 2005, 2006).

6.5.4.2
Managing environmentally induced migration

If migration flows cannot be halted, strategies to control environmentally induced migration can help prevent conflict. Environmental migrants need not inevitably represent a security risk if receiving countries take appropriate measures.

SUDDEN ENVIRONMENTALLY INDUCED MIGRATION AFTER EXTREME EVENTS

- *Accompanying measures:* The first step in the management of temporary environmentally induced migration after extremes of weather is to put disaster relief measures in place; in particular, this involves setting up emergency settlements and refugee camps. In order to reduce the potential for conflict, it is important to ensure that migrants are adequately provided with food and medical care. In addition, steps should be taken to prevent the infiltration of rebels and the spread of weapons in refugee camps. These should be erected as far away from conflict regions as possible and at sufficient distance from national borders, in order to avoid being affected by the overspill of existing conflicts. Conflicts can arise if different ethnic groups have differential access to aid and locally available resources; aid measures should therefore be adapted to local needs, with the resident population being involved in the relevant decision-making processes (OECD-DAC, 2003).

Extremes of weather, in particular, can suddenly trigger the migration of large numbers of people, who exceed the absorptive capacity of the region in which they arrive. Such situations call for multilateral strategies, such as agreements under which other countries temporarily take in environmental migrants. In the case of developing countries, which will normally be dependent in such cases on the financial support of international donors, the financing and implementation of such strategies can – as in the past – be coordinated under the aegis of the UN, in particular the UNHCR.

- *Government-induced environmentally induced migration:* If the threat posed by sudden environmental changes in a region is so severe that protective measures are no longer politically or economically justified, the state itself may encourage migration. In high-risk areas such as coasts exposed to tornadoes and storm floods, national resettlement programmes prepared for such a contingency would ensure that migration takes place in more orderly fashion than might be the case if there were a sudden mass exodus after an extreme weather event. They could also ensure that the conditions awaiting displaced people in the destination areas are conducive to integration (WBGU, 2006). However, experience gained from previous government resettlement programmes (such as the relocation of the settlements of Garzweiler and of Großgrimma in Germany to make way for opencast lignite mining, or the 'Transmigrasi' project in Indonesia) shows that – despite careful and circumspect planning, the efforts of decision-makers to ensure social acceptability, and the acceptance by displaced persons of the need for resettlement – such strategies can have considerable conflict potential. Conflict may be triggered by, for example, the level of government compensation payouts, the subjectively perceived favouring of some sectors of the population, or general frustration over the loss of one's home. The possibility of heightened conflicts over resource use with the population in the receiving region must also be considered (Scholz, 1992; Berkner, 2000, 2001; Hansen, 2003; WBGU, 2005; ADB, 2007). It is therefore essential for government resettlement programmes to strike a careful balance between security risks and the needs of the local population (Section 9.3.4).

Environmentally induced migration as a result of gradual environmental degradation

- *Integration of environmental migrants in the target regions:* In managing environmental migrants who migrate permanently from their regions of origin, measures to ensure their integration in their destination region are crucial. Such measures include, for example, legal access to employment. Dialogue with local communities aids the early recognition of economic and social tensions and helps to defuse them. Evidence also indicates that violence tends to erupt where there has been a failure to acknowledge cultural differences between the local population and the immigrants and to take account of them in formulating an integration policy. Governments can play a part here too: conflict can be reduced through the introduction of measures – such as the facilitating of access to public services and social activities – aimed at preventing discrimination against immigrants. The creation of a legal framework that enables immigrants to acquire citizenship in the medium term also aids integration. Last but not least, the risk of ethnic tensions can be reduced by safeguarding minority rights and adapting the electoral system and federal structures to take account of minorities. Such steps must be accompanied by measures to encourage integration and to promote tolerance among the local population – that is, they must be actively promoted politically.

- *Multilateral coordination:* Where environmentally induced migration takes place across national borders, steps must be taken to prevent illegal immigration. Political liaison between the countries of origin and arrival can help to defuse possible tensions. Through international cooperation, strategies such as the imposition of take-up quotas can be developed.

- *Repatriation agreements with environmental migrants:* National and international policy must not neglect the repatriation of environmental migrants. In particular, rights of ownership and use must be regulated by the government authorities, especially when periods of many years elapse before environmental migrants return to their regions of origin. In this situation there is frequently a need for dispute settlement mechanisms. Such systems should aim primarily at achieving a consensual resolution of disputes through the mediation procedures that are put in place; access to the courts should be available to the parties involved as a last resort. Where people return to areas that have been the scene of conflict, repatriation agreements drawn up in collaboration with the former parties to the conflict can help to counteract specific factors that could trigger conflict.

6.5.4.3
Supporting developing countries

Migration currently takes place mainly within and between developing countries; it seems that this continues to be the case where environmentally induced migration is concerned (Clark, 2006). On account of inadequate infrastructure, the lack of political and economic stability and limited economic capacity, developing countries must also be regarded as particularly susceptible to conflict. Implementation of the measures described above for managing environmentally induced migration in these countries therefore requires international donors to play a part in coordination and funding. An appropriate forum for discussion of these issues might well be the International Dialogue on Migration set up in 2001 by the International Organization for Migration (IOM, 2001). The Dialogue could also serve as a platform for organizing a system by means of which costs can be fairly and efficiently shared between affected and unaffected countries (Section 10.3.4.3).

Particular challenges arise in dealing with environmental migrants who are classed as Internally Displaced Persons (IDPs) in their country of origin. Compared with cross-border refugees, IDPs have only limited protection under international law, even when they are victims of internal unrest and armed conflict (Geissler, 1999; Phuong, 2004). In such cases the international community can only complement the work of national bodies when affected governments are themselves unable to ensure the welfare of IDPs. The leading role in caring for IDPs is played by the UNHCR, with the support and cooperation of other UN organizations and relevant NGOs (in particular the International Red Cross and the International Organization for Migration) (Phuong, 2004; OHCHR, 2006; UNHCR, 2006a). There is, however, as yet no procedure for the binding allocation of responsibilities under international agreements. In order to deal appropriately in future with the pressing problems of internal environmentally induced migration, these omissions in the provisions of international agreements should be remedied.

Economic and social development is in itself an effective strategy for preventing environmentally induced migration, because it strengthens countries' ability to adapt to environmental changes, increases the capacity of institutions to regulate migration movements and thus reduces susceptibility to security risks arising from environmentally induced migration. In addition, the closer networking of development, environmental and migration policy should be pursued; this would facilitate the identification of related problems and the utilization of synergy effects in selecting appropriate instruments (WBGU, 2005; Section 10.3.4.3).

6.5.4.4
Instruments of international law

It is likely that growing numbers of people will be affected by environmentally induced migration and migration movements will more and more frequently take place across national borders; in view of this it is crucial that the problem of environmentally induced migration is dealt with at the level of international law (Section 10.3.3.3). Because environmental migrants are not covered by the Geneva Refugee Convention and because the current international refugee regime should not be dismantled, the required measures need to be effected outside the existing refugee regime. The most appropriate course of action would be to seek to regulate the legal position of environmental migrants by drawing up a separate cross-sectoral multilateral convention.

A future regime designed to protect environmental migrants should at a minimum cover the following aspects, each of which should involve the entire international community: acknowledgement of environmental damage as a cause of environmentally induced migration; protection of environmental migrants through the granting of at least temporary asylum; establishment of a formula for the distribution of environmental migrants which ensures that among potential host countries no individual states are overburdened; establishment of an equitable formula for the distribution of the costs of receiving refugees; equalization of the financial burdens of climate-related environmental degradation.

Hotspots of climate change: Selected regions

Building upon the analysis of global warming impacts performed in Chapter 5, the following chapter identifies a set of regions in which, due to their special ecological, demographic or socio-economic features, climate change will present particularly major challenges in the next decades. The assessment of problem-solving capacity and of susceptibility to conflict in the individual regions uses the findings of others chapters which reviewed the body of research on environmental conflict and the causes of war and identified past hotspots with an environmental element of causation (Chapter 3), explored the effects of state fragility and a changing international system upon conflict dynamics (Chapter 4) and constructed conflict constellations arising from anticipated climate change (Chapter 6).

The synopsis of conflict research performed in Chapter 3 yielded several key findings. First, environmental degradation is always only one of several factors which can amplify or even trigger conflict. There is a complex interplay among these factors. Whether environmental degradation translates into destabilization, collapse of order and finally into conflict depends upon specific aspects of the social setting. Second, the environmental conflicts of the past did not transcend the local scale, nor did they compromise international stability and security. Third, it became clear that the problem-solving capacity of states – but also of stakeholders in society – plays a key role, both in the emergence and in the resolution of conflicts. States and societies are especially prone to violent conflict when several of the following attributes apply:

- They are in a process of political transition (i.e. they are neither clearly democratic nor clearly autocratic in constitution).
- They are at a low level of economic development and are marked by wide social disparities.
- They have a large population or a high population density.
- They are characterized by rough terrain or border on a neighbouring country in which a violent conflict is being waged.
- They have themselves experienced violent conflict in the very recent past on their own state territory.

The above attributes are particularly likely to lead to conflict if they arise in societies in which state structures and capacities are weak and the problem-solving capacity of social actors is small.

Furthermore, the discussion of state fragility, governance issues and probable turbulence in the international system makes it plain that national and international actors are ill-prepared for the challenges ahead (Chapter 4). In particular, it must be feared that growing climate-induced pressures will overwhelm the capacity of states that are already considered weak or fragile today to perform their central governance functions. In this situation, the present architecture of global governance lacks the necessary capacity to counter these problems effectively.

The conflict constellations explored in Chapter 6 illustrate how conflicts can develop out of the interplay among diverse environmental, political and socio-economic factors. Such conflicts have an inherent potential to endanger human security, to contribute to the destabilization of societies to the point of state collapse, to increase the tendency to use violence both within and among states, and to engender cross-border insecurity. In this context, WBGU has identified the following conflict constellations as likely for the future: climate-induced degradation of freshwater resources, climate-induced decline in food production, climate-induced increase in storm and flood disasters, and, finally, environmentally induced migration.

These conflict constellations may arise with different frequency and intensity in the various regions of the world. The following chapter therefore sets out climate change impacts specific to particular regions, together with their consequences for the biosphere and human society. This is followed by a discussion of the political and socio-economic situation in each region, leading on to an assessment of that region's problem-solving capacity and susceptibility to conflict.

It is worth stressing here once more that WBGU considers climate-induced conflicts conceivable from roughly the mid-2020s onwards if climate change continues unabated. The social sciences lack the tools to predict societal developments reliably over such long periods. The following analysis therefore concentrates on present developments in the selected regions, and then uses the findings to identify, in a form of risk analysis, plausible development trends. WBGU makes no claim that the trends outlined here must necessarily materialize in these forms. Nonetheless, to take decisions for the future, preventive security and development policy needs knowledge which, while it should be as robust as possible, may well be subject to a degree of uncertainty.

The selection of regions presented here is not complete. It does, however, identify those particularly at risk in view of anticipated climatic changes and concrete regional developments. WBGU by no means implies with this selection that climate change will have no security-relevant impacts in the future in other regions. On the contrary – in several regions not dealt with here, major social impacts of climate change are to be expected. It does not appear likely with present knowledge, however, that these impacts will generate security problems that escalate into violent conflict within the coming 20–30 years, at least in the 'Western' industrialized countries (e.g. in Australia, the USA or Central Europe). In contrast, it is evident that climate-related stressors will make it yet more difficult to ease political tensions in notoriously conflict-prone regions such as the Near and Middle East, or the Horn of Africa. However, the aim of this report is to shift the focus of security policy discourse towards those regions that have received insufficient attention until now.

7.1
Arctic and Subarctic

7.1.1
Impacts of climate change on the biosphere and human society

In the Arctic, climate change is giving rise to major warming, resulting in thawing of the permafrost soils of the Arctic tundra and melting of the glaciers and sea ice.

Geographically the Arctic is not clearly delimited, but it is usually defined as the region north of the polar circle. Alternatively, the region's limits are often defined in terms of criteria relating to climate (10 °C July isotherm) and vegetation geography (tree line). For our purposes, the region is taken to consist of the Arctic Ocean and parts of the adjoining seas – including the numerous islands – and the neighbouring parts of the North American, European and Asiatic continents. The Subarctic comprises the climatic zone of boreal coniferous forests that adjoins the Arctic to the south.

The rise in annual mean temperature anticipated for the Arctic in the year 2100 under the A1B scenario is 2.8–7.8 °C (mean 5 °C); this is almost twice as high as the global mean (ACIA, 2005; IPCC, 2007a). Warming in winter, at 4.3–11.4 °C, is likely to be considerably more marked than in summer (1.2–5.3 °C). The anticipated temperature rise accords well with the temperature trends measured in the second half of the 20th century. The anticipated warming will be accompanied by an increase in annual precipitation of 10–28 per cent (IPCC, 2007a).

The effects of this major climate warming are already observable and, if climate change continues unabated, they will intensify further in coming decades (ACIA, 2005). Climate change is further accelerated by the thawing of the permafrost soils of the Arctic tundra, the shrinking ice cover of the Arctic Ocean and the melting of the glaciers and ice sheets. This acceleration is due to the increased absorption of solar radiation by ice-free surfaces and the potential release of methane from permafrost soils and from methane hydrate deposits under the seabed. In addition, the melting of the Arctic glaciers and ice sheets contributes to the rise in sea level (IPCC, 2007b). Thus WGBU estimates (WBGU, 2006) that if global warming is stabilized at 3 °C above pre-industrial levels, the Greenland ice sheet will contribute 0.9–1.8m to sea-level rise by the year 2300.

The retreat of the Arctic sea ice is leading to increased erosion of the coasts by wave action and storms – processes amplified by the rise in sea level and thawing of permafrost soils. This puts coastal settlements at risk. In addition, there is increased risk of rock avalanches and mud floods. Warming also causes vegetation zones to shift northwards, although this is limited by the Arctic Ocean. This shift has an impact on many Arctic animal and plant species. The traditional ranges of important fish stocks are also affected by the warming of the Arctic (ACIA, 2005; WBGU, 2006). Significant warming could ultimately cause lasting damage to the boreal coniferous forests as a result of heat and water stress, the spread of pests and more frequent forest fires (IPCC, 2007b).

7.1.2
Political and economic situation in the region

The largest land areas of the Arctic extend over northern Scandinavia, Russia, the US state of Alaska,

Canada, the larger islands such as Greenland, Iceland, Spitzbergen and Novaya Zemlya and the numerous islands of the Canadian archipelago. Politically the Arctic is divided into a number of sectors to which Russia, the USA, Norway and Denmark lay claim. Some four million people live in the region. Around 10 per cent of these are members of indigenous ethnic groups; in some parts of the region – such as in parts of the Canadian and Greenlandic Arctic – this proportion rises to more than 50 per cent (ACIA, 2005). Overall, the region is very thinly populated. In the Arctic regions of Alaska, Canada and Greenland there are only a few large settlements, although in Scandinavia and Russia there are several bigger cities, such as Murmansk and Norilsk in Russia and Tromsø in Norway. The Icelandic capital of Reykjavík is also an important urban centre for the region.

The Human Development Index (HDI) of the United Nations Development Programme (UNDP) classes all the countries of the region – with the exception of Russia – as highly developed (UNDP, 2006). In the Arctic land areas there are development deficits and in some cases governance problems associated with one-sided economies that focus on resource extraction, but overall the political situation in the region is stable. The most important economic activities in the region are fishing and the extraction of minerals. The Arctic has extensive deposits of important resources; those that are extracted include oil and gas, iron ore, nickel, zinc, coal, uranium, tin, diamonds, gold and cryolite.

The thawing of permafrost soils has a considerable impact on the structure of settlements, their supply and waste management, and on the mining and transport sectors. Transport becomes increasingly difficult on the softer ground. Thus in Alaska the number of days on which driving on the tundra is allowed has fallen in the last 30 years from 200 to 100 (ACIA, 2005). Coastal industrial facilities such as harbours – which are crucial for the export of minerals – are increasingly endangered by coastal erosion. Economic activities could therefore be increasingly restricted as a result of climate change.

However, climate change also presents the region with opportunities. If the polar ice continues to retreat, the extraction of natural resources will become easier and new shipping routes will become viable. The possibility of expanding or improving areas of land used for agriculture is also often discussed, although limits are imposed by the poor quality of the soil, precipitation trends, inadequate infrastructure, the small scale of the local market and the large distances to potential larger markets (ACIA, 2005). New opportunities can, however, turn out to be potential sources of conflict. For example, increased access for shipping brings with it the threat of ecological problems, such as the increased risk of tanker accidents. Because of the slow rate at which oil slicks break down in the Arctic, such accidents cause relatively greater damage here than they do elsewhere. Experts also warn of new social and security risks arising from potential conflicts over newly accessible minerals (Herrmann, 2006). The disputes between Russia and Norway over the Barents Sea, which date back to Soviet times, demonstrate that the risk of resource conflicts is real (du Castel, 2005).

The population of the region is therefore affected by climate impacts in many ways. This is particularly true of the indigenous population. By no means the least important aspect for them is the fact that hunting in traditional ways becomes more difficult on account of the melting of the sea ice cover, changes in the animal populations, and increased weather variability. This has a serious impact on the cultural identity of the indigenous peoples who live there (Bangert et al., 2006).

7.1.3
Conclusions

The Arctic and Subarctic are among the regions particularly severely affected by climate change. Major warming induces the thawing of the permafrost soils and melting of the glaciers and sea ice, which has negative impacts not only on the environment but also on the economy and society. The resettlement of inhabitants of endangered coastal regions and international disputes over claims to land and minerals could lead to future conflicts. However, since the affected regions are very sparsely populated and lie for the most part in politically and economically stable countries, and since various forms of regional cooperation already exist, WBGU estimates that the significance of the Arctic in security terms is low.

7.2
Southern Europe and North Africa

7.2.1
Impacts of climate change on the biosphere and human society

The principal consequence of climate change in the Mediterranean area is increasing aridity. In some regions the impact on agricultural yields can be considerable; in Egypt, for example, soya production could fall by almost 30 per cent by 2050 (IPCC, 2007b). It is thought likely that, as a result of climate change, agricultural production in North Africa will decline by an amount corresponding to 0.4–1.3 per cent of GDP by the year 2100 (IPCC, 2007b).

In southern Europe and the coastal areas of North Africa, the climate is Mediterranean. The subtropical high-pressure area ensures hot, dry summers, while in winter the region receives cooler, wetter weather as the west wind zone shifts southwards. Areas of North Africa away from the coast are affected by the north-east trade wind and have a hot, dry desert climate. According to model calculations based on the A1B scenario, temperatures in the Mediterranean area will rise by 2.2–5.1 °C (mean 3.4 °C) by the end of the century compared to the baseline period of 1980–1999; this is likely to be slightly above the average global warming (IPCC, 2007a). Model calculations identify the greatest warming as taking place in summer, although with considerable temperature fluctuations; temperature variability in winter will be smaller (Giorgi and Bi, 2005). Sea-level rise could impact on the Nile Delta and on people living in the delta and other coastal areas.

Over the last 50 years, winter precipitation in the Mediterranean area has declined; this is ascribed to a trend in the North Atlantic Oscillation (Xoplaki et al., 2004; Scaife et al., 2005). It is anticipated that the greatest decline in precipitation will be in summer. By the end of the century the average annual precipitation rate will likely fall by 4–27 per cent in southern Europe and by on average 20 per cent in North Africa. The higher temperatures will increase the rate at which water in the soil evaporates, exacerbating the summer water shortages that are already a feature of many countries in southern Europe and North Africa (IPCC, 2007b). Rivers, too, are affected: the seasonal differences in the amount of water they receive are becoming bigger, with higher water levels in winter and lower levels in summer, when demand for water is greatest (Santos et al., 2002). This increasing water scarcity has a particularly negative impact on agricultural and forestry yields and on the generation of electricity by hydropower (IPCC, 2007b).

The anticipated climate trends will not only accelerate soil desiccation in southern Europe; they will also increase the risk of wind erosion and lead to fires, including forest fires, which will further compromise the vegetation cover. The result could be a gradual vegetation shift from forest to bush (Mouillot et al., 2002). Such a change in the ecosystem would in turn impact on the quality of the soil and its water retention capacity, the carbon cycle and the local climate. The increase in summer aridity and consequent reduction in soil moisture will further exacerbate soil degradation caused by unadapted agriculture, forestry and dairy farming (Section 5.4). As a result, Mediterranean soils will suffer further from overuse unless more sustainable methods of agriculture are introduced. Plans to increase productivity through irrigation as a response to drought and future periods of extremes must be viewed with great caution: water resources are already scarce and irrigation can result in salination. Soils in large areas of Spain and some regions of Italy and Greece are already salinated (MA, 2005a). If this trend continues it can lead to desertification, with the land then being lost for agricultural use.

The initial situation of the North African countries bordering the Mediterranean is already worse than that of their southern European neighbours: in large parts of Algeria, Libya and Egypt and certain regions of Morocco and Tunisia soil salination is already widespread and at risk of increasing. As a result of overgrazing, deforestation and non-sustainable irrigation practices, the soils are seriously eroded. The vegetation cover is very patchy and likely to decrease further. The subsequent consequences are even lower air humidity and less precipitation – a self-reinforcing cycle. Using fossil groundwater to irrigate agricultural land is not a sustainable way out of this vicious circle. According to the IPCC-SRES scenarios, surface run-off in North Africa will fall significantly by 2050 (Arnell, 2004). It is likely that the speed of soil degradation – which is already classed as high in Egypt – will continue to increase throughout North Africa (Oldeman et al., 1991). The risk of desertification is present in all countries, but particularly on the northern fringes of the Sahara (USDA, 1998; MA, 2005a). Even today, land use is not sustainable; these facts show that it will come under even more pressure from future climate change.

7.2.2
Political and economic situation in the region

There are clear differences between southern Europe (that is, Portugal, Spain, France, Italy and Greece – the Balkan region is excluded here) and North Africa (Morocco, Algeria, Tunisia, Libya and Egypt) in respect of their vulnerability and problem-solving capacity. The southern European countries are consolidated democracies with functioning governance structures and are, in addition, members of the European Union. North Africa, by contrast, is dominated by autocratic regimes that are currently undergoing a phase of political transition (Jacobs and Mattes, 2005; Kaufmann et al., 2006). According to Schneckener (2004), all the North African countries are weak states – that is, states that still, to a large extent, have a monopoly on the use of force but which in some cases have significant deficits in their welfare and rule-of-law functions (Section 4.2). On top of this, they have a history of virulent conflict (e.g. civil war in Algeria, intra-community conflicts in Morocco), suffer periodic social unrest, and harbour radicalized fundamentalist religious movements and an opposition that is ready to use violence. This situation creates multiple problems for people's security and for the stability and problem-solving capacity of the North African countries bordering the Mediterranean.

These regional discrepancies are also reflected in economic and social capacity. In the southern European EU countries, annual per capita income is US$19,000–29,000; in the countries of North Africa, in contrast, it is only US$4,000–7,500 (in purchasing power parities; World Bank, 2006a). The countries of the European Union have established market economy structures, social security networks and supranational equalization systems such as the European Structural Funds. Many countries have established state mechanisms that can provide know-how and financial resources in the event of drought or flooding; there is also an extensive private insurance network and the possibility of European subsidies, for example via the European Disaster Relief Fund.

The countries along the southern shore of the Mediterranean present a different story. Here there are major economic development deficits. Although some countries, such as Algeria, have valuable deposits of fossil resources, North African societies are characterized by poverty, high youth unemployment, wide social discrepancies and scanty state social security networks. Only in the case of Libya does the Human Development Index reveal a better picture (UNDP, 2006). Regional attempts at integration between the North African states – all the states are members of the Arab League – have so far generally been unsuccessful, so that no strong institutionalized structures exist. Within the Euro-Mediterranean Partnership (Barcelona Process), too, issues of climate change and its effects on the region have been almost entirely neglected to date (Brauch, 2006).

In both southern Europe and North Africa, the economy is dependent to a relatively large extent on climate-sensitive sectors. In Greece, for example, agriculture accounts for 7–8 per cent of GDP (World Bank, 2006a). Another water-intensive sector is tourism, which in Spain accounts for 9 per cent of GDP (Chatel, 2006). In addition, hydropower plays a relatively important role in electricity generation in southern Europe. For this reason, the European economies, with their resource-intensive agricultural systems and existing ecological problems, are vulnerable to the anticipated climate effects, too (WWF, 2006; IPCC, 2007a). Nevertheless, loss of pasture and arable land as a result of water scarcity, soil erosion and salination is a far more serious problem for the countries south of the Mediterranean. For example, in Morocco and Egypt a far larger proportion of the population is directly employed in agriculture (44 per cent and 28 per cent respectively) and the contribution of the agricultural sector to GDP is higher, at 15 per cent (World Bank, 2006a). Tourism in these countries accounts for almost 8 per cent of GDP. In addition, increasing water scarcity exacerbates competition for its use between tourism, agriculture and drinking water provision. Resources and institutions that would facilitate protective and adaptive measures, for example more efficient water management, are virtually non-existent.

In combination with the demographic trends, climate-induced environmental changes become a particular problem. Estimates assume that the population in North Africa and the eastern Mediterranean will grow by 40 per cent by 2025 and that the total population will increase by 95 million, of whom 31 million will live in coastal areas (Brauch, 2006). Hence by 2025 agricultural production will be stagnating while the population steadily increases. A factor further amplifying the population pressure is the migration of people from the Sahel region. The North African countries are both a destination and a transit region for migrants, a situation that has already led to violent conflicts in the past (Section 6.5).

The serious overuse of crucial resources such as water and soil is already having a detrimental effect on the region's food security; droughts caused by climate change could in future further accelerate the decline in food production. In 1984, droughts and failed harvests in Morocco, combined with an increase in bread prices, led to violent unrest that was put down by the military (Brauch, 2006). Urban agglomerations such as Cairo, where high population pressure is combined with non-sustainable agriculture, are particularly vulnerable to future climate events. For example, a 50cm sea-level rise in the Mediterranean would enable salt water to penetrate 9km into the coastal aquifers of the Nile Delta. This would have consequences for the agricultural sector (salination) and the whole economy (Brauch, 2006). Furthermore, Egypt is dependent on the Nile for approximately 95 per cent of its drinking and industrial water supply. Climate-induced changes in precipitation and the allocation of water could lead to water conflicts between the ten states that border the Nile, particularly between Egypt and Ethiopia. Previous crises have been resolved cooperatively through the setting up of the Nile Basin Initiative (Stroh, 2005). Increased pressure due to climate-induced problems could, however, overtask these conflict resolution mechanisms (Brauch, 2006; Stern, 2006).

7.2.3
Conclusions

Southern Europe and North Africa are severely affected by climate change through rising temperatures, precipitation variability and rising sea-levels combined with existing environmental problems. However, there are clear differences between the two major regions with regard to their vulnerability and problem-solving capacity. In southern Europe, environmental changes such as droughts and heatwaves are unlikely to lead to outbreaks of violence in the foreseeable future. Relatively high economic and social capability and support from the EU will mitigate the impact and make long-term adaptation possible (Brauch, 2006). The North African countries, by contrast, are far more vulnerable, because their environmental situation is significantly worse than that of southern Europe and their societies have less problem-solving capacity. The interaction of high population growth, the critical importance of agriculture for the economy and society, weak governance capacities and existing conflicts within society make these states, caught up as they are in political reform processes, particularly vulnerable to the effects of climate change. As usable land and water resources become increasingly scarce and use of non-sustainable methods of agriculture continues, desertification will cause further impoverishment and the risk of water- and land-related conflicts at regional and local level will increase throughout North Africa (Brauch, 2006).

However, the risk of destabilization applies not only within these countries; the situation can have consequences for the stability of the entire region. One result of climate change will be further emigration from rural areas to cities and migration via the countries of North Africa to the EU countries. Migration issues will therefore become increasingly sensitive; in southern Europe this could trigger potentially violent conflicts (e.g. the youth riots in France in 2005). In the long term – i.e. after 2025/30 – the possibility of water conflicts between Egypt and individual countries of the Nile basin cannot be excluded if Egypt's water supply is significantly reduced by the actions of countries further upstream, thereby endangering Egypt's viability. In extreme cases, these conflicts might involve violent interstate clashes (Brauch, 2006). The consequences of such an escalation in political and security terms might be felt far beyond the region.

7.3
Sahel zone

7.3.1
Impacts of climate change on the biosphere and human society

While current climate models predict definite warming of the Sahel zone, there are no reliable forecasts of the future change in average precipitation in the region. However, there is a heightened risk that there will be a significant increase in aridity and hence an expansion of the areas affected by drought.

The term Sahel zone describes the semi-arid strip of land running from west to east that divides the Sahara desert to the north from the savannah to the south; it passes through the countries of Senegal, Mauritania, Mali, Burkina Faso, Niger, Nigeria, Chad and Sudan. Even today, the Sahel is one of the regions of Africa most frequently affected by drought. Between 1950 and 1980 it witnessed a marked trend towards aridity. In some places this trend then reversed, although without a return to pre-1950 precipitation levels. This phenomenon was known as 'greening of the desert'. In the model results of Held et al. (2005), however, this situation simply reflects an element of natural variability within a long-term anthropogenic trend towards increased aridification.

For such biogeophysical conditions, population density is relatively high and thus exerts pressure on land use. In many places this has already led to overgrazing, significant soil degradation and desertification. The vegetation cover has been reduced or destroyed, increasing the albedo (i.e. ability to reflect light). The degradation or absence of vegetation cover reduces evaporation, thus leading to a reduction of water vapour in the atmosphere and hence to less precipitation (Schlesinger et al., 1990).

According to climate simulations under the A1B scenario, the anticipated degree of warming by the end of the century could be 2.6–5.4 °C (mean 3.6 °C), somewhat above the average global rise in temperature (IPCC, 2007a). However, current climate models provide no reliable prediction of the future change in mean precipitation in the Sahel zone. For the Western Sahara, in particular, the models yield contradictory results; some anticipate considerable further desiccation of the region, while others predict increasing moisture and an advance of vegetation into the Sahara (IPCC, 2007b). Nevertheless, scenarios involving a future trend towards greatly increasing aridity, and thus leading to an expansion of the areas affected by drought, must be considered (IPCC, 2007b). In addition, models for the Sahara show increased interannual variability in precipitation, leading to an increase in both very dry and very wet years (IPCC, 2007b).

The anticipated increase in annual temperatures brought about by climate change combined with increasing precipitation variability leads one to expect that in future the vegetation cover will reduce further, regeneration or the establishment of new vegetation will be more difficult, and soil erosion and desertification will increase (MA, 2005a). The risk of failed harvests is therefore likely to rise, particularly if no adaptation takes place in agriculture.

Further losses of potential arable land are anticipated not only for the Sahel zone, but for the entire region south of the Sahara, resulting in a deterioration of the food production situation in many countries by 2080 (IPCC, 2007b). According to the MA (2005a), the number of people affected by water scarcity will in future triple in each generation, leading to severe crises in food production and other problems. This could result in increased internal migration as well as migration beyond the affected regions (Brauch, 2006).

7.3.2
Political and economic situation in the region

The Sahel's vulnerability to climate change is amplified considerably by the susceptibility of the region to socio-economic crises. All the countries of the Sahel zone have a high proportion of people living in absolute poverty; they are all among the Least Developed Countries and lie towards the bottom of the Human Development Index (UNDP, 2006). The Sahel states are also characterized by poorly developed transport and communication infrastructures, weak markets, high population growth and weak governance capacities (Karim and Gnisci, 2004). As a result, the volume of international direct investment in the region is very low in international terms. Many countries have no sea ports of their own; in view of the weak transport infrastructure within Africa, this represents a major obstacle to development and the increased transport costs raise prices of imported goods.

The inefficacy of government action is reflected even now in many of the Sahel countries in poor physical security (Grimm and Klingebiel, 2007). Often the state has no functioning monopoly on the use of force. In most outbreaks of violence, a major part is played by non-governmental participants such as rebel groups, warlords and traditional authorities. However, the complete collapse of government structures remains an exception (Somalia) in the Sahel zone, although many countries of the region are structurally unstable. Almost all large, sparsely-populated countries, such as Sudan, are unable to maintain their monopoly on the use of force throughout their territory. Even in relatively stable countries, retention of the government's monopoly on the use of force is not guaranteed in rural areas. The economic and political capability of the Sahel states also depends to a significant extent on external support through development cooperation (Grimm and Klingebiel, 2007).

Some countries in this zone have been dominated for years by severe crises, civil wars and government-supported violence. In the Sudanese province of Darfur, a violent conflict has been raging for years with over 200,000 dead and 2.2 million refugees so far (as at the end of 2006; UNHCR, 2007). In Chad a number of rebel groups are battling for control of the country. In Senegal the struggles of the Casamance for autonomy have resulted in permanent conflict. Darfur is not the only crisis area in Sudan in which ethnic conflicts take violent form. Disputes also erupt between countries; an example is the border conflict between Chad and Sudan, each of which accuses the other of supporting rebel groups (HIIK, 2006). Their neighbour, Somalia, is also affected by major crises and civil wars. The political and economic fragility of the region is reflected in the large number of refugees: Sudan (over 690,000 people) and Somalia (over 390,000 people) are currently among the countries from which most refugees originate (UNHCR, 2006b). The number of internally displaced persons is considerably larger: within Sudan alone, more than 840,000 people are currently in flight (UNHCR, 2006b).

As well as violence and civil war, the extreme droughts of 1972/73 and 1984/85 also gave rise to much movement of refugees in the region, involving hundreds of thousands of people (Richter, 2000). Many individual examples (Box 6.3-1; Section 6.2.2.3) demonstrate that environmental degradation has, in the past, already contributed to the volatility of the region, particularly by giving rise to disputes over the use of ever-scarcer resources. The principal problem is desertification through overuse. This primarily 'home-grown' problem is amplified by droughts. Research has not investigated whether climate change was a significant driver of the degradation of the natural environment experienced in the Sahel in the 1980s and 90s, thus contributing to present conflicts in the region. Some voices, however, describe climate change as the principal cause of the Darfur conflict (Faris, 2007).

The agricultural sector that dominates the economies of the Sahel zone is particularly vulnerable to climate change. The main threat is to the food security of the population, most of whom raise cattle and grow food crops. Farmers grow millet and sorghum, and sometimes groundnuts and sesame, primarily on a subsistence basis. The most important agricultural export is cotton. The second pillar of agriculture in the region is cattle-rearing, which increases in importance in the drier areas (north of the arable farming limit). Population growth has reduced the amount of land available for agriculture; in the past this has repeatedly led to conflicts of use between nomadic animal keepers and arable farmers. Though nomads have evolved and handed on adaptive strategies for dealing with droughts (roaming cycles, adapted stock densities, etc.), it is unlikely that these strategies will be adequate for dealing with the future consequences of climate change – in many places adaptive capacities are already overstretched by the increasing scarcity of natural resources that has arisen as a result of desertification and population growth.

After the catastrophic droughts at the beginning of the 1970s the supranational Comité Inter-Etats de Lutte contre la Sécheresse dans le Sahel (CILSS) was set up; almost all the Sahel states are members. CILSS is a regional competence centre for food security, natural resource management and combating desertification. The CILSS countries see themselves as a community of solidarity, not only in emergencies but to an increasing extent also in regional development. Intergovernmental cooperation is still, however, in its early stages. Alongside CILSS, the Sahel and West Africa Club of the OECD is the second important platform for the Sahel countries; it is an informal alliance that brings together the Sahel countries and the most important donor nations. The OECD club also strives to negotiate peacemaking measures in the region.

7.3.3
Conclusions

Current climate models do not permit of any reliable prediction of the future change in mean precipitation in the Sahel zone. For the Western Sahara, in particular, the models yield contradictory results; some indicate considerable further desiccation in the region, while others predict increasing moisture and an advance of vegetation into the Sahara. Scenarios involving markedly increasing aridity as a possible future trend must therefore be considered. In addition, it must be assumed that variability in precipitation will increase in the Sahara. The Sahel – which is already affected by aridity – could therefore suffer further from droughts and desertification and thus face an increased risk of food crises.

The high vulnerability of the Sahel zone in the face of climate change is amplified by the large number of weak and fragile states in the region, as well as by violent conflict within society and the region's susceptibility to socio-economic crises. There is often no functioning state monopoly on the use of force; most of the countries in the region are among the world's least developed countries. As well as violence, civil war and extreme poverty, extreme drought and environmental degradation are causes of migration. Internal migration continues to predominate, although in recent years there has been increased international migration, including illegal immigration into southern Europe. Overall, food production is particularly vulnerable to climate change. It is anticipated that even warming of 2°C or less compared to pre-industrial levels will cause a fall in production; dependency on food imports is therefore likely to increase throughout the region (Section 6.3). For countries without their own sea ports, this reliance on imported food increases the risk of supply bottlenecks as a result of seasonal disruption of transport links (for example, during the rainy season). Such countries will also be affected by high additional transport costs. Overall, therefore, it is to be feared that in the Sahel zone climate change will lead to increasing regional destabilization, including increased potential for violent conflict.

7.4
Southern Africa

7.4.1
Impacts of climate change on the biosphere and human society

The high vulnerability of the African continent to climate change also includes southern Africa. In particular it is anticipated that climate change will result in changes in the availability of water and will increase water stress in parts of southern Africa. This is likely to further weaken food security in the region, which is in any case susceptible to crises (IPCC, 2007b).

The region of southern Africa lies in the tropical and subtropical climate belt and comprises the countries of Angola, Botswana, Lesotho, Madagascar, Malawi, Mozambique, Namibia, South Africa, Swaziland, Tanzania, Zambia and Zimbabwe. The climate in the northern part of southern Africa ranges from moist tropical (in the north of Angola) to dry tropical arid, with long dry periods and irregular rainfall (Kalahari, Namib). The Cape region has a subtropical Mediterranean climate with an arid summer period and winter precipitation. Projections show that there will be a noticeable decline in precipitation during the months of the southern winter, particularly in the extreme south-west of the region. In percentage terms this decline will be greatest between June and August. These are, however, the dry months of the year, so absolute changes will be small. Around half of the decline in precipitation – averaged over the year – will occur in spring (September–November); this will have the effect of delaying the start of the rainy season (IPCC, 2007a). In addition, the poleward shift of convection over the South Atlantic and the Indian Ocean will increase daytime temperatures and the duration of dry periods in the region (Tadross et al., 2005; New et al., 2006). Finally, sea-level rise will affect the coastal states of southern Africa and the flat coast of Angola with its many lagoons. Fast-growing coastal cities such as Dar es Salaam, Cape Town and Maputo will be particularly at risk.

Decreasing precipitation and increasing temperatures affect the availability of water and food production in the region and thus impact on the living conditions of the population; in addition they affect ecosystems that are already weakened by various further factors. Parts of southern Africa are regarded as biodiversity hotspots that are being put under stress as a result of climate change (Cape Floristic Region, Succulent Karoo, Maputaland-Pondoland-Albany). Even where species diversity and the associated performance of the ecosystems in the region are considered to be remarkably robust, there is an increase in symptoms indicating a heightened threat from climate change (MA, 2005b; AMCEN and UNEP, 2006).

This is also true of soil degradation – most of it anthropogenically caused – which by international standards is already far advanced in southern Africa and is proceeding at a rapid rate, particularly in the east of the region (Oldeman et al., 1991). It is due in the main to overgrazing, deforestation and non-sustainable agriculture. Different models predict that throughout Africa arid and semi-arid areas will increase by 5–8 per cent, or 60–90 million ha, by 2080 (Fischer et al., 2005). It is anticipated that even well below the 2 °C warming guard rail the arid and semi-arid areas of Africa will see forest give way to savannah and savannah give way to desert (Scholz and Bauer, 2006).

The use of water in agriculture in most countries of southern Africa is currently regarded as almost balanced or slightly non-sustainable (USDA, 1998; MA, 2005b). Except in Malawi, Mozambique and Zambia, where water supply has until now exceeded demand, further negative consequences for people and the environment are predicted – due to decreasing precipitation combined with increased need for water – unless an adaptive water management system is introduced. Even without climate change, Arnell (2004) estimates that the number of people in southern Africa exposed to water stress (defined as less than 1,000m^3 of water per person per year) will rise from 3.1 million in 1995 to 33–38 million in 2025, 50–127 million in 2055 and 50–188 million in 2085. In particular it must be assumed that in future there will be only limited opportunities for irrigation agriculture if current regional water supplies are not to be overstretched. Overall, falling groundwater tables and increasing salination will make it more difficult to find solutions to the primarily socio-economically and politically induced problems of water provision (Scholz and Bauer, 2006).

It is likely that the simultaneous occurrence of desertification, salination and regional water scarcity will cause grain harvests to fall throughout southern Africa. Fischer et al. (2005) calculate that by 2080 climate change will render an additional 11 per cent of the total area of southern Africa unsuitable for growing cereals. In particular, wheat farming could by then have disappeared completely from the African continent (IPCC, 2007b).

Declining grain harvests go hand in hand with increasing population pressure as a result of population growth and migration. It is likely that population will have doubled by 2050 in all the countries of southern Africa, with much of the increase occurring in cities (UNDP, 2005b; Swatuk, 2007). Exceptions to the general trend are Botswana, Lesotho, South Africa and Zimbabwe, where the population will stagnate or even decline, primarily as a result of the HIV/AIDS pandemic. Regional population growth is fuelled by migration, involving both labour migrants and refugees. The Republic of South Africa attracts an increasing number of labour migrants from West and Central Africa, while many people flee from unstable neighbouring states and regions affected by civil war – particular around the East African lakes – to the comparatively more stable countries of the south such as Malawi, Tanzania and Zambia (Tull, 2004; Swatuk, 2007).

7.4.2
Political and economic situation in the region

The political and economic situation in southern Africa is to a large extent dominated by the end of the apartheid regime, the effects of which have determined the development of the region for decades. The Republic of South Africa is an upcoming economic and political power and today plays a major role in the region. Neighbouring states view this situation with a mixture of great hope for the future of regional development and concrete fear of a neighbour who is overpowerful in every way.

Both the Republic of South Africa and southern Africa as a whole are still characterized by widespread poverty and social inequality as well as weak political and economic structures. Although recognizable progress towards political liberalization and good governance has been made in the last 10–20 years (except in Zimbabwe, which is experiencing increasingly autocratic rule under Robert Mugabe), state institutions remain relatively ineffective and little socio-economic improvement has been achieved. With the exception of South Africa, Namibia, Botswana and the small state of Swaziland, all the countries of the region are among the 30 lowest states on the Human Development Index, which covers 177 countries in total (UNDP, 2006). In the somewhat better-placed states just mentioned, however, unemployment is high and national income very unevenly distributed. In terms of Gini coefficients, which measure the inequality of income distribution, Namibia is at some distance from the others in bottom place, while Botswana, Swaziland and South Africa are all in the bottom dozen (UNDP, 2006; Swatuk, 2007).

On the other hand, the region – at least in comparison with Africa as a whole – can be considered stable and peaceful: Botswana is generally regarded as the only established multi-party democracy in continental Africa. Namibia and Mozambique have achieved noticeable consolidation in recent years. In addition, the Republic of South Africa acts as an economic magnet; this and its political weight contribute to regional stability. In particular, regional cooperation among the 'front-line states' that came together in the armed liberation struggle against the apartheid regime has now developed into a force for integration, led by South Africa, working primarily within the context of the Southern African Development Community (Hofmeier, 2004). In relation to the protection of regional and global environmental resources, various trans-border nature conservation and resource management initiatives are also contributing to an improvement of the cooperation climate in the region. Examples of this are the Gaza-Kruger-Gonarezhou Transfrontier Conservation Area in the border area of Mozambique, South Africa and Zimbabwe, the Four Corners Initiative between Botswana, Namibia, Zambia and Zimbabwe and the OKACOM River Commission of the countries bordering the Okavango (Angola, Botswana and Namibia) – although the acceptance and effectiveness of these and other initiatives in the region varies (Swatuk, 2005; Lindemann, 2006). In view of the often violent history of the region, the importance of institutionalized intergovernmental cooperation of this sort cannot be overestimated.

Nevertheless, with regard to the security of southern African societies, the superficial description of stable state structures does not tell the whole story. For many structurally disadvantaged groups, the state in southern Africa plays an almost irrelevant role; in many cases the state represents an additional problem rather than a guarantor of security and wellbeing (Booth and Vale, 1997; Swatuk, 2007). The enormous social inequality, combined with the importance of subsistence agriculture and the predominance of land use issues, harbours considerable potential for conflict within southern African societies. Even if the escalation of the land issue in Zimbabwe is viewed primarily as the consequence of poor governance, it is nevertheless an example that should serve as a warning. It is by no means certain that the fundamentally comparable conflicts of interest in Namibia or Botswana will be permanently and peacefully resolved, particularly as racial and xenophobic attitudes towards long-established settlers, foreign workers and minority groups are virulent in those countries (Swatuk, 2007). It is, however, foreseeable that the usable land area in these countries will diminish further as a result of climate change.

7.4.3
Conclusions

The anticipated medium- to long-term effects of climate change point towards an increasing worsening of general living conditions in southern Africa, interspersed with scattered and dynamic pockets of wealth. With the burden of widespread poverty, social inequality and weak governance, the societies of the region already face major challenges – such as control of the HIV/AIDS pandemic – even before the serious consequences of climate change are taken into account. Without a sustained economic revival, of which there is at present little prospect, improvements in living conditions are unlikely. It is more likely, instead, that the effects of global warming in the region will compound existing problems by having a seriously detrimental effect on regional food production, water supplies and accelerating soil degradation. As a result, rural-urban and internal migration will increase further.

Quite apart from the question of whether the capacity for action exists at all, it is not yet apparent that government and civil society decision-makers in southern Africa recognize the need to focus pre-emptively on these causal linkages and adopt a proactive policy of adaptation. Whether and when this changes will depend to a large extent on further developments in the most promising countries of the region, such as Namibia, Mozambique and Tanzania, the behaviour of the leading country in the region, South Africa, and the type and extent of external support.

From the perspective of human security, preventive climate policy adaptation measures must be regarded as essential in southern Africa, even though climate change will not necessarily lead to large-scale escalating conflict. However, since tensions in society along ethnic and social lines are already virulent and are likely to increase further, it is probable that there will be an increase in local and potentially violent resource conflicts resembling the familiar 'environmental conflicts' (see Sections 3.1 and 3.2).

7.5
Central Asia

7.5.1
Impacts of climate change on the biosphere and human society

Climate change will have a major impact in Central Asia. The substantial temperature rise, an increase in periods of drought and – in the long term – the melting of the glaciers will endanger the provision of freshwater in the region (IPCC, 2007b).

Central Asia is defined here as comprising the countries of Kazakhstan, Kyrgyzstan, Tajikistan, Turkmenistan and Uzbekistan and the autonomous Uyghur region of Xinjiang in the People's Republic of China. Xinjiang, however, is touched upon only briefly. Central Asia is characterized by high mountain ranges and large basins with no outflow; in some places there are great changes in altitude over very short distances. The climate is continental with noticeable differences in temperature between summer and winter months. While the mountain ranges – depending on their exposure – receive annual quantities of precipitation varying from over 400mm to over 600mm in places, the more densely populated lowland basin regions receive no more than 150–200mm per year; desert and semi-desert therefore predominate in these areas (Giese and Sehring, 2006). The climate is therefore characterized by aridity in addition to its continental nature.

In recent decades the climate of the region has changed markedly: since the beginning of the 1970s the recorded air temperature has risen by 0.3-0.4 °C per decade – more than twice the global mean. No tendency towards change in annual precipitation levels has been identified during this period (Giese and Mossig, 2004). Corresponding with these observed climatic changes, models project a rise in annual mean temperature of 3.7 °C (with a range of 2.6 °C to 5.2 °C) by the year 2100, based on the A1B scenario (IPCC, 2007a). A slight fall of 3 per cent (-18 per cent to +6 per cent) in the annual precipitation quantity is also anticipated, with precipitation increasing slightly in winter but falling slightly during the summer dry period. This will further exacerbate the existing imbalance in the annual distribution of precipitation.

In the markedly continental climate of Central Asia, grassland systems and steppes dominate the landscape. Temperature and availability of water limit plant productivity in the areas used predominantly as pasture (UNEP, 2002a; Nemani et al., 2003). Rising temperatures in the summer months – which are already warm and dry – increase evaporation rates, thus leading to a rise in water consumption. The soils of arid areas fail to benefit from increasing winter precipitation because they cannot retain sufficient moisture (Barnett et al., 2005). More rain and less snow result in more surface run-off in winter but are of no help during the periods in which the vegetation has the greatest need for water. This will further increase the already marked seasonality of plant productivity and make agriculture more difficult. Climatic changes will therefore have a negative impact on future agricultural usage. These trends will have serious consequences for the Central Asian republics that depend to a significant extent on the cultivation of water-intensive plants such as cotton and rice; in these countries agriculture accounts for 90 per cent of water use.

Moreover, poorly managed agricultural irrigation has led to the salination of large areas of Central Asia, particularly in Turkmenistan and Uzbekistan (USDA, 1998). The clearing of forests, overgrazing and non-sustainable agriculture are resulting in significant soil degradation (MA, 2005a). This process will be accelerated in future by the anticipated periods of very high temperature (almost every summer by the end of the century) and low precipitation. It can therefore be anticipated that Central Asia's vulnerability to desertification – which is already high – will increase further (USDA, 1998; MA, 2005a). These environmental changes also have an impact on food production. Model calculations show that harvest yields in Central Asia could fall by up to 30 per cent as a result of climate change, even if the positive physiological effects of the rise in CO_2 are taken into account (IPCC, 2007b).

A further important effect of the higher air temperatures is the melting of the glaciers in the mountains of the region that has been observed since the beginning of the 20th century (Dikich and Hagg, 2004). The intensity of this melting process has increased noticeably since the beginning of the 1970s. If one assumes that summer temperatures will rise by 5 °C in the next 100 years and that annual precipitation levels will remain constant, the result will be, for example, that by 2050 around 20 per cent of the glaciers in the Kyrgyz part of Tienshan – particularly smaller glaciers – will have disappeared and that glacier volume will have shrunk by around 32 per cent

(Giese and Sehring, 2006). In summer, the proportion of glacier water in the rivers can be as much as 75 per cent; this process could therefore have far-reaching consequences, because without an inflow of water, irrigation agriculture in the foothills would no longer be viable. With the melting of the glaciers and the trend towards more frequent heavy rain events, there is, moreover, an increased risk of rock falls, landslides and mud and scree avalanches.

7.5.2
Political and economic situation in the region

Central Asia is an area prone to conflict. All the countries of Central Asia are characterized by a major democracy deficit, autocratic and paternalistic forms of government and weak governance structures (Kaufmann et al., 2006; Grävingholt, 2007). The post-Soviet political institutions lack legitimation and fail to operate in accordance with the rule of law. Inefficient public administration and widespread corruption hinder economic and social development. Unmet material and participative needs fuel discontent among the population, leading increasingly to internal political conflict. Political tensions usually have ethnic or religious undercurrents, because the region is home to a variety of ethnic groups and Islamic opposition movements are growing in strength (Lüders, 2003). Arbitrary border-drawing between the areas occupied by different ethnic groups further exacerbate the situation. In consequence, the internal stability of these countries has been repeatedly shaken in the recent past by terrorist campaigns (Uzbekistan), civil war (Tajikistan) and criminal penetration of politics. On account of their poor governance capacities, many countries are regarded as so weak and fragile that it would take little in the way of critical events for the state to collapse.

The unstable internal situation is compounded by global developments: the geostrategic importance of the region has increased as a result of the 'war on terror' (Halbach, 2002). On account of its proximity to Afghanistan, Central Asia is regarded, moreover, as a hub of the international drugs trade. In addition, in connection with the securing of global resources and energy supplies, this resource-rich region is increasingly caught up in the potentially conflicting interests of powers such as the USA, Russia and China (Lüders, 2003; Amineh, 2006).

The region is characterized in some places by great poverty. For example, the proportion of the population living below the poverty line of less that US$2 per day is 16 per cent in Kazakhstan, 21 per cent in Kyrgyzstan and as high as 43 per cent in Tajikistan (World Bank, 2006e). The Human Development Indices of theses countries are close together, all lying in the lower part of the middle range (UNDP, 2006). The sole exception is Kazakhstan, which scores better than its neighbours. A large proportion of the population of Central Asia is affected by unemployment and struggles for opportunities to earn a living. Kazakhstan, Uzbekistan and Turkmenistan are the only countries in which the state fulfils its welfare obligations, and then only at a very low level (Schmitz, 2004). At the same time, there are extremely wide social differences in these three countries, with a large proportion of the profits from gas and oil exports benefiting only a small, affluent minority.

Economic structures depend to a large extent on natural resources. In addition, these countries still have to contend with inefficient management structures created for a planned economy; this is particularly significant for the supply of water. Agriculture forms the basis of existence for large sectors of the population and accounts for up to 40 per cent of GDP. The proportion of agricultural land requiring irrigation is 75–100 per cent (Bucknall et al., 2003; Giese and Sehring, 2006). The non-sustainable monoculture has far-reaching consequences, including soil salination, declining availability of pasture and arable land and contamination of water with fertilizers and pesticides. In addition, the large-scale abstraction of water from the Syr Darya and Amu Darya rivers has led to a gradual silting up of the Aral Sea, with disastrous consequences for climate and environmental conditions in the region and the health and socio-economic situation of the population (WBGU, 1998). Health problems caused by impure drinking water and sandstorms, high unemployment and impoverishment have fuelled social unrest and migratory movements in the region. The allocation of scarce water supplies, combined with controversial measures such as the building of the Golden Century reservoir, places strain on relationships between Uzbekistan and Turkmenistan and between Uzbekistan and Kazakhstan. Because of the anticipated affects of climate change in the Aral Sea region, the potential for destabilization is therefore particularly high (Giese and Sehring, 2006).

In addition to agriculture, another important economic sector is the generation of hydroelectricity for both the domestic and – increasingly – the export market. With regard to the use and distribution of water, there is sometimes a divergence of interest between countries adjoining the upper and lower reaches of rivers that flow through different states: for example, the requirements of electricity generation in winter (Kyrgyzstan, Tajikistan) as opposed to agricultural irrigation in summer (Uzbekistan, Kazakhstan, Turkmenistan). Since the Central Asian republics gained political independence, these differences of interest between riparian countries has resulted in more interstate disputes over water throughput quantities. Increased worsening of the water supply situation in summer would significantly increase the existing potential for conflict and far overtask current regional water management structures such as the Interstate Commission for Water Coordination, which was later included in the International Fund for Saving the Aral Sea (Section 6.2.3.2).

A part of Central Asia that is particularly prone to conflict is the Fergana basin. This is the most important area of agricultural cultivation and the most densely populated part of the region. It is situated in Uzbekistan, Kyrgyzstan and Tajikistan. Since the end of the 1980s there has repeatedly been conflict over access to resources that has erupted along ethnic lines. On account of social impoverishment and existing ethnic tensions, there is a high security risk attached to the anticipated consequences of climate change – that is, the probable increasing loss of valuable arable land, the risk of landslides and the growing scarcity of usable water resources in summer. This applies not only to the valley itself, but to the entire region.

In the neighbouring autonomous Uyghur region of Xinjiang in the People's Republic of China, too, scarce water and land resources are already the principal cause of conflict between ethnic groups. The tensions between immigrant, controlling Chinese on the one hand and the local population (Islamic Turkic ethnic groups: Uyghurs, Kazaks, Kyrgyz and Buddhist-Lamaist Mongols) on the other have increased in severity in the recent past and have erupted openly. Further developments in the province could have a destabilizing effect on the whole of China (Section 7.7).

7.5.3
Conclusions

Central Asia is a region severely affected by climate change. An increasing shortage of water is already noticeable. The above-average warming of Central Asia and the increasing variability of precipitation will exacerbate the situation. Combined with increasing demand for water, this will lead to further water scarcity in lakes and rivers that are shared between countries, siltation of inland lakes and desertification. Although the melting of the glaciers will increase the flow of water in the short to medium term, it will further exacerbate the water shortage in the long term. Water is both a key resource for agriculture (ensuring the survival of the population) and a strategic resource (electricity generation), and the region is already characterized by political and social tensions, the growth of Islamic movements, civil war and resource disputes; there is therefore considerable additional potential for conflict.

The socio-economic consequences of global warming are borne in particular by small farmers. Environmental degradation affects primarily regions that are politically and economically marginalized and whose problems are not a priority for those in power (Giese and Sehring, 2006). This can lead to unrest or to the escalation of existing tensions, particularly if the problems are instrumentalized ethnically or nationalistically, as for example in the Fergana valley or Xinjiang.

Climatic changes could also affect the strategic interests of countries through which waterways flow. Declines in agricultural exports (cotton) and disputes over water throughput quantities affect the economies of some countries. The countries themselves have very weak capacity for dealing with the consequences of environmental problems and climate change. This applies both to the implementation of national and international agreements and to domestic political reforms. Water management problems are further compounded by the geostrategic and economic interests of powerful countries in the region and by geographical aspects such as proximity to Afghanistan. The further development of the region could depend to a significant extent on resolving this critical complex of socio-economic and ecological problems.

7.6
India, Pakistan and Bangladesh

7.6.1
Impacts of climate change on the biosphere and human society

Climate change affects the South Asian countries of India, Pakistan and Bangladesh particularly severely. Its consequences include a rise in sea level, threatening areas such as the densely populated Ganges delta, changes in the monsoon rains that are so important for agriculture, the melting of the glaciers in the Hindukush-Karakorum-Himalaya region whose meltwaters are crucial for the water supply in the dry seasons, and the foreseeable increase in heavy rain events and intensity of tropical cyclones (IPCC, 2007b).

For the Indian subcontinent, comprising the countries of India, Pakistan and Bangladesh, climate models project warming of 2-4.7 °C, with the most probable level being around 3.3 °C by the year 2100 (A1B scenario; IPCC, 2007a). Warming is expected to be more marked in the winter half of the year (3.6 °C) than in summer (2.7 °C), and it is stronger in the north than in the south. Most models project a decrease in precipitation quantity during the winter dry period (mean change -5 per cent, range -35 per cent to +15 per cent) and an increase for the rest of the year (mean change +11 per cent, range -3 per cent to +23 per cent). At the same time, an increase in heavy rain events is probable, particularly in the north of India, in Pakistan and in Bangladesh.

The summer monsoon is crucial to the annual precipitation total of the Indian subcontinent (Lal et al., 2001). The effect that global warming will have on the Indian monsoon is still unclear, but increased variability in the monsoon rains is probable (Section 5.3.2; IPCC, 2007a). In addition, the strength of tropical cyclones, which represent a threat to the eastern coast of India and to Bangladesh, could increase (IPCC, 2007a). The risk to these areas will be aggravated by the rising sea level (WBGU, 2006).

The impacts of climate change in India vary from region to region. The effect of possible changes in the intensity of the monsoons will be particularly sensitive, because large parts of India receive the majority of their annual precipitation during the summer monsoon rains, which already vary noticeably in different regions (Lal et al., 2001). Agriculture is highly dependent on the monsoon and therefore sensitive to changes in monsoon intensity. In the past, variable monsoon rains have already led to harvest losses as a result of droughts or heavy rain. The east coast of India, which lies in the path of tropical hurricanes from the Gulf of Bengal, is particularly at risk of being damaged by storms and floods (Emanuel, 2005; IPCC, 2007a).

Pakistan is affected both by increasing precipitation variability and accelerated melting of the glaciers in the Himalayas. The increased frequency of heavy rain events anticipated in the north of the Indian subcontinent will lead to an increase in flooding. At the same time, the shrinking of glaciers caused by the rise in temperature presents a great challenge to Pakistan's water supply, because the country's great rivers are fed by glacier meltwater (IPCC, 2007b).

Bangladesh will be particularly severely affected by the effects of climate change. This is a result of its coastal location on the Gulf of Bengal and its geography, in particular the extensive delta of the Ganges and Brahmaputra rivers. Large areas of the country are predominantly flat and lie only a few meters above sea level. It is already prone to flooding; the risk will be heightened by the rise in sea level and by more intensive tropical cyclones over the Gulf of Bengal (Ali, 1999; IPCC, 2001; IPCC, 2007a).

In the greater part of India, deforestation and subsequent non-sustainable land use is the principal cause of soil degradation. In the northern regions overgrazing also plays a part. Soil salination occurs primarily in the north-west of India, but also in various regions of Pakistan. In both areas, overuse of the land is the main cause of degradation (Section 5.2). In Bangladesh, too, agriculture is responsible for soil degradation. In addition, soils become salinated through the intrusion of seawater. Soil degradation is proceeding moderately fast in India and slowly in Pakistan and Bangladesh (Oldeman et al., 1991). There is a very major risk of desertification in western Pakistan, north-western India and central India. In all these areas, agricultural irrigation is often not practised in a sustainable fashion (USDA, 1998; MA, 2005a). It must be assumed that agriculture and food production in the entire region will suffer as result of climate change and that the decline will be exacerbated by anthropogenic degradation. Studies show, for example, that in Bangladesh production of rice and wheat could fall by 8 per cent and 32 per cent, respectively, by the middle of the century; in India a rise in winter temperatures of 0.5 °C would reduce the wheat harvest by 0.45t per hectare (IPCC, 2007b).

7.6.2
Political and economic situation in the region

In this region climate change affects states with precarious political structures, major economic and social problems and a range of existing intra- and interstate conflicts. High population figures and rapidly increasing urbanization are typical of the region. The population of India is forecast to be around 1.4 thousand million by 2020, while Pakistan's will reach 212 million and that of Bangladesh 181 million (UN DESA, 2006). Population density in Bangladesh is already extremely high at 1,079 people per km^2 (2004); there is correspondingly high pressure of migration to the cities and to other regions. The urban population is growing faster than the rural population (UNPD, 2005). The coastal megacities of Chennai (2005: population 6.9 million), Dhaka (12.4 million), Karachi (11.6 million), Calcutta (14.3 million) and Mumbai (18.2 million) lie generally only a few metres above sea level. The high population pressure combined with increasing cyclone intensity and sea-level rise as a result of climate change will put millions of people at risk of being hit by storm and flood disasters (Ali, 1999).

As the largest democracy in the world, India has considerable experience of efficient conduct of elections, regular changes of government, forming of coalitions and the complex interaction between national and state governments. The judiciary exercises considerable influence in individual cases, for example in the environmental field. There are, however, considerable variations in governance between the different Indian states. The reputation of politicians is marred by frequent corruption scandals and criminal influences. The complexity of the system makes decision-making slow and difficult. Overall, though, the country can be regarded as politically stable.

The situation is different in Pakistan, where short spells of democracy are repeatedly interspersed with military coups. The government of Pakistan is now caught between the need to prove itself an ally in the worldwide 'war on terror' and attempts to stand up to radical Islamic parties and groupings. The government stabilizes the situation through negotiations and compromises, but also resorts to repressive measures. Since the elections of 2002, the military continues to hold crucial power and uses both the large people's parties and the small religious groups for its own interests. This strengthens the structures and influence of radical groups and increases the risks of sectarian conflict. In some parts of the country, such as the Federally Administered Tribal Areas on the eastern border with Afghanistan, the government is unable to maintain its hold on the state's monopoly on the use of force.

For a long time, political stability in Bangladesh was threatened by the hostility between the two major parties that crippled the political system. Since 2001 militant Islamic forces have had increasing influence (International Crisis Group, 2006). The country's problems include corruption, the use of violence within and outside politics and weaknesses in the organs of justice and security. These are countered by the stabilizing factors of a free and active press, a lively civil society and a tradition of liberal secularism. Effective NGOs have helped bring about improvements in social indicators such as the reduction of child mortality, enabling Bangladesh to achieve a better than average position among poor developing countries in such respects (Houscht, 2003; World Bank, 2006d).

Despite the economic growth of recent years, around 410 million people still live in poverty in South Asia. There are, moreover, large regional inequalities within countries. In terms of social development South Asia, despite significant progress, has not kept up with other regions such as Southeast Asia and East Asia (UNDP, 2006). In India, poverty is concentrated primarily in the states of northern India and in the particularly conflict-prone north-east. In 1999 the national average poverty rate in India (less than US$1 per day) was 26.1 per cent, but in Bihar it was 42.6 per cent, and in the north-eastern states of Assam and Meghalaya 36.1 per cent and 33.9 per cent respectively (ADB, 2006a). Poverty and social disadvantage are predominantly rural phenomena. The percentage contribution of agriculture to GDP is declining; in 2005 it was around 20 per cent in all three of these countries (World Bank, 2006a, c). Nevertheless, agriculture continues to form the livelihood of a large proportion of the population – 42 per cent in Pakistan and around 60 per cent in Bangladesh and India (Südasien Info, 2006). Pakistan, with its low precipitation, is particularly dependent on irrigation agriculture (which accounts for a quarter of GDP, two-thirds of employment and 80 per cent of exports) and on the Indus river system (World Bank, 2005a). As a result of climate change, therefore, there is likely to be a further increase in internal political conflict.

Security trends in the region are significantly influenced by the conflict between India and Pakistan over Kashmir. The tension between the two countries persists despite the commencement of a dialogue process. The situation is not helped by the fact that both countries now have nuclear weapons. Moreover, in connection with relationships between India and Pakistan, the geostrategic interests of the USA are an important factor influencing the regional security of South Asia. Additional security problems include cross-border extremism, the drugs trade in Afghanistan and Myanmar and the simmering civil wars in Nepal and Sri Lanka (South Asia Intelligence Review, 2003). In India the growing influence of militant communist groups (Naxalites) is seen as the largest internal political threat (South Asia Intelligence Review, 2007).

In South Asia there are organizations that concern themselves to varying degrees with trans-border environmental problems and climate impacts. These include regional organizations such as the South Asian Association for Regional Cooperation (SAARC) and the Bay of Bengal Initiative for Multisectoral Technical and Economic Cooperation (BIMSTEC), and international NGOs such as the International Centre for Integrated Mountain Development (ICIMOD). In addition, there are intergovernmental agreements that have so far been successful in resolving water conflicts. One such pact is the Indus Water Agreement of 1960 between India and Pakistan, which was mediated by the World Bank. This regulates the allocation of Indus river flows in Kashmir and was used recently in settling the dispute over the Baglihar dam (BBC report 13.2.2007). For a long time there were conflicts between India and Bangladesh over the allocation of water from rivers that flow between the countries (Ali, 2006; Tänzler et al., 2006). The conflict over India's unilateral extraction of large quantities of water from the Ganges (Farakka dam), which caused enormous economic and ecological damage to Bangladesh, was only resolved through a water allocation agreement in 1996 (Lailufar, 2004). The management of climate impacts could prove to be a test case for these intergovernmental agreements.

At national level there is a range of unresolved political disputes that are related to environmental issues. In India and Pakistan, for example, there are water conflicts between the individual states or provinces that have not yet been resolved and that repeatedly give rise to outbreaks of violence. As a result of poverty, natural disasters, land shortage and soil degradation, Bangladesh has for many years witnessed internal migration and migration to neighbouring Indian states, where native population groups repeatedly resort to violence against immigrants from Bangladesh (Pathania, 2003; Reuveny, 2005; Section 6.5.3.1).

7.6.3
Conclusions

South Asia will in many ways be especially severely affected by climate change – notably in the Ganges delta, which is not only densely populated but also the granary of an entire region. The region will be subjected to severe pressures as a result of sea-level rise, the effects of changes in yield and pattern of the monsoon rains on agriculture and forestry, the rapid melting of the glaciers with consequent drastic water scarcity during the dry seasons, the impact of higher temperatures and weather extremes on agriculture and the anticipated increase in the intensity of tropical hurricanes. These consequences will affect a region that is already among the most crisis-ridden in the world (World Bank, 2006a) and whose state institutions and intergovernmental capacities are weak. It is therefore foreseeable that climate change will overwhelm political structures and will further exacerbate economic and social problems.

The consequences of climate change are likely to affect primarily the poorer sectors of the population such as subsistence farmers and slum dwellers in the megacities. There are indications that these groups tolerate circumstances less patiently than previous generations (Imhasly, 2006). In combination with awareness of increasing social inequality and inadequate access to social services, this will increase the existing susceptibility – arising in part from ethnic and religious tensions – to violently expressed conflict.

The effects of climate change on the economy will vary: Pakistan's irrigation agriculture could be seriously affected. Severe floods could overwhelm the capacity of Bangladesh to deal with disasters and ultimately its capacity to provide food for the population. As a result of the threatened loss of arable and residential land through flooding, it is likely that migration towards India will increase and that existing conflicts will escalate. The instability in the east of the subcontinent is likely to increase and relationships between the countries will come under additional strain.

In India it is likely that the poorer states in the north and north-east will be most affected by climate impacts. Economic and social inequality within the country is likely to increase further. Decreasing water availability in the dry seasons will lead to increased incidence of usage conflicts both locally and between countries. Not least, water conflicts between South Asian countries and China arising from the reduction in the flow of glacier water from the Himalayas represent a potential problem.

7.7
China

7.7.1
Impacts of climate change on the biosphere and human society

China is a country particularly affected by climate change. Models project that China will have an above-average rise in temperature accompanied by increases of periods of drought and heavy rain events. This will result in an increase in soil degradation and desertification, particularly in the north of China, and – amplified by the melting of the glaciers in north-west China – endanger the water supply to large areas of land (IPCC, 2007b).

For the year 2100, models based on the A1B scenario of the IPCC show a rise in mean temperature of 2.3–4.9 °C in eastern China and a rise of as much as 2.8–5.1 °C for the Tibetan highlands (IPCC, 2007a). This warming, particularly in eastern China, will be stronger in winter than in summer. It is projected that annual precipitation levels will increase by 2–20 per cent in eastern China in the same period; this increase will be evenly distributed over the seasons. It is very probable that heavy rain events will occur more frequently, as will heatwaves and periods of drought; precipitation variability will thus increase markedly (Gao et al., 2002). In western China, on the other hand, it is expected that there will be more precipitation in winter and less in summer (IPCC, 2007a). However, uncertainty attaches to predictions of precipitation on account of the complex topography of the mountainous regions of western China and lack of clarity concerning the trend of the Asiatic monsoon and the behaviour of El Niño/Southern Oscillation (ENSO). Overall, it is primarily the north of China that is threatened by increasing drought (Section 5.1.2).

If present warming continues, the area covered by glaciers in the Tibetan highlands could decline by more than 60 per cent. Combined with significantly earlier snowmelt, this would lead to a considerable seasonal reduction of runoff (IPCC, 2007b), thus further jeopardizing in the long term water supplies for this already very arid region. There are already 200 million people in China who have no access to clean drinking water. In view of environmental pollution and the highly uneven regional distribution of water resources (rivers at risk of drying up in the north; rivers with abundant water in the south), the most populous country in the world, with its 1.3 thousand million inhabitants, is particularly sensitive to temperature changes, glacier melt and precipitation variability. In regions affected by decreasing precipitation at the same time as rising temperatures, the availability of water for human use will decline further. It is assumed that by 2015 it will be possible to meet only around 85 per cent of the water needs of the rural population (ADB, 2005; Oki and Kanae, 2006).

In addition, the coast of China is threatened by sea-level rise (WBGU, 2006). A rise of 30cm could inundate 8,000km² of the densely populated and highly industrialized Chinese coastal region (IPCC, 2007b). Moreover, the frequency and intensity of strong typhoons have been mounting since 1950 (IPCC, 2007b). Should that trend continue, it will, in combination with sea-level rise, present a major additional threat to the coastal region.

This wide variability in climatic conditions and their future dynamics is set against the backdrop of already major problems of soil degradation. Erosion, salination and also increasing desertification (Section 5.2) and loss of arable land as a result of industrialization and pollution are presenting serious environmental threats. Almost a third of China's land area (27.3 per cent) is already affected by desertification, particularly in Inner Mongolia (Nei Mongol) and around Beijing (USDA, 1998; MA, 2005a), and this tendency is increasing. A consequence of desertification is frequent sand and dust storms (IPCC, 2007b). Soil degradation is due to non-sustainable land use. In the northern, rather dry regions of China, overgrazing is the principal cause of the deterioration in soil quality, which may extend to loss of the upper soil horizons. In the southern, wetter regions, on the other hand, deforestation is the principal cause of soil degradation. The degradation rate is particularly high in the south-east of the country (Oldeman et al., 1991; Oldeman, 1992). It can be assumed that this trend will be further amplified by more frequent extreme precipitation events, but also droughts. In addition, northwestern China – where deserts are already a feature of the landscape – faces increasing problems from soil salination as a result of inappropriate, non-sustainable irrigation; these problems are also to be observed in the east of the country, in the provinces around Beijing. In this area, in which rainfed agriculture predominates, corresponding negative effects on harvest yields are therefore to be expected (Tao et al., 2003). A rise of 2 °C in the global mean temperature could lead to a 5–12 per cent reduction in the rain-fed rice yield in China as a whole (IPCC, 2007b).

7.7.2
Political and economic situation in the region

Against the background of existing and – as a result of climate change – increasing environmental problems and the economic, ecological and social contrasts within the country, the capacity of the Chinese state for management and action can be described as follows: marked regional, social and economic disparities are a feature of China's development (Alpermann, 2004; Heberer and Senz, 2006b). Since the introduction of its reform policies and increased openness in 1978/79, economic growth has proceeded rapidly, with GDP growing by more than 9 per cent annually in real terms; per capita income (in US$) quadrupled between 1981 and 2003. In the same period, the proportion of the population living in extreme poverty (less than US$1 per day) fell from more than 60 per cent to just below 10 per cent (World Bank, 2006a; Chaudhuri and Ravallion, 2007). At first glance, therefore, China appears to be a major economic success story.

However, the population participates very unequally in the economic dynamism, resulting in social polarization and wide income disparities. Economic development takes place almost exclusively in the urban centres of the coastal region; many inland regions, by contrast, continue to resemble typical developing countries. Around 45 per cent of China's workforce is still occupied in agriculture (Heberer and Senz, 2006b; Winters and Yusuf, 2007b). This disparity between and within regions is reflected in the Human Development Index: in 2005, Shanghai was on the same level as Portugal, while Tibet was level with Gabon. In addition, income disparities in China are among the highest in the world (UNDP, 2005a). A new urban middle class contrasts with a majority of some 770 million rural dwellers who have no adequate access to education and health care provision. Since the 1990s unemployment, income disparities and the need for cheap labour in China's coastal provinces have led to a large-scale exodus from the rural regions. The number of migrant workers in China is currently around 120–180 million. They form a growing stratum of impoverished urban dwellers and highlight the imbalance in China's impressive socio-economic development (Heberer and Senz, 2006b).

The poorest and least developed regions are the areas in which ethnic minorities live; these make up around 60 per cent of Chinese territory. These districts are dependent on agricultural yields and are therefore particularly affected by deforestation (Tibet, Yunnan, Sichuan), desertification (Xinjiang, Inner Mongolia) and glacier melt (Tibet, Xinjiang, Qinghai). In the autonomous Uyghur region of Xinjiang, for example, scarce water and land resources are already the principal cause of disputes between different ethnic groups (Giese and Sehring, 2006). In addition, the Chinese government's modernization policy for the autonomous regions based on economic development and resource extraction has bred political and ethnic conflict that continues to simmer (Heberer and Senz, 2006b). Social disparities among the population, increasing extraction of resources to underpin the economic upswing, and the effects of climate change threaten to further exacerbate these internal Chinese conflicts and beget the risk of growing political instability.

In addition to the conflicts in the interior of the country, the regional security situation is becoming increasingly tense. Tensions are growing as a result of migration from China to neighbouring countries such as Russia and Mongolia, and as a consequence of the resource and raw materials policy pursued by China and by Chinese companies in adjoining states (e.g. Mongolia, Russia, Central Asia) (Barkmann 2006; Sidikov 2006; Umbach 2006). China's need for resources also forms the background to territorial disputes over the islands of the South China Sea and disagreements between China and Japan (Heberer and Senz, 2006a). In addition, there is significant crisis potential in China's relationship with Taiwan. The pursuit of national interests in dealings with neighbouring countries can, in the long term, serve to exacerbate conflict. This effect is heightened if the 'nationalist card' is perhaps played in an attempt to stabilize the state and maintain the Communist Party's power (Klenke, 2006).

A central challenge to China's problem-solving capacity lies in the need to tackle not only social problems but also the environmental consequences of economic growth. The loss of arable land, the decline in soil quality and pollution of air and water are already giving rise to costs amounting to around 8 per cent of GDP (Economy, 2004). Every year millions of people fall victim to natural disasters; many of these are dependent on state aid. The health consequences of environmental pollution (e.g. 'cancer villages') are serious, and the number of protests is growing. The Chinese government has so far succeeded in presenting itself as capable of solving these problems. In fact, its position and legitimacy is built on its promise to develop and modernize the country, to solve the pressing social and ecological problems and to ensure political and social stability (Heberer and Senz, 2006b).

The government's goals and options for action in the face of these challenges are ambivalent; Chinese politics has broken away from Marxist-Leninist ideology and is based on pragmatism. The new innovative and technocratic political elite continues to focus primarily on economic development. The Chinese leadership, however, is to an increasing extent including ecological goals among its aims, particularly through the environmental authority SEPA. It has started to measure 'green GDP', for example, and has introduced measures to increase energy efficiency and expand renewable energies (SEPA, 2006; Schumann, 2007; Richerzhagen, 2007). The 'Communist party state' of the People's Republic of China is not a homogenous structure but represents a system of 'fragmented authoritarianism' (Lieberthal and Lampton, 1992). Within this system, a range of stakeholders exert influence on political decisions and their implementation – central government, the provinces, local government, the military, new social strata, new social organizations, etc. In many respects, the goals of central and local government conflict. For example, the implementation of environmental legislation often founders in the hands of administrative bodies at the lower level that evade the regulations in the interests of local economic development (Lan et al., 2006; OECD, 2007a). Awareness of regional and global responsibility for environmental issues appears to be slowly growing (Wenk, 2007; Schumann, 2007).

Negotiation and less coercion are now important tools of political consensus building and decision-making within China. A civil society movement is slowly emerging and giving rise to China's own environmental NGOs. On account of the prevailing political structures, however, they are still fragmented, often locally oriented and not entirely independent of the government in the western sense. Provided that they do not pursue explicitly political goals, the state supports such organizations because they often contribute to solving concrete social problems. Overall, though, the party state is set against a splintered and weak civil society that is at present unable to challenge it effectively.

7.7.3
Conclusions

The major environmental changes to be expected in China as a result of climate change are increasing soil degradation and the loss of arable land, an increase in extreme weather events and periods of drought and increasing water scarcity. The effects will, however, vary in different parts of China. Nevertheless, existing ecological and social problems in the cities and rural areas could escalate, undermining the dynamism of the economy and ultimately affecting the political stability of the country.

The south will be confronted with an increase in flood disasters, severe weather events, storms and landslides. The main problems in the north will be water scarcity in summer and an increase in droughts – with subsequent harvest failures – as a result of glacier melt and precipitation variability. The rural population will bear the brunt of climate impacts. Compounded by ethnic tensions, poverty and social disadvantage, these regional developments could have a destabilizing effect on the entire country and on neighbouring states. Since rural exodus as a result of environmental degradation is already being observed, it is very likely that internal migration will increase further and will represent one of the central challenges of the coming decades (Heberer and Senz, 2006b). Further significant risk and conflict potential arises from the concentration of the economic infrastructure in the densely populated river deltas of the east coast. The rising sea level and an increase in flood disasters, possibly also driven by more powerful typhoons, could cause major damage both to industrial facilities – thus striking at the heart of the Chinese economy – and to residential areas in which millions of people live. The potential for conflict would be considerable.

The anticipated problems present a major challenge to the political structures and the legitimacy of the Chinese government both internally and externally. Externally China, with its growing economic strength and global political weight, is not only becoming a serious competitor of the United States but is also being seen increasingly as a key player in matters of international climate policy. Tough international disputes are emerging over responsibility for the reduction of greenhouse gas emissions; such disagreements put the international system under pressure and can lead to tension in foreign policy matters. At home, the Chinese leadership runs the risk of being overstretched by the combined effects of heightened environmental problems, social polarization and economic and political liberalization. The political elite has recognized this problem and begun to develop mechanisms to protect against and deal with environmental problems and climate impacts. However, the government's capacity to control and manage the situation has struggled to keep pace with modernization and the dynamism of economic growth (Richerzhagen, 2007). As regards preventing internal conflicts from escalating into violence, much will depend in future on whether the current balance between economic capability, the state's ability to govern, and social acceptance or the 'willingness to suffer' of parts of the population continues to prevail (Heberer and Senz, 2006b). The ability to adapt to the anticipated climate impacts could thus be a crucial factor in China's further development.

7.8
Caribbean and the Gulf of Mexico

7.8.1
Impacts of climate change on the biosphere and human society

As a result of climate change, the countries in the Caribbean and the Gulf of Mexico are likely to be increasingly affected by major hurricanes. This could further exacerbate social and political tensions in the region.

The Gulf of Mexico and the Caribbean are tropical marginal seas connected to the Atlantic Ocean. Countries in the region include many island states and the coastal countries of North, Central and South America. For the Central American mainland, regional climate models based on the A1B scenario project a temperature rise of 1.8–5 °C by 2100; the most probable increase under this scenario is 3.2 °C (IPCC, 2007a). This rise is higher than the global mean. The anticipated rise in temperature will be particularly marked in the spring months of March to May. For the Caribbean islands, a lower rise in temperature of 1.4–3.1 °C with a most probable level of 2.2 °C is projected. According to most models, the amount of annual precipitation will decrease in both parts of the region. In addition, there is likely to be an increase in periods of drought and noticeably increased water stress, particularly on the islands (IPCC, 2007a).

Tropical cyclones are a weather phenomenon particularly relevant to the region. Climate warming leads to an increase in the sea surface temperatures, and high surface temperatures provide the energy source for tropical cyclones. An increase in the intensity of such cyclones while the climate is warming is therefore plausible, and a correlation between sea surface temperature and tropical cyclone intensity has already been observed (Emanuel, 2005; WBGU, 2006; Section 5.1.3). Because tropical cyclones often generate flood waves several metres high, this puts lower-lying coastal areas at major risk from floods – especially in view of global sea-level rise (IPCC, 2007b).

Hurricanes already frequently cause considerable loss of life and property; a single event can often result in more than 1,000 deaths and cause damage of over US$1,000 million. Examples are Hurricane Mitch in Nicaragua and Honduras, which claimed over 18,000 lives, the regular storm and flood disasters in Haiti and Hurricane Katrina in the USA, which took 1,322 lives (CRED, 2006). Such disasters have a major impact on the development of the affected regions; some have also triggered conflict (Section 3.2.2).

In addition, hurricanes represent a major threat to the oil and gas infrastructure in the United States part of the Gulf of Mexico. There are over 800 manned drilling platforms in the area, connected by an elaborate pipeline network to the refineries on the coast of the United States. The refineries on the Gulf of Mexico are also used to process foreign oil of inferior quality. The region therefore plays a crucial role in the United States' supply of fuel and gas. Past natural disasters, in particular the hurricane season of 2005, caused major damage to the oil and gas infrastructure and led to losses of extraction, transport and refinery capacities that lasted for many months. As a result, the USA had to resort to using strategic oil reserves in order to sustain the country's energy supply (Bamberger and Kumis, 2005; Energy and Environmental Analysis, 2005).

The sea-level rise will particularly affect the flat coastal areas of the United States and Mexico along the Gulf of Mexico (IPCC, 2007b). In contrast to the situation in the Pacific, sea-level rise does not threaten the very existence of island states in the Caribbean, but it will have serious consequences for both coasts and islands in the region. These effects will include flooding of coastal plains and increased coastal erosion.

In the Caribbean it is likely, in addition, that precipitation variability will increase markedly, with periods of drought and intense precipitation events representing the two extremes. The effects of intense precipitation events can be as destructive as those of hurricanes: during such an event in Venezuela in 1999, 30,000 people lost their lives as a result of landslides and floods (CRED, 2006). Droughts may also become a more significant problem in parts of the region, and may, due to non-sustainable land use – particularly in Honduras, Costa Rica, Nicaragua, Cuba and the Dominican Republic – result in increased soil degradation and eventually desertification (USDA, 1998; MA, 2005b). The consequences in many places will include additional agricultural losses and increased risk of disasters (Section 6.4.2.1). Grain harvests could fall by up to 30 per cent by 2080 (IPCC, 2007b).

7.8.2
Political and economic situation in the region

According to the Human Development Index, the USA, Barbados, Costa Rica, Saint Kitts and Nevis, the Bahamas, Cuba, Mexico, Panama and Trinidad and Tobago are highly developed (UNDP, 2005b). Almost all other countries in the region lie in medium positions on the index. Haiti is the only country in the region with a low index.

Economically the region is very diverse: the USA is the largest economy in the world with a very high per capita income. Neighbouring Mexico is an emerging country with relatively strong economic power. Venezuela's affluence is due largely to its major oil reserves. For the majority of countries in Central America, the export of clothing is of great economic importance. The proportion of export revenue attributable to this sector is 28 per cent in Guatemala, 50.2 per cent in El Salvador and 59.9 per cent in Honduras. Costa Rica and Mexico also export large quantities of electronic goods (WTO, 2006). The USA is the most important trading partner for almost all countries in the region; it is likely that economic linkages will increase further under existing and future free trade agreements.

In some Caribbean island states (Dominican Republic, Barbados, Saint Kitts and Nevis and others), tourism has become a crucial economic factor. Nevertheless, agriculture remains extremely important in most Caribbean and Central American countries, providing the majority of the population with a basic income or a livelihood through subsistence production. In the Central American countries, in particular, wealth is very unequally distributed: according to their Gini coefficients, Honduras, Mexico, Nicaragua, Colombia and Panama have a high inequality of income distribution (CIA, 2006).

For many inhabitants of the region, emigration is an attractive economic option. In many countries remittances from migrants have become an important economic factor. In Haiti such remittances account for more than half of GDP, while in Jamaica and Honduras the figure is 17 per cent and 16 per cent respectively. Mexico is the largest recipient of remittances from abroad in the world, receiving US$21.8 thousand million annually (Fajnzylber and López, 2006). The most important destination region for Latin American migrants is still the USA. The USA's increasing attempts to close its borders to migration from Latin America, for example by erecting a border wall to deter illegal migrants, fuel discord between it and its neighbours (HIIK, 2006).

Various conflicts, some of them armed, are taking place in the region. Colombia has for many decades witnessed violent conflict between left-wing rebel groups, the central government and right-wing paramilitaries. Intense armed conflict has been taking place in Haiti for more than 20 years; the central government has permanently lost control of large parts of the country and since 2004 a UN peace mission has been attempting to stabilize the situation there. In Belize there were violent clashes between police and demonstrators at the beginning of 2005. In 2006, internal political tensions in the Mexican state of Oaxaca erupted violently, resulting in many deaths.

Since 2000, Venezuela has been locked in an internal ideological conflict in which the populist left-wing government of President Hugo Chávez and the established middle class confront the large landowners. As one of the last remaining Cold War-era system conflicts, the relationship between Cuba and the USA is also important for security in the region. However, violent escalation of this situation appears very unlikely.

In addition, there are a number of other conflicts that, on the whole, are taking place non-violently. These include disputes between Costa Rica and Nicaragua and between Colombia and Venezuela. In Guatemala and Mexico there are also considerable internal political tensions (HIIK, 2005).

7.8.3
Conclusions

The analysis shows that the region will be exposed to considerable social and political risks. The anticipated increase in severe tropical cyclones is likely to have a negative impact on all the countries in the region, particularly those of Central America. After Hurricane Mitch in 1998, it was estimated that the event had set back economic development in the most severely affected developing countries by several decades (IDA, 2006). More frequent incidence of such extreme events is therefore likely to lead to permanent economic destabilization in many parts of the region. This is all the more probable because, with few exceptions (Cuba, USA), the countries of the region have few effective disaster prevention measures in place.

Moreover, in many parts of the region there is a latent risk of conflict that erupts repeatedly in armed disputes. Since weak governance structures are a feature of most of the countries of Central America and the Caribbean, it can be assumed that climate-induced environmental changes will markedly increase the existing risk of conflict. The vulnerability of the oil and gas infrastructure in the Gulf of Mexico represents a further important factor in regional and global crisis susceptibility. Short-term disruptions of oil and gas production already have significant economic and sometimes political consequences. Increased frequency of severe hurricanes in the Gulf of Mexico could therefore have global economic and political consequences.

The economies of many countries in Central and South America are stabilized by migration to the USA and the remittances sent back to families in the migrants' countries of origin. Therefore the USA's attempts to limit migration already entail a risk of social upheaval. In the wake of climate change and its anticipated effects on the Caribbean and Central America it can be assumed that migration flows towards the USA will increase, and that within Central America they will also intensify. In particular, people will leave those regions that are especially severely affected by extreme events and ecosystem degradation. Urban centres will be the primary destinations for migration. In view of the weak governance capacities in the region and the limited employment opportunities in the cities, it can be assumed that here, too, latent conflicts are likely to be exacerbated.

7.9
Andes region

7.9.1
Impacts of climate change on the biosphere and human society

The most important climate impacts for the countries of the Andes include major warming and the resulting rapid glacier melt. These factors, combined with increasing precipitation variability, have serious impacts on the water supply and on agriculture, which is also affected by increased soil degradation (IPCC, 2007b).

The Andes extend along the west coast of the South American subcontinent; they pass primarily through Colombia, Ecuador, Peru, Bolivia and Chile, but also parts of Venezuela and Argentina. Climate models project warming in excess of the global mean by the end of this century; temperatures are projected to rise by 1.8–5 °C (mean 3.2 °C) in the northern Andes and by 1.7–3.5 °C (mean 2.5 °C) in the southern Andes, with the increase being distributed relatively equally across the seasons (A1B scenario, IPCC, 2007a). In recent decades a rise in temperature has already become apparent in the region; this is responsible for progressive melting of the Andes glaciers. For example, the glaciers in the Peruvian Andes have lost 25 per cent of their area in the last 30 years alone (Barnett et al., 2005; IPCC, 2007b).

In the Andes the anticipated precipitation changes vary considerably from region to region. It is assumed that annual precipitation will decrease in the southern part of the Andes (Chile and Patagonia), while the northern Andes – in particular the north of Peru – will benefit from a slight increase. It should, however, be borne in mind that the climate models used in control experiments tend to overestimate precipitation along the Andes (IPCC, 2007a). Uncertainties are introduced especially by difficulties in modelling the interaction between ocean and atmosphere in the tropics as well as the atmospheric circulation, and by the morphology of the high, narrow mountain range of the Andes.

The majority of the local population is dependent on meltwater from the Andes glaciers, either directly for household and industrial water supplies or indirectly via the generation of electricity. Accelerated melting of the Andes glaciers, such as is already measurable (Barnett et al., 2005) and expected to continue (IPCC, 2007a), will have a severely detrimental effect on the regional water budget and thus also on the availability of water for the population. Water resources for people and the environment are used very differently in different regions, with impacts on both agriculture and industry (Liniger et al., 1998; Mark and Seltzer, 2003). Ecuador, the coastal area of Peru and parts of Bolivia and Argentina are already characterized by non-sustainable agricultural irrigation practices – that is, the sustainable water yield is less than the amount consumed by agricultural irrigation (MA, 2005a).

An additional problem is soil degradation, which is rated as medium to high in large areas of the Andes countries (Oldeman et al., 1991). The reason for this is non-sustainable land use – particularly deforestation, but also overgrazing. In some locations, for example on the west coast of Ecuador, salination is the primary type of soil degradation. The salination problem might initially be improved by accelerated melting of the Andes glaciers (if more water is available, salination can be counteracted by methods such as targeted over-irrigation), but in the long term the more or less complete disappearance of the glaciers will cause a general shortage of water as a resource for agriculture. Combined with higher temperatures, this will increase the risk of humanly induced desertification – which is already high to very high – for the west coast of Ecuador and for parts of Chile and Peru.

7.9.2
Political and economic situation in the region

The effects of climate change depicted above affect countries that currently have only limited capacity for adaptation and conflict resolution. While all the countries in the region are formally constituted as democracies, with the exception of Chile they are politically unstable, not democratically consolidated, more than averagely corrupt and characterized by significant deficits in the rule of law (Hagopian and Mainwaring, 2006; Mainwaring et al., 2006; Sample and Zovatto, 2006). In Colombia the armed conflict between the drugs mafia, paramilitary groups, guerrillas and officers of the state has, in addition, led to erosion of the state monopoly on the use of force. Against this background it is currently hard to conceive of political reforms or a sustainable formulation of policy, especially where these run counter to the interests of powerful groups.

The democratization processes of the late 1970s and 1980s in Bolivia, Ecuador and Peru, combined with economic reforms, have triggered major socio-economic distribution conflicts in already strongly divided societies. Groups that had previously been economically, politically or ethnically disadvantaged have used their newly acquired rights to demand removal of privileges of traditional elites, thereby exacerbating and diversifying these distribution conflicts. Due to continuing socio-economic distortions and inequalities that from time to time favour circumstances resembling civil war in Peru (Sendero Luminoso, or Shining Path), there has been an erosion of political parties and a renewed increase in populist rule since the 1990s (O'Neill, 2005; Roberts, 2007). A similar process has developed in Colombia and Venezuela. In both of these countries, non-transparent patronage and clientele networks had led to the formation of entrenched political structures that massively disadvantaged large sections of the population. With the base level of social conflict already high, radical political and economic liberalization measures triggered the erosion of the functionality of the state in both cases.

Despite macroeconomic stabilization and at least moderate overall growth, attempts to reduce poverty have been unsuccessful and economic distributive justice has not gone far enough. The societies of the Andes region therefore have some of the highest income disparities in the world and meeting basic needs is beyond the reach of large sections of the population around 60 per cent in Bolivia, 50 per cent in Ecuador, Peru and Colombia and around 20 per cent even in relatively affluent Chile are poor (ECLAC, 2006). The deficits in the political systems are therefore to a significant extent responsible for the unresolved social problems of the region. From the point of view of environmental policy and the ability to adapt to anticipated climate impacts, it is worrying that, in order to solve their problems of legitimation and governance, governments are focusing primarily on rapidly increasing agricultural exports and intensifying resource extraction while having very little in the way of coherent strategies for sustainable resource management or being unable to implement appropriate strategies.

The resurgence of development strategies based on the export of resources particularly affects the Amazonian regions of the Andean countries. Rapid extraction of resource reserves creates short-term macroeconomic stability and growth and secures the privileges of powerful interest groups. Exports in all the Andean countries are based primarily on resources (mining, oil, gas) and – depending on climatic conditions – on agricultural products; the importance of processing industries has declined during the economic reforms of past decades. This trend, as well as having negative environmental consequences, has amplified social disparities in the regions concerned and increased the potential for conflict. For example, it has given rise to conflict between indigenous groups, migrants, private and nationalized companies and government bodies in the lowlands of Ecuador and Peru. The Peruvian government, in particular, is attempting to force up agricultural exports and follow the Chilean model; to this end the coastal regions, which already have a relatively limited water supply, are earmarked for the expansion of Peruvian agriculture. Major infrastructure projects (such as diversion of rivers in the Lambayeque region of northern Peru) take very little account of environmental aspects or integrated water management criteria. This results in inefficient use of scarce water resources and is likely to increase allocation conflicts between water-rich and water-poor parts of the country in the medium term.

Nevertheless, for the Andes region as a whole, a gradual increase in awareness of environmental issues is discernible; this has led, for example, to establishment of environment ministries and agencies and inclusion of environmental principles in countries' constitutions (for example in Colombia). In addition, government development cooperation and the activities of NGOs are helping to put environmental issues on the agenda and bring about improvements in environmental legislation.

A central hurdle for a sustainable policy that would enable forward-looking climate adaptation steps to be taken is the incoherent administrative structure of the countries in question. Currently observable conflicts of relevance to the environment do not affect the entire country, but local or regional areas. The relevant authorities in the region (central government, provinces, municipalities), however, lack clear administrative responsibilities and appropriate resource endowment (Faust, 2006; Faust and Harbers, 2007). This makes it more difficult to design and implement coordinated local and national adaptation measures for a forward-looking environmental policy that would prevent conflict. A coherent decentralization strategy oriented towards local authorities cannot be identified in any of the Andes countries with the exception of Chile; instead, such strategies are hampered by the political conflicts associated with them (Falletti, 2005; Daughters and Harper, 2006). In addition, the fact that political and legal responsibilities are often unclear and laws are not effectively implemented encourages disputes between different levels of government, as a result of which conflicts proceed in a less predictable and less focused manner.

In the case of future water scarcity in the Lima region caused by glacier melt accelerated by climate change, the problems arising from a situation in which conflict management is rarely process-oriented are already apparent (Section 6.2.3). It is true that Lima can exert considerable pressure on central government on account of its political and economic significance. However, it is questionable whether the individual municipalities within Lima can solve the anticipated conflicts effectively between them. It is equally doubtful whether an equitable solution can be found to the allocation between the Lima region and neighbouring regions that have less negotiating power. Even for the allocation of water beyond provincial boundaries there is as yet no institutional structure in Peru; this encourages politicization of anticipated water conflicts in connection with the planned intensification of the agricultural export industry.

In addition, there are problems of cooperation between states in solving environmental problems that involve more than one country. Since the 1960s, there have been moves to develop cooperation and integration in the Andes region, but these efforts have been repeatedly undermined by the particular interests of individual governments and a heightened nationalism among the people. Numerous long-standing border conflicts – such as those between Ecuador and Peru, Peru and Chile, or Chile and Bolivia – amplify the difficulty of cross-border cooperation in matters of environmental policy or sustainable resource management.

7.9.3
Conclusions

The countries of the Andes are severely affected by climate change as a result of significant warming, the rapid glacier melt that this causes and increasing precipitation variability. Above all, serious impacts on the population's water supply and on agriculture are to be expected. The societies concerned already face major political and socio-economic challenges and are characterized by political disputes and conflicts over the allocation of scarce resources. The (partial) destruction of the rainforest and water scarcity – amplified by largely non-sustainable water usage and population growth – have so far resulted mainly in conflicts that are limited to local and regional areas. Even though it is likely that there will be a gradual shift in awareness among the population and some sections of the political elite as a result of the increasing pressure of problems in the region, countries other than Chile are comparatively poorly equipped to meet the additional challenges arising from climate change effectively. As a result of political instability and economic pressure, political planning is extremely short-sighted; there are serious defects in the administrative structure of the state, deficiencies in the rule of law and a low level of cooperation between countries. Hence, in the near term, iterative learning processes are possible, but the development of a long-term adaptation strategy is unlikely. It must therefore be assumed that climate change and the environmental changes that it causes will in future exacerbate existing political tensions and conflicts in the Andes region.

7.10
Amazon region

7.10.1
Impacts of climate change on the biosphere and human society

The Amazon rainforest is the largest contiguous expanse of tropical forest in the world, harbouring a significant proportion of all terrestrial plant and animal species (IPCC, 2007b). The Amazon basin extends over eight Latin American countries; around 60 per cent of the basin lies in Brazil, while the remaining 40 per cent is divided among Bolivia, Peru, Colombia, Venezuela, Ecuador, Suriname and Guyana. Around 50 per cent of precipitation in the Amazon region is generated by evapotranspiration in the region (Schubart, 1983; Salati, 1987). The most major problem in the Amazon region is continuing deforestation. If present trends continue, 30 per cent of the Amazonian forest could have disappeared by 2050. The resulting regional climatic changes could lead to 'savannization', particularly of the eastern Amazon area; this will be significantly amplified by global climate change. The transformation of tropical rainforest into dry grassland savannah would lead to the extinction of a significant number of species (IPCC, 2007b).

For the Amazonas region of northern Brazil, regional climate projections based on the A1B scenario predict a rise in temperature by 2100 of 2.6–3.7 °C against a 1990 baseline; such a level of warming is 30 per cent above the global mean (IPCC, 2007a). These model projections are robust and accord well with the warming already recorded in the 20th century. The seasonal change in temperature distribution shows a trend towards more marked warming in the months of June to August compared to the months of December to February, thus reducing the annual temperature range (IPCC, 2007a).

Changes in the regional distribution of precipitation are always hard to predict, and in the Amazon region the difficulty is increased by the very important interactions between vegetation and climate, and by the effect of the high but narrow mountain range of the Andes. Both factors are poorly depicted in current models. A further factor contributing to the level of uncertainty is the future trend of the El Niño phenomenon, which leads to significant droughts in the Amazon region. At present, therefore, it is impossible to make any statement about future changes in mean precipitation (IPCC, 2007a). However, droughts will occur in future as a result of the significant warming of the Atlantic and the associated changes in atmospheric circulation, irrespective of the El Niño phenomenon (Cox et al., 2004; Shein et al., 2006).

In the Amazon region, the year 2005 was notable for its extraordinary dryness. This unusual event could be a herald of the drought years that, according to climate projections, will occur ever more frequently in the region. According to model calculations of the Hadley Centre (HadCM3), from 2050 onwards the Amazon region will be able to absorb less and less carbon from the atmosphere. Higher air temperatures combined with increasing dryness will reduce carbon fixing by the rainforest; the effect will be amplified by a further decrease in the area of the rainforest. In the models of Cox et al. (2000), this reduction of carbon storage in the Amazon region results in the terrestrial biosphere becoming a global source of carbon in future. More recent predictions assume that, in the extreme case, 65 per cent of the Amazon forest area will disappear by 2090 as a result of increasing incidence of droughts (Cox et al., 2004; Hutyra et al., 2005). The transformation of the Amazon basin into a savannah landscape (IPCC, 2007a) would release additional carbon dioxide into the atmosphere, which would in turn accelerate climate change (Section 5.3.4).

Clearing of the Amazon rainforest (Nepstad et al., 1999) reduces air humidity because there is less transpiration from vegetation; there is then less precipitation. The air over the deforested areas heats up more markedly than that over the forest, which in turn influences the local climate and thus the neighbouring vegetation. Moreover, the complete removal of biomass after clearing (including the burning of roots and vegetation remaining after harvesting) releases a noticeably larger quantity of CO_2 than is taken up by subsequent field crops (Morton et al., 2006). In addition, clearance results in considerable fragmentation of ecosystems and within a very short time leads to increased erosion and to soil degradation.

Rising temperatures, increasing droughts and soil degradation have serious consequences for agriculture. In tropical regions it is generally assumed that warming of as little as 1–2 °C has a negative impact on grain production. As further warming takes place, all crops in tropical countries will be affected by falling harvests (IPCC, 2007b). In the case of the Amazon region, it is calculated that even moderate warming will reduce yields of wheat and maize by around 30 per cent and 15 per cent, respectively, as a result – in particular – of heat stress and water shortage. In the case of typical market crops, simulations of the effects of moderate warming predict varying results: while the areas suitable for growing coffee are likely to shrink considerably in extent, soya yields are forecast to rise – at least temporarily – by around 25 per cent (IPCC, 2007b).

Inland fishing would also be affected. This would have negative consequences for the rural population, because fish is their primary source of animal protein (Waichman et al., 2002). Fish stocks are already under threat as a result of overfishing. With increasing dryness and rising temperatures, the natural habitat of many fish species dwindles. The drying up of channels linking inland lakes and rivers interrupts the cycles of reproductive migration; unable to migrate, an excessive number remain in the lakes, where they suffocate as water levels fall. In drought years the reproduction rate therefore falls drastically. In addition, the low water level can cause agricultural pesticide residues to become concentrated in rivers and lakes to toxic levels.

A further aspect is disruption of transport routes. The water level in the rivers may sink dramatically, with the water turning to mud; in many areas the result is that the only transport routes become impassable. Forest and bush fires occur frequently in drought years, often affecting large areas and severing important land transport routes. The negative consequences affect not only the transport of goods but also the medical care of the population. The effects of sea-level rise on the Amazon delta have not yet been studied.

7.10.2
Political and economic situation in the region

Brazil is the most important country for the development of the Amazon region. In the government's view, the region functions as a hinterland that has yet to be opened up. Although the Amazon region constitutes 58 per cent of the land area of Brazil, it is home to only 8 per cent of the population and contributes only 5 per cent of GDP (IBGE, 2007). The Amazon region's resources play an important role in the expansion strategies of private and public stakeholders. Resource use is already an issue with major potential for conflict in the region.

The Amazon region was opened up in the 1960s through the construction of cross-country highways. Site development was financed by public funds and served the economic valorization of the area's productive resources (Mahar, 1988). The land was first offered to large businesses for cattle-rearing; then the rural population from the north-east, an area plagued by drought, was to be settled there. Since the 1970s, ore deposits have been mined, large hydropower plants have been built and the settling of small farmers has been extended. The result has been a growth in population and a 17 per cent reduction in the area of forest (INPE, 2007).

Since the 1990s, the deforestation dynamic has developed independently of public investment. Cross-country highways are increasingly being built by financially powerful sawmill owners, cattle rearers and soya farmers. Cattle rearers and soya farmers have an eye on major export markets, because Brazilian meat and soya exports have increased substantially in the wake of the BSE crisis in Europe. Brazil has the largest cattle herd in the world (200 million head in 2003; IBGE, 2007) and is the world's second-largest soya producer (52 million t in 2006); whereas the Amazon region currently remains marginal as a production location for exports. The savannahs in the south of the Amazon region have already been opened up for soya cultivation through soil improvements, and large areas in the dry region around Santarém in the central Amazon region are now being cleared. The vast expansion in production of bioethanol from sugar cane in the south and southeast of the country is in part replacing the cultivation of soya in the Amazon region. The Amazon region is also being considered for the production of fuel from biomass.

In terms of GDP, Brazil is one of the leading 13 economies in the world. Between 1995 and 2005

GDP grew on average by 2.1 per cent annually. Real growth was particularly strong in agriculture, where it was driven primarily by animal production. Economic growth has improved the income situation of all population groups; nevertheless, income disparities in Brazil are among the largest in the world (UNDP, 2005b). Poverty in Brazil is concentrated regionally in the dry north-east and the Amazon area (Brazil, 2004). The situation is worst for the indigenous population. The expansion of large cattle and soya farms has led to a noticeable increase in violent conflict and in expulsions of small farmers in the Amazon region (CPT, 2007). But also the majority of large landowners have no title to the land. Clearance is usually carried out without regard to statutory requirements, but the majority of these offences go unpunished. These conflicts over land and resources may increase as farmland is further expanded and as a result of the effects of climate change.

Tensions are also to be expected in the Andean countries, because significant migration from the mountain regions to the Amazonian lowlands is taking place, bringing with it the familiar consequences of deforestation, minoritization and displacement of indigenous peoples. This migration is directly or indirectly encouraged by governments, for example by stripping protected areas retrospectively of their status in order to develop them for economic use. Since the source areas of the most important Amazonian rivers lie in the Andes, the destruction of the forest in this part of the Amazon region has particularly drastic consequences, influencing as it does the water supply of the entire area, including the Brazilian lowlands.

While the measures taken by the Brazilian government to protect the Amazon region from over-exploitation of natural resources have improved, there are still conflicts of goals between environmental plans (protected areas) and plans for the energy and transport sectors (building of hydropower plants, gas pipelines, roads). After the deforestation rate rose very sharply in 2002 and 2003, special measures for tackling deforestation were agreed. However, only the monitoring measures of the environment ministry have so far been implemented. Improvements in the awarding of land titles and in environmental impact assessment have not yet been achieved. The range and effectiveness of the government's measures is significantly compromised by general governance weaknesses in the areas of rule of law and control of corruption (World Bank, 2005b). However, the government has for some years been attempting to strengthen the state's monopoly on the use of force and improve adherence to the law in the region.

With the publication of the IPCC's Fourth Assessment Report, climate change has become an international political issue. Brazil participated proactively and constructively in international climate negotiations, and the environment ministry subsequently took up the issue with great resolve. Brazil is also involved in regional cooperation on environmental policy and is the strongest member of OTCA (Organização do Tratado de Cooperação Amazônica), which focuses on cooperation to ensure the sustainable development of the Amazon region. With the transfer of the permanent secretariat to Brasilia in 2003, Brazil has assumed responsibility for the further development of this organization.

7.10.3
Conclusions

If the changes anticipated for the Amazon region as a result of global warming occur, life in the region will become more and more difficult, not only for small farmers. It will no longer be possible to use the Amazon region as a location for expansion of large-scale agriculture or as a destination area for poverty-induced migration. At the same time yields in the traditional, better developed agricultural regions in the south, the centre and the south-east of Brazil will fall as a result of the reduced availability of water and higher temperatures. This may lead to a crisis situation developing in the agricultural sector, because soya and coffee are the third and fifth most important export goods respectively. In consequence, struggles over the allocation of land could intensify and the already high potential for violence could increase further. Scope for conserving biodiversity in the Amazon region would narrow even more.

In addition, the role of Brazil as a stabilizing factor and leading regional economy in Latin America could be weakened. Brazil is one of the few countries that perceives itself as a democracy and, despite internal tensions, it can ensure a relatively high degree of political stability. It is an important partner of the OECD in developing global solutions to problems and driving forward global regimes. If the country is taken unawares by the effects of climate change, it will have significantly less capacity for attending to projects of regional and global governance in addition to dealing with internal crises. This is particularly important because other Latin American countries will also be significantly affected by climate change and might turn to Brazil with requests for support (Section 6.9). If Brazil were to give lower priority to the needs of the Amazon region and world climate policy than to short-term national trade and energy interests, this would have a negative impact on its ability to prevent or handle crises in the Amazon area.

Climate change as a driver of social destabilization and threat to international security

8

The discussion in Chapters 5 to 7 has shown that climate change is altering the natural systems that support human societies worldwide. These transformation processes translate into the four conflict constellations set out in Sections 6.2–6.5, in a manner moderated by the specific geographical and social conditions. Those conflict constellations show how climate change can become a driver of social destabilization in several regions of the world. The following analysis in Section 8.1 highlights structural attributes of these climate-induced conflicts that clearly differ from the conflicts surrounding environmental degradation and resource scarcity explored in Chapter 3.

Given the importance of climate policy for international security, Section 8.2 discusses the prospects for effective climate policy in the coming two decades – the time remaining for the international community to avert dangerous climate change. Forecasting international policy processes is extremely difficult. Hence three scenarios are presented that outline possible and conceivable futures. This instrument makes it possible to gain a glimpse of the bifurcations in development trajectories facing climate policy in the coming years. Finally, Section 8.3 examines the possible impacts upon the international system of the social destabilization dynamics previously analysed. The analysis reveals that, above and beyond the regional crises that climate change will probably induce, challenges to the global governance system will arise that may undermine international stability and security.

8.1
Climate-induced conflict constellations: Analysis and findings

8.1.1
Key factors determining the emergence and amplification of conflicts

The analysis of the four conflict constellations driven by climate change, undertaken in Chapter 6, reveals the processes of social destabilization induced by climate change and the associated patterns of conflict, and thus provides the basis for an assessment of the various influences and key factors in terms of their conflict relevance (Table 8.1-1). Some of the key factors can be influenced by political action, which makes them important starting points for the recommendations for action presented in Chapter 10.

STATE CONSTITUTION, POLITICAL STABILITY
The constitution of a state (democratic, autocratic) does not, per se, appear to have any direct impact upon its coping capacities in situations of crisis or conflict. Nonetheless, societies in transition from authoritarian to democratic forms are particularly susceptible to crisis and conflict. Moreover, analysis of the 'Climate-induced decline in food production' and 'Climate-induced increase in storm and flood disasters' conflict constellations shows that disasters regularly heighten dissatisfaction with a government perceived to be illegitimate, especially in situations where there is a pressing need for action. Environmentally induced social challenges or even destabilization can thus become catalysts for processes of delegitimation of governments, and may therefore contribute to an escalation of conflict to the point at which political order collapses. This finding is relevant to African countries, most of which are in such phases of transition; these countries will be affected by climate change and their societies will come under pressure to adapt. This nexus may also become relevant to China.

GOVERNANCE STRUCTURES
The factor of governance is paramount in all conflict constellations. Whether climate-induced pressures upon societies erupt into crisis and conflict depends primarily upon the performance and problem-solving capacity of the states in question. A confluence of weak state institutions with processes of social overstretch induced by climate change thus harbours great potential for conflict. Building and strengthening national and international governance structures (developing viable administrative structures,

Table 8.1-1
Key factors determining the emergence and amplification of conflict constellations.
DCs = developing countries, ICs = industrialized countries
Source: WBGU

Conflict constellation / Key factors	State constitution, political stability	Governance structures	Economic performance and distributional equity	Social stability and demographics	Geographical factors	International power distribution and interdependency
Degradation of freshwater resources	General relevance	Effective national and international water management prevents crisis	High conflict potential in DCs due to great dependence upon agricultural sector	High risk of water crises due to rising demand in conjunction with stagnating supply	High risk of local conflict in catchment areas. Risk of regional destabilization ('spillover')	High conflict potential due to disparate national interests and local needs
Decline in food production	General relevance	High conflict potential where land-use rights are inequitable	High conflict potential in DCs due to great dependence upon agricultural sector	High risk of food crisis due to drought and population growth and density	Risk of regional destabilization ('spillover')	Major relevance of world market conditions in cases where agricultural production declines
Increase in storm and flood disasters	General relevance. Acute pressure to act compromises government legitimacy	Effective disaster risk management prevents crisis	High conflict potential in DCs. Elevated conflict potential in ICs due to dependence upon complex infrastructures	High conflict potential due to high population density and weak institutions	Risk of regional destabilization ('spillover')	General relevance
Migration	General relevance	Effective migration management prevents crisis	High conflict potential in DCs	High conflict potential due to changes in or instrumentalization of ethnic composition	Elevated conflict risk due to mounting resource competition in destination country	General relevance
General relevance for all conflict constellations	Lack of stability acts as catalyst. High conflict risk in periods of state transition (e.g. democratization)	Effective performance of state governance functions forestalls crisis and conflict	Poverty and socio-economic disparities increase susceptibility to crisis and potential for conflict	High risk of conflict where civil society structures are weak	Increased potential for conflict with neighbouring countries	Divergent impacts of economic integration and world market conditions. Cooperative foreign policy and action forestalls conflict

strengthening the judiciary, improving disaster management) are therefore extremely important. This is particularly apparent in the case of functioning national and international water resources management systems, which are pivotal in helping to prevent crisis. The 'Climate-induced increase in storm and flood disasters' and 'Environmentally induced migration' conflict constellations also confirm this finding. Here, too, disaster management and refugee management essentially determine whether critical developments arise at all. The finding applies not only to the specialized institutions responsible for crisis management; the fact is that the better and the more effectively a state as a whole performs its regulatory and steering functions, the less likely it is that a crisis will escalate into violent conflict. Climate change will impact especially hard on those regions of the world in which states with weak governance and problem-solving capacity already predominate. Climate change could therefore lead to a further increase in the number of weak and fragile states and thus to a greater likelihood of violent conflict.

ECONOMIC PERFORMANCE AND DISTRIBUTIONAL EQUITY
The economy of a country is particularly vulnerable to climate impacts if the agricultural sector accounts for a large proportion of employment and value creation. The 'Climate-induced degradation of freshwater resources' and 'Climate-induced decline in food production' conflict constellations illustrate that agriculture-based developing country economies will be hit particularly hard by the impacts of climate change. Because many of these countries lack the financial capacity to import food or to adapt water management in order to compensate for mounting resource scarcity, social crises are likely. Moreover, the high economic costs can further impede devel-

opment. This can entrench development blockages and poverty, and thus increase the societies' susceptibility to conflict. The constellations also show that good economic performance does not always insulate against the impacts of climate change. Extreme weather events are also a threat to the complex infrastructures of industrialized and newly industrializing countries, particularly in coastal areas. All the conflict constellations confirm that pre-existing socioeconomic disparities and inequities in the allocation of resources such as land or water – even if such inequality is merely a matter of perception – can contribute significantly to the further escalation of a conflict. Distributional equity and poverty reduction therefore play a major role in efforts to prevent and manage environmentally induced conflicts.

SOCIAL STABILITY AND DEMOGRAPHICS

Demographic development plays an important role in all conflict constellations. The combination of high population density and growth with increasing drought and water stress, for instance, harbours a particularly high risk of food crisis. In the same vein, the combination of high population density and weak state institutions greatly heightens the potential for damage and crisis in the event of storm and flood disasters. Ethnicity is a factor of relevance mainly in the 'Environmentally induced migration' conflict constellation. This constellation is particularly likely to heighten conflict if, for example, immigration changes the ethnic balance in the destination region and if ethnic differences are instrumentalized for political purposes. In general, it seems that when a society is already significantly destabilized and civil society structures are weak, the danger of conflict being induced by climate change grows.

GEOGRAPHICAL FACTORS

All the conflict constellations show that a destabilization of regions through 'spillover' effects is possible. Migration can heighten the risk of conflict in neighbouring countries or regions, for instance due to growing resource competition in combination with high population density or rebel incursions. The danger of spillover can arise in relation to water resources use as well, as river catchments and groundwater reservoirs often extend across national borders. Conflicts that were initially limited to the local or national level can thus destabilize neighbouring countries. As the social impacts of climate change are transboundary in nature, they can easily cause regions of crisis and conflict to expand.

INTERNATIONAL POWER DISTRIBUTION AND INTERDEPENDENCY

The distribution of power among states can play a role in the de-escalation or escalation of a conflict. For instance, the analysis of the 'Climate-induced degradation of freshwater resources' conflict constellation found that states which may have an incentive to wage war over water resources are primarily those with high political and military capacity and without democracy. World market conditions can impact in various ways on the development of a conflict constellation: In terms of food security, access to world markets in situations of declining agricultural production is an important factor that may be decisive in easing tensions. As the freshwater conflict constellation shows, however, integration within international trade systems can also have an adverse effect, for instance if transnational economic interests oppose local development interests. In general, the following applies to all the conflict constellations: (1) If the ability and willingness of a state to engage in intergovernmental cooperation, participate in supranational forums and join international decision-making bodies are high, it is more likely that agricultural production losses, freshwater crises or migration-related crises can be mitigated and interstate conflicts prevented. (2) A cooperative international setting and well-performing international organizations can help improve the prospects of managing the impacts of climate change and supporting national actors.

8.1.2
Reciprocal amplification of conflict constellations

The problems associated with climate change are particularly likely to develop their critical impact when several adverse causal complexes act in concert (Fig. 8.1-1). Societies and regions are particularly vulnerable where unfavourable biogeophysical conditions (e.g. coastal or arid areas), poverty, high population density, weak social and political institutions and political instability converge. As indicated in the characterizations of the individual conflict constellations (Sections 6.2–6.5), there are, however, also numerous interdependencies and feedback effects among the four constellations. Interactions hard to predict can thus occur, causing policy challenges to grow exponentially.

INTERACTIONS AMONG THE 'FOOD', 'FRESHWATER' AND 'MIGRATION' CONFLICT CONSTELLATIONS

The linkages between the conflict constellations surrounding food production and freshwater availability are readily characterized (Sections 6.2 and 6.3). Freshwater is fundamental to food security and pov-

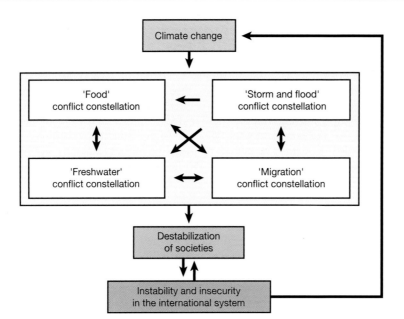

Figure 8.1-1
Conflict constellations as drivers of international destabilization.
Source: WBGU

erty reduction. 70 per cent of freshwater is used for farming. Unfavourable climate-related changes in water availability, for instance as a result of glacier melt or greater variability of precipitation, can have severe impacts upon the conditions for agricultural production and thus upon the livelihoods of rural populations. Such trends are amplified by competition with other forms of resource use, e.g. hydropower and bioenergy, and by pre-existing environmental problems such as soil degradation. Where the key political and institutional preconditions for functioning water resources management are absent and water crises emerge, food crises can be triggered or worsened. Conversely, non-sustainable agricultural practices can play a part in causing water stress and conflict by overexploiting water resources, or contaminating them with fertilizers and pesticides and thus rendering them unusable. Pre-existing soil degradation, water scarcity, non-sustainable utilization methods (e.g. expansion of unsustainable irrigation farming) and resource competition (e.g. due to the construction of dams) in combination with poverty, weak institutions and political instability can thus concentrate and amplify impacts regionally, thereby overwhelming institutions and causing crises to escalate.

Finally, in extreme cases, drought or water stress will force people to leave their homelands because they can no longer cultivate the land and have no alternative way of making a living. They might also flee war and violence once ecological problems have induced a social crisis. Migration can then, in turn, become the cause of various types of conflicts in the transit and destination regions. A growth in population and thus in demand could, for instance, over-whelm local water management or lead to further demographic pressure, e.g. in coastal conurbations at risk of storm surges. Through these mechanisms, poor or absent migration management can adversely affect water availability and food production as well as disaster risk management structures in the transit and destination regions.

It can be expected that climate change impacts will lead to a considerable increase in the number of migrants. These migration movements will occur mainly between developing countries, but the developed world will also be a target destination of environmental migrants. An increase of the associated conflicts is therefore likely. Here, too, there are signs of self-reinforcing feedback among the conflict constellations, and their concentration in the developing world.

The interactions described here can already be observed today. In future they will occur more frequently and in new regions. For instance, the population of the Sahel zone already suffers drought and hunger today. Further contraction of farmland due to drought and water shortage and an expansion of the drought zones towards the North into the Mediterranean region and towards East and West Africa will amplify the potential for crisis and the probability of migration (Fig. 8.1-2). North-western Brazil is also at risk of becoming a new drought-prone area. The Andes and Himalayan regions will be particularly exposed to the problem of glacier melt and resultant water shortage in the dry months (IPCC, 2007b).

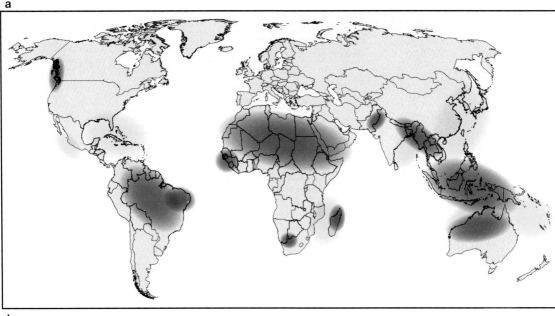
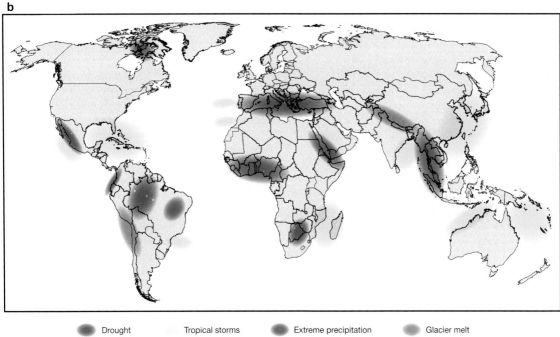

Figure 8.1-2
a) Climate status: Regions in which extreme climatic conditions already prevail today.
b) Climate future: Regions which could be put at risk in future by unabated climate change. The various climate impacts are set out in detail in Chapter 5.
Source: WBGU

INTERACTIONS DRIVEN BY THE 'STORM AND FLOOD' CONFLICT CONSTELLATION

As a result of climate change, weather extremes such as heavy rains will increase, as will the intensity of tropical cyclones. Combined with deforestation in the upper catchments of rivers, land subsidence in urban agglomerations and the growing concentration of cities and settlements in coastal zones, disaster risks are mounting considerably. In some regions of the world, coastal zones and entire island groups will be inundated. This will erode the coasts, destroy infrastructure and wreak economic damage, for instance in the fisheries and tourism sectors. Large numbers of people will be forced to flee abruptly. This will place major pressures upon disaster risk management systems and upon the coping capacities of the states and regions affected. Moreover, increasing extreme weather events lead not only to sudden disasters. They also trigger gradual processes of environmental degradation (WBGU, 2006). When the sea level rises, so too can the groundwater, and saltwater may intrude into aquifers and rivers. This can lead to the waterlogging and salination of soils, culminating in large-scale losses of agricultural and settlement areas. Storm and flood disasters thus have the potential to worsen the availability of drinking and irrigation water as well as soil quality over the long term, and, in conjunction with other factors, to induce water and food crises. These processes, too, force large numbers of people to move away because of the gradual deterioration of environmental conditions.

It is mainly in the densely populated river deltas of South Asia, East Asia and South-East Asia that this type of interaction will play a key role (IPCC, 2007b). However, as Fig. 8.1-2b shows, a number of other regions will suffer such effects due to tropical storms, e.g. in the Gulf of Mexico, the Pacific Islands, the west coast of Australia and the east coast of South America (e.g. Guyana). A number of larger cities in Africa will also be put at risk by sea-level rise.

8.1.3
The new quality of conflicts induced by climate change

Research on the environmental degradation–conflict nexus has shown that environmental degradation has been an important factor in the complex causal web of conflict emergence and amplification in recent decades, but has rarely been the key factor (Chapter 3). The environmentally influenced conflicts observed until now have generally remained at the local or intrastate level. In countries where problem-solving capacity is high or where there is international cooperation, such conflicts can be contained. Thus no large-scale escalation of violent conflict triggered by local environmental changes such as land degradation, desertification, water shortage or biodiversity loss is presently to be expected (Section 3.4).

The analysis in Chapters 5–7 does show, however, that climate-related regional risks of destabilization and violence threaten to gain a new dimension previously unknown to human civilization if climate change is not abated within the next 10–15 years through effective climate protection policy. Far-reaching impacts upon national societies, entire continents and the international system are then to be expected. This results from the new quality of climate change in terms of the globality of causation and impacts, the speed and scale of climate-induced environmental effects, the large numbers of people affected, and the reciprocal dynamic amplifications of conflict constellations.

The two maps in Figure 8.1-2 provide a global synopsis of the regional distribution of extreme climatic conditions today and in the event of unabated climate change. These maps depict climatic parameters as discussed in Chapter 5. Figure 8.1-3 further takes account of the social, political and economic conditions that determine the emergence of a conflict constellation in a given region; this map summarizes the security risks arising from climate change for a set of regions selected as examples. It is based on the regions examined in Chapter 7 and their susceptibility to the conflict constellations analysed in Chapter 6. As reliable regional information on climate change impacts is still lacking in many instances, a complete global overview cannot be presented here.

The particular volatility of this issue for international relations results from the fact that
- the impacts of climate change are escalating worldwide and simultaneously (Fig. 8.1-4);
- climate change can only be managed by means of international cooperation, whereby regional and local strategies need to be deployed in concert;
- there are major disparities between the per capita emissions of industrialized countries and those of developing and newly industrializing countries, and these disparities are perceived by the latter as an 'equity gap'.

WBGU has identified the following eight attributes that encapsulate this new quality:
- *Concentration* of conflict constellations in certain types of countries. Not all countries are exposed equally to the conflict risks arising from climate change. Developing countries, and especially the fragile states among them, will be hardest hit. The low problem-solving capacities of this group of states makes a high conflict potential probable. The largest concentration of fragile states is to be found in sub-Saharan Africa, where the effects of climate change must be expected to include a sharply rising number of people affected by water shortages (IPCC, 2007b).
- *Cumulation* of different conflict constellations in one region, and resultant feedback effects and interactions. Conflict constellations can reinforce each other; this is exemplified by the interaction between increasing freshwater scarcity and declining food production (Section 8.1.2). In the drylands of South America, for instance, increased soil salination and desertification are to be expected, leading to a heightened risk of food insecurity (IPCC, 2007b). Similar feedback effects are anticipated in sub-Saharan Africa, where models indi-

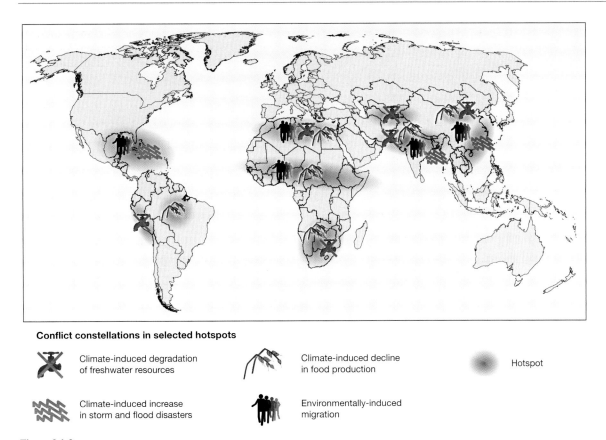

Figure 8.1-3
Security risks associated with climate change: Selected hotspots. The map only shows the regions which are dealt with in this report and which could develop into crisis hotspots.
Source: WBGU

cate a decline in freshwater availability together with a decline of farmland and shortened crop maturation periods caused by heat stress (IPCC, 2007b). In the Himalayas, increased glacier melt will cause more mudflows and landslides, destroying regional infrastructure and greatly hampering the external provision of supplies to this mountain region.

- *Scale:* The potential for crisis and the number of people affected will both reach unprecedented levels worldwide. Global population amounts to around 6.6 thousand million people today, and they are exposed in different ways to the consequences of climate change. Coastal communities, for instance, are affected particularly: In 1995 some 60 million people lived within the 1-metre elevation zone and 275 million within the 5-metre zone above mean sea level. Projections of population growth indicate that these figures will rise by the end of the 21st century to 130 million (1-metre zone) and 410 million (5-metre zone) (WBGU, 2006). The densely populated major delta regions of South, East and South-East Asia (e.g. Ganges-Brahmaputra or Zhujiang) will be exposed to a particularly high flood risk (IPCC, 2007b).
- *Recurrence* and a greater frequency of extreme events can undermine all development efforts and progress in a region. Models now indicate that in the cities of North America already affected in the past by heatwaves, the frequency of such events will rise to such a degree that the over-65s in particular will be exposed to an increased mortality risk (IPCC, 2007b). A greater frequency of floods and drought will pose particular risks to people dependent upon subsistence farming in lower latitudes (IPCC, 2007b), i.e. especially in Africa, parts of Asia and South America.
- *Propagation* of conflict constellations beyond the affected region: Environmental migration is a key mechanism here. It can be assumed that the number of migrants moving away from areas particularly affected by climate change will increase in future and that this may destabilize neighbouring regions. Moreover, the area immediately affected by the consequences of climate change can also expand. A poleward shift in hurricane activity is to be expected, as the higher sea temperatures

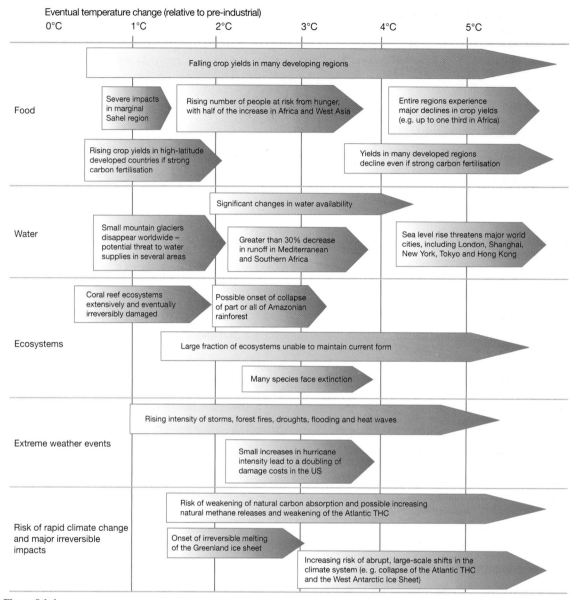

Figure 8.1-4
Consequences of climate change for ecosystems and economic sectors at different levels of warming.
Source: Stern, 2006

which favour hurricane formation are becoming established across larger parts of the oceans (Section 5.1.3). The existing toolbox for conflict prevention, conflict management and reconstruction of post-conflict regions will not suffice in future to cope with the problems if climate-induced conflicts threaten to engulf entire regions.

- *Intensification* of crisis situations: The new quality of problems resulting from climate change as set out above can engender a new dimension of conflict risks. This results from the interplay between increasing frequency and regional propagation of conflict constellations.
- *Disputes* and conflicts of interest between states will become more probable through an altered global balance of power in combination with the new problems driven by climate change. A world in transition to a multipolar order is likely to be less stable than today's world (Section 4.3), so the risk of confrontation among key players will rise. This fragile political setting would greatly hamper environment and development policy.

- *Time lag:* The time dimension presents particular challenges in efforts to address these problems. Resolute action must be taken to mitigate climate change within the next 10–15 years in order to reduce the risks of major socio-economic disruption. If climate policy fails, this will increase social destabilization, crisis, instability and violence in the particularly affected and most vulnerable regions of the world from roughly the mid-century onwards. Present political and economic systems are poorly equipped to respond to such global, long-term effects.

8.2
International climate policy scenarios and their long-term implications

There is no longer any doubt that humankind must greatly intensify its mitigation efforts if dangerous climate change (i.e. a rise in globally averaged near-surface air temperature of more than 2 °C relative to the pre-industrial value) is still to be avoided. To meet this goal, WBGU considers that strong climate policies must be adopted and energy systems transformed within the next 10-15 years to achieve the necessary 50 per cent reduction in global greenhouse gas emissions by 2050 compared with the 1990 baseline (WBGU, 2007). Yet there is a widening gap between the action that is urgently needed and current climate policy, as is evident from the international community's faltering efforts to implement the Kyoto Protocol and develop it further.

The particular challenge facing international climate policy is that according to current knowledge, the climate impacts resulting from inaction or inadequate action today will only be felt to their full extent in the next decades, by which time they will be severe and irreversible. Yet these impacts are still being widely ignored or are impossible for the public to imagine. The following 'Green Business As Usual' scenario is therefore intended to illustrate the impacts to be expected if international climate policy continues with the same lack of dynamism that has been witnessed in recent years. Unless there is significant climate policy impetus in one direction or another, this type of 'Green Business As Usual' scenario seems entirely plausible.

Besides the 'Green Business As Usual' scenario, WBGU aims to demonstrate the possible implications if the international community were to drop back behind the Kyoto commitments, for example ('International Policy Failure' scenario), or, alternatively, adopts the measures which current scientific knowledge indicates are both necessary and appropriate ('Strong Climate Policy' scenario).

What all these scenarios have in common is the underlying premise that based on the given path dependencies within emissions-intensive sectors and the delayed response of many climate parameters to greenhouse gas emissions, climate change over the next 10–15 years will be almost impossible to influence. However, relatively minor differences in the climate policy decisions adopted during this period will determine how the future maps out between various emissions pathways (Fig. 8.2-1), thus leading to substantial variations in the medium- to long-term climate impacts.

Figure 8.2-1
Emissions reductions required in order to avoid global warming of more than 2 °C: The red and green curves show the emissions pathways along which there is a 50 resp. 75 per cent probability of achieving this target. The various SRES scenarios developed by the IPCC are shown for the purpose of comparison. These do not allow for any explicit climate change mitigation policies.
Source: WBGU, based on Meinshausen, 2006

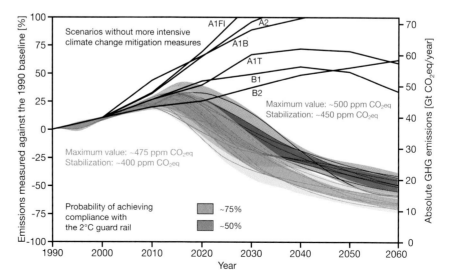

8.2.1
'Green Business As Usual' scenario: Too little, too late, too slow

Based on the outcomes of the 12th session of the Conference of the Parties to the United Nations Framework Convention on Climate Change (COP 12) held in Nairobi in November 2006, expectations of the progress to be achieved in the climate process in the coming years are relatively clear. As in the past, many of the small steps agreed in the negotiations and pointing in the right direction were hailed as successes even though no one could seriously claim that any real 'breakthroughs' were achieved in terms of solving the problem.

In Nairobi, the national delegations agreed to make appropriate preparations in 2007 for negotiations on more comprehensive reduction commitments for the industrialized countries for the Kyoto Protocol's second commitment period. In view of the political starting point, this can indeed be viewed as an – albeit modest – diplomatic success. Given the urgency of the climate problem, however, it is minor progress at best. Dangerous climate change can only be averted through a comprehensive reversal of global emissions trends, which requires a global transformation of energy systems. This means that the USA, Australia and the major newly industrializing countries must be integrated into climate protection as a matter of urgency (WBGU, 2007).

However, the 'Green Business As Usual' scenario outlined below assumes that the current trend – which can be summed up as 'slow drilling through hard boards' – will continue in future. The following developments are conceivable in this context:

- The majority of the developed countries commit to a 30 per cent reduction in greenhouse gas emissions by 2020; however, the USA – the world's largest emitter of greenhouse gases – and Australia do not enter into any binding reduction commitments.
- Various exemptions, e.g. in relation to land use changes, the inclusion of 'hot air' and the overly generous use of flexible mechanisms mean that even those developed countries which have undertaken reduction commitments in effect only reduce their emissions by 10-15 per cent.
- The USA increases its greenhouse gas emissions by around 10 per cent to 2020; at best, its emissions rates remain constant.
- China and other newly industrializing countries show a growing willingness to adopt voluntary climate measures and make efforts to curb the increase in their emissions. However, they balk at binding commitments, invoking the developed countries' responsibility and their own right to pursue 'catch-up' economic development.
- The developed countries make major concessions to the newly industrializing countries on technology transfer, so that with this support, the increase in the latter's emissions is indeed curbed compared with the reference pathway. For example, China's emissions increase by 'only' 150 per cent to 2020 instead of the expected 200 per cent against the 1990 baseline, while the emissions produced by the other newly industrializing and developing countries rise by just 100 per cent instead of 150 per cent.

Overall, then, global greenhouse gas emissions continue to rise and in 2020 exceed the 1990 figure by more than 30-40 per cent. The trajectory continues and no reversal of the trend is achieved. In this scenario, emissions develop along a pathway on which stabilization could still be achieved at 550ppm CO_2eq but not at 450ppm CO_2eq. This certainly cannot be regarded as the requisite reversal of the climate and energy policy trend. Instead, it would put humankind on an extremely precarious course which would be likely to result in global warming of 3 °C by the end of the century, even though the mitigation measures adopted appear to be reasonably ambitious compared with the climate policy status quo.

As this scenario will delay many of the transformations that must take place in the global economy and production in the interests of climate protection, especially the shift away from fossil fuels and hence the radical restructuring of energy and transportation systems, lock-in effects will occur. The failure to initiate radical systemic reform will entrench an emissions-intensive infrastructure (e.g. fossil fuel-based power generation systems) for decades, dragging out the already protracted process of developing low-emission alternatives. Due to the delay in achieving the requisite reversal of global emissions trends, the bar for safeguarding compliance with the 2 °C guard rail will be raised ever-higher: The longer the delay, the more swiftly emissions reductions would have to be achieved at a later date (i.e. through technology conversion) in order to reach the target on schedule.

In this scenario, the 2 °C guard rail is exceeded. The consequences described in detail in this report – notably in relation to the freshwater supply (Section 6.2), food production (Section 6.3), the greater frequency of storm and flood disasters (Section 6.4) and an increase in environmentally induced migration (Section 6.5) – become more likely. The corresponding climate-induced security risks increase as a result, and the associated security policy implications become ever more relevant.

It can also be assumed that constructive management of the anticipated difficulties by the international community will not become any easier if climate change exceeds critical limits. On the contrary, mutual recriminations – e.g. among the major powers in a multipolar world – cannot be ruled out, posing a great strain on international relations and global policy-making and permanently blocking a problem-solving-oriented system of global governance.

8.2.2
'International Policy Failure' scenario: Collapse of the multilateral climate regime

In light of the future problems associated with the 'Green Business As Usual' scenario, it would be extremely reckless and risky for the international community to drop back behind its multilateral commitments and to fail to live up to the expectations associated with them. Yet this type of scenario is by no means unrealistic.

A failure of the multilateral climate regime would lend considerable weight to the arguments of the pessimists who, in recent years, have claimed that multilateralism is in serious and permanent crisis. In this case, in addition to dangerous and no longer controllable climate change, it is likely that the old and new great powers would pursue unilateralist approaches which would become the norm in world politics, and that this unilateralism would be mutually reinforcing in light of the major problems faced by the world. Indeed, a reversion to a global confrontation between separate power blocs is conceivable, albeit under different conditions from those prevailing at the time of the East-West conflict. Here too, mutual recriminations and the violent escalation of proxy wars are possible. The United Nations would be at risk of losing all its political significance outside its formal institutional structures, while the EU could either be absorbed by one of the power centres or forfeit its capacity to act as a global player. In such a scenario, the constructive management of the world's numerous problems, for which international cooperation is indispensable, would be a vain hope.

8.2.3
'Strong Climate Policy' scenario: Compliance with the 2 °C guard rail

In its latest policy paper, WBGU has described in detail the form that an international climate and energy policy must take in order to achieve a scenario in which dangerous climate change can still be averted (WBGU, 2007).

If the German Federal Government were to champion the implementation of the measures proposed by WBGU, this would result in significant but by no means utopian reforms compared with the pathways outlined in the 'Business As Usual' scenario. If, for example, the developed countries were to implement their reduction commitments in such a way that a 30 per cent emissions reduction by 2020 compared with the 1990 baseline were genuinely achieved in practice, and furthermore, if more substantial concessions could be secured from the USA and China through adroit climate diplomacy and altruistic pledges on technology transfer, with mechanisms to curb emissions from land use changes (especially deforestation) also taking effect, then it is entirely realistic to stay within the 2 °C guard rail, thus indeed preventing dangerous climate change.

A further factor in favour of this scenario is that the relevant investments in climate change mitigation are worthwhile in economic terms, as the costs of effective climate protection are likely to be far lower than the costs of 'Green Business As Usual'. The fundamental principle which applies here is that the longer the delay in initiating action on climate protection, the more expensive it becomes. From WBGU's perspective, the requisite global transformation of energy systems is technically feasible, leads away from fossil fuels towards renewable energies, and requires rapid use of high efficiency potentials.

In such a scenario, a key step which is difficult to negotiate but which is by no means utopian is to achieve an international consensus on establishing the 2 °C guard rail within the framework of the UNFCCC, which means stabilizing the concentration of greenhouse gases below 450ppm CO_2eq (Fig. 8.2-1). Moreover, a progressive development of the Kyoto Protocol would create effective incentives for the global transformation of energy systems. On the one hand, the industrialized countries would have to commit to ambitious emissions reductions in this context; on the other hand, mechanisms for differentiated and progressive integration of the newly industrializing and developing countries into a commitment regime would need to be established. In order to facilitate this latter step, the issue of adaptation would have to be prioritized, and this must include firm pledges from the industrialized countries on financing and technological cooperation.

Almost inevitably, the G8 would have to play a key role in achieving this type of strong climate policy scenario. Ideally, the faltering climate process would receive the necessary impetus before the end of 2007; for example, the leaders of the G8 countries and the five major newly industrializing countries – Brazil, China, India, Mexico and South Africa – could announce an 'innovation pact on decarbonization'.

Not least, the European Union would utilize the opportunity to demonstrate global leadership in the climate process and expand and strengthen its pioneering role in international climate protection. In order to prove its credibility and seriousness on a lasting basis and also remain a respected and relevant partner for the key global actors in a multipolar world order, the EU must achieve – or, better still, exceed – its agreed emissions reduction targets. By demonstrating the viability of pro-active climate protection and generating worldwide impetus for an efficiency revolution and a shift towards renewables, the European Union could become the driver of strong climate policy and thus underscore its own significance and status in world politics.

In principle, other actors could drive such policy as well. If the forces of inertia emanating from industrial interest groups and other actors in Europe and North America remain so strong that governments dispense with responsible climate policy beyond 'Green Business As Usual', it is quite possible that the Group of 77 (G-77) – a caucus of developing countries – will take on the climate challenge under Chinese and Indian leadership.

Prompted by traditional political sensitivity to the poorest and least developed countries in the G77 which are hardest hit by climate change, a recognition of its own responsibility in relation to global warming and an awareness of its increasing global capacity to take action and its own vulnerability to the impacts of climate change, the People's Republic of China, for example, could emerge as a driving force in international climate policy. Simply on account of the sheer size of the Chinese population and economy, a resolutely implemented Chinese climate protection policy could make a major contribution to the reversal of the global trend, thereby massively increasing the pressure on the hesitating developed countries to take action, especially given that these countries would otherwise be threatened not only with the economic costs of inaction but substantial political costs as well.

However, non-state actors could also send out important climate policy signals which could generate the dynamics on the scale necessary for a bottom-up 'climate turnaround'. The US economy – which, historically, has always been open to innovation, technologically sophisticated and, not least, keen to generate export revenue – could seize the opportunity and, regardless of the decision-makers in Washington, initiate an energy revolution with global impacts. The example of the US state of California – one of the world's ten largest economies in its own right, and with its own ambitious climate goals – suggests that these need not be utopian ideals. Not least, civil society actors could also intensify the pressure on industry and governments to take action. By making widespread use of their consumer power, they could force industry, energy suppliers, airlines and others to develop climate-safe corporate strategies. This could mean that individual conduct which is incompatible with climate goals, such as short trips to the Caribbean or the use of gas-guzzling SUVs in urban areas, would become socially unacceptable. This trend would be most likely to emanate from the consumption-oriented and increasingly climate-aware industrial societies of Western Europe, but certainly need not be restricted to them: The urban middle classes in cities such as Mexico City, Rio de Janeiro, Johannesburg, New Delhi, Mumbai, Hong Kong or Beijing could also follow suit.

Moving beyond the prospect of resolute action by the EU and G8, a compelling and positive strong climate policy scenario is most likely if there are mutually reinforcing dynamics, e.g. between the governments of the newly industrializing countries, private-sector interests in the United States, and a 'revolution in consumption patterns' in the EU. In any event, WBGU's message is clear: Dangerous climate change can still be prevented if strong climate policies are adopted pro-actively, consistently and above all without delay.

8.3
Climate change as a threat to international security

The analysis of the conflict constellations that are likely to be initiated or intensified in many societies as a result of climate change has shown that in developing countries in particular, political, economic and social adaptation strategies must be developed in order to avoid future destabilization processes and conflicts (Chapter 6). At present, the transformation of climate change impacts into social crises is only just beginning. To date, climate change has resulted in temperatures rising by around $0.8\,°C$ above the pre-industrial level. Without more intensive mitigation measures, a $2–7\,°C$ increase in global temperatures above the pre-industrial level must be expected by the end of the 21st century, with the precise increase depending on the amount of greenhouse gases emitted and uncertainties in the climate system.

In the event of climate change mitigation efforts failing, WBGU anticipates major destabilizing effects on societies from around the mid 21st century onwards. The greater the climate change, the greater the likelihood that climate-induced social destabilization processes and conflict constellations will impact not only on individual countries or subregions in the coming decades but will also affect the global governance system as a whole. If an effective climate policy could be implemented in the next two decades which limits the rise in near-surface air temperature to a maximum of 2 °C relative to the pre-industrial value, a climate-induced threat to international security would likely be averted. Viewed in this light, climate policy is preventive security policy.

The analysis of the interactions between the conflict constellations (Section 8.1.2), the description of the world regions especially affected by climate change (Chapter 7) and the overview of possible impacts of rising temperatures (Fig. 8.1-4) illustrate the potentially explosive nature of climate change for the international community. Global temperature increases from 3–5 °C – which are extremely likely unless greenhouse gas emissions are drastically reduced – would lead to major changes in the natural environment worldwide and radically alter the living conditions of hundreds of millions of people. In some regions of the world, food production could collapse on a massive scale. Extreme weather events could threaten the major coastal cities of China and India, but also London, New York or Tokyo, and climate change could deprive up to a thousand million people of their access to potable water. The Stern Review observes: 'The latest science suggests that the Earth's average temperature will rise by even more than 5 or 6 °C if emissions continue to grow and positive feedbacks amplify the warming effect of greenhouse gases (e.g. release of carbon dioxide from soils or methane from permafrost). This level of global temperature rise would be equivalent to the amount of warming that occurred between the last ice age and today – and is likely to lead to major disruption and large-scale movement of population. Such 'socially contingent' effects could be catastrophic, but are currently very hard to capture with models as temperatures would be so far outside human experience' (Stern, 2006).

WBGU's assessment of the future security impacts of climate change at local, national and regional level shows that the rise in global temperatures is unlikely to lead to classic interstate wars. A more probable scenario is the proliferation of processes of destabilization and collapse in countries and regions which are especially hard hit by climate change, which overstretches the political and economic capacities of states and societies. The breakdown of law and order and the erosion of social systems in climate crisis areas could reinforce the trend towards 'new wars and conflicts' which has been observed since the 1990s, and whose characteristics include violent intra-societal conflict, state collapse and lawlessness, and cross-border conflicts over resources, accompanied by increasing migration. What starts out as local and national crises will ultimately have an impact on the international system as well. In light of current knowledge about the social impacts of climate change, WBGU identifies six key threats to international security and stability which could be triggered by global warming. Climate change will

– accelerate the proliferation of the 'fragile state' phenomenon,
– jeopardize global economic development,
– trigger international distributional conflicts between the main drivers of climate change and those most affected,
– undermine fundamental human rights and lead to crises of legitimacy in the countries which cause climate change,
– trigger migration flows and crises,
– overstretch classic security policy.

The interplay between the threats that unabated climate change poses to the international system would overstretch the capacities of the existing global governance system.

8.3.1
Possible increase in the number of destabilized states as a result of climate change

Weak and failing states already pose a major challenge to international security policy (Section 4.2). This view is endorsed by the European Security Strategy, which highlights the links between state failure, cross-border civil wars, transnational terrorism and trafficking in humans and weapons (EU, 2003). So far, however, the international community has failed to summon the political will or the necessary financial resources to safeguard lasting stability in the world's 30 or so fragile states. What's more, the mechanisms available through development cooperation and military policy to deal with the problem of weak states are contentious in both conceptual and political terms, although the concern about the threat that state failure poses to regional and global stability is widely shared.

The impacts of climate change, such as the threat of food crises, water scarcity, extreme weather events and ensuing migration, will expose many of the already weak states, especially in southern Africa, to additional pressure to adapt (Fig. 8.1-2b). However, climate change could overstretch countries' problem-solving capacities in other regions of the world too:

- Hurricanes gaining in destructive force as a result of rising global temperatures could overwhelm the economic capacities of states and societies, especially in Central America (Section 7.8).
- Protracted droughts or even a collapse of the Amazon rainforest (Sections 5.3.4 and 7.10) would present northern Brazil and neighbouring regions

in Latin America with unprecedented challenges and heighten the distributional conflicts between the poor Amazon regions and relatively affluent southern Brazil.
- From the mid 21st century onwards, sea-level rise could confront the farming regions of the Ganges Delta – home to as many as 200 million people – with grave socio-economic problems.
- The melting of the glaciers in the Andean and Himalayan regions would jeopardize water supply and trigger agricultural crises.

The risk that economic capacities, political systems and societies will be overstretched by climate change therefore increases as the climate problem intensifies. As the effects of climate change do not stop at national borders, 'failing subregions' could emerge, consisting of several simultaneously overstretched states.

The international community should thus be prepared for a situation in which the global problem of weak and overstretched states will become even more pressing in future if climate change is not stopped in time. A proliferation of fragile states is unlikely to trigger major military conflicts; rather, it will cause the diffuse erosion of international stability and security and the spread of failing regions. The unstable peripheries of the international system could expand, with a widening of the 'black holes' in world politics that are characterized by the collapse of law and public order, i.e. the pillars of security and stability. There is no sign, at present, that the international community would have the capacity to halt this process of erosion effectively if climate impacts intensify. Climate change therefore triggers and amplifies international insecurity and widening instabilities by overstretching the capacities of states.

This overstretch is relevant to global climate change mitigation in economic terms as well: Relatively low mitigation costs presuppose that cost-effective mitigation options can be implemented in developing and newly industrializing countries as well. However, with the proliferation of state fragility, some countries are facing the loss of the institutional structures that would enable them to carry out mitigation measures efficiently and effectively in the global interest.

8.3.2
Risks for global economic development

Climate change alters the conditions for regional production processes and supply infrastructures, e.g. by causing regional water scarcity or variability in water availability (Section 6.2), drought and declining soil productivity (Section 6.3), or storms and flooding of coastal locations and infrastructures (Section 6.4). These climate impacts force companies to relocate, either spontaneously or at best on a planned basis, and lead to the closure of production sites. People abandon their home regions in coastal or arid regions because under the changed climatic conditions, they no longer have adequate employment and income generation opportunities, or perhaps even because their previous living and working environment has become hostile to life (Section 6.5).

Climate change thus leads to the destruction and devaluation of economic capital as well as the loss of skilled and productive workers through environmentally induced migration and an increase in climate-induced diseases and malnutrition. Furthermore, economic resources which would normally be channelled directly into the production process instead have to be spent on adaptation measures, e.g. preparing for extreme events, or on reconstruction or the delivery of additional health services. The impairment of international trade routes as a result of changed climatic conditions may also mean that the benefits and growth stimuli resulting from the ongoing international division of labour are capitalized on to a lesser extent.

The impacts outlined above contribute – each according to their form and intensity – to a slowing of economic growth processes and/or the stagnation of or even a drop in the affected countries' gross domestic product (GDP). These negative economic impacts may initially be offset in some regions by the limited economic benefits of changed climatic conditions, such as more moderate temperatures and increased precipitation, which could prove advantageous for some sectors of agriculture, for example. Technological innovations could also reduce the economic pressure of adaptation and stimulate growth. From a global perspective, however, these regionally limited effects will not compensate for the overall negative trend. Rather, the drops in growth and prosperity are likely to be very substantial if climate change continues unabated and causes greatly intensified climate impacts.

In terms of the range of economic impacts, which may even include global economic crisis (Stern, 2006), a key factor is the importance of the worst affected regions in global economic relations. It is notable that the developing countries would bear the main burden of the impacts of climate change, yet play a relatively insignificant role in the global economy. However, the major newly industrializing countries such as China and India will become ever more important economic players in the coming decades, also as trade partners for the export-oriented industrialized countries (Section 4.3.2; Goldman Sachs, 2003) – and these countries are highly exposed to the threat of major climate impacts (Chapter 5 and Sections 7.4, 7.6, 7.7 and 7.10). Significant impairment of the global economy is therefore a distinct possibility.

Another factor of relevance in terms of avoiding global economic crisis is the level at which stabilization of the concentration of greenhouse gases can be achieved – and therefore which climate impacts are likely to be felt in the global economy. According to Stern (2006), inaction in climate policy and the consequent unabated climate change would put global GDP on a far lower development pathway, costing between 5 per cent and 20 per cent of annual global GDP now and into the future, compared with a hypothetical development pathway without climate impacts (Kemfert and Schumacher, 2005; Stern, 2006). Even though comparisons with the world economic crisis of the 1920s and 1930s (Stern, 2006) appear overstated, it is nonetheless the case that any inaction or delay in pursuing resolute and ambitious climate protection policies will jeopardize the growth prospects of the newly industrializing/developing countries and the industrialized countries alike. This in turn will steadily reduce the economic scope for action at both national and international level on urgent challenges such as poverty reduction, demographic change, control of major diseases, growing energy and resource scarcity, and environmental protection.

8.3.3
Risks of growing distributional conflicts between the main drivers of climate change and those most affected

Because climate change will cause substantial damage and consequential costs worldwide, conflicts and discussions about compensation can be expected in future between the main drivers of climate change and those countries whose role in causing climate change is negligible but which will be hard hit by its effects. These distributional conflicts will probably arise primarily between those countries which are mainly responsible for CO_2 emissions from fossil fuels and those which sustain especially high levels of damage from climate change. Admittedly, CO_2 emissions from large-scale deforestation and methane/nitrous oxide emissions are contributory factors in climate change (Box 8.3-1), but these emissions are not increasing anything like as sharply as CO_2 emissions from fossil fuels. Emissions from fossil fuels are also a key issue in the distribution debate because they are particularly bound up with economic development. Industrialization and the economic prosperity of the industrialized countries are based, not least, on massive use of fossil fuels.

So it is the industrialized countries which have been and continue to be primarily responsible for CO_2 emissions from fossil sources, whereas it is the developing countries which are bearing the main burden of the rising costs associated with climate impacts. The greater the burden of adaptation in the South, however, the more intensive the conflicts between the main drivers of climate change and those most affected will be. The worst affected countries are likely, with good reason, to invoke the 'polluter pays' principle and demand not only support for adaptation measures but compensation as well. Furthermore, because the industrialized countries' sophisticated economies are based on energy-intensive industrialization processes which have caused the massive increase in CO_2 emissions since the 19th century, it is clear that the prosperity of the industrialized world has ultimately been achieved at the expense of developing regions, which are being damaged by the ensuing climate change. The compensation claimed on this basis could reach a scale far greater than the fairly modest investments in international development cooperation to date. The countries damaged by climate change could point to the US state of California, which announced in 2006 that it was suing the US automobile industry for allegedly contributing to global warming in order to establish its liability for the environmental damage caused. The international community is ill-prepared for this type of conflict, however. At present, no institution exists at international level with a mandate to rule on climate-damaged countries' compensation claims. Indeed, it seems very likely that a global compensation regime for climate change will prove highly controversial at international level. Unabated climate change could thus herald the onset of a diplomatic and foreign policy 'freeze' between the main drivers of climate change, particularly the USA, the EU, Japan and Australia, and a substantial number of developing countries which are especially hard hit by its impacts. A worsening of the North-South conflict would be likely to occur, with aggression and aversion against the affluent industrialized countries intensifying.

Beside today's industrialized countries, the major ascendant economies whose emissions have also increasing substantially since the end of the 20th century, notably China but also India, Brazil and Indonesia, could also be called to account by the majority of developing countries in future (Box 8.3-1), even though their per capita emissions are still well below those of the industrialized countries (Table 8.3-2). Distributional conflicts over compensation payments are likely to play an increasingly important role in relations between these groups of countries as well. Against this background, it is clear that a key line of

Box 8.3-1

The major newly industrializing countries' possible future share of global greenhouse gas emissions

In 2000, China accounted for around 12 per cent of the world's total greenhouse gas emissions. India's share was 4.5 per cent, Brazil's 5.3 per cent, and the OECD countries' share was 37 per cent (CAIT WRI, 2007). The emissions can be broadly categorized as CO_2 emissions from land use changes, CO_2 emissions from fossil fuels, and emissions of other greenhouse gases (non-CO_2 emissions, e.g. methane, nitrous oxide).

In 2004, around 20 per cent of global greenhouse gas emissions consisted of CO_2 resulting from land use changes (Fig. 8.3-1). Among the countries listed, this especially affects Brazil: Its CO_2 emissions from land use changes (mainly from deforestation in the Amazon region) are currently four times higher than its energy-related emissions and account for around two-thirds of its total emissions (CAIT WRI, 2007). In India and China, by contrast, emissions from land use changes are negligible. Projections of CO_2 emissions from land use changes are extremely unreliable, but it is expected that emissions from deforestation in the tropical regions will remain high in the medium term (IPCC, 2007c). In view of the strong growth in total emissions, however, the relative share of emissions from land use changes will probably decrease.

Non-CO_2 emissions accounted for 23 per cent of global emissions in 2004 (IPCC, 2007c). Projections from the US Environmental Protection Agency (EPA) show that global non-CO_2 emissions will rise by around 30 per cent to 2020, but the relative shares of China, India and Brazil in these emissions (currently 16 per cent, 6 per cent and 6.5 per cent respectively) are likely to remain fairly constant (EPA, 2006).

The main driver behind the expected relative increase in the significance of emissions from the newly industrial-

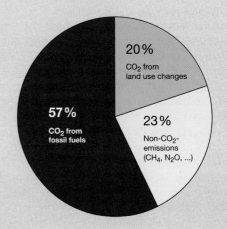

Figure 8.3-1
Rough breakdown of global greenhouse gas emissions in 2004.
Source: WBGU, based on IPCC, 2007c

izing countries is therefore fossil fuel use. CO_2 from fossil sources not only accounted for the major share of current emissions in 2004, i.e. 57 per cent; these emissions are also increasing at a disproportionately high rate (IPCC, 2007c). There are numerous scenarios for these emissions which all show an increasing relative share for the newly industrializing countries. For illustrative purposes, Table 8.3-1 shows the percentage increases of selected newly industrializing countries and the OECD countries in global energy-related CO_2 emissions based on the IEA's Alternative Policy Scenario (IEA, 2006c). The absolute level and the rapid growth of emissions from the newly industrializing countries make it clear that climate protection policy cannot succeed in future without the participation of the newly industrializing countries.

Table 8.3-1
Global energy-related CO_2 emissions and selected countries'/groups of countries' shares in these emissions based on the IEA's Alternative Policy Scenario.
Source: IEA, 2006c

	1990	2004	2015	2030
Global energy-related CO_2 emissions [Gt CO_2]	20.5	26.1	31.6	34.1
China's share [%]	11.2	18.3	23.1	25.8
India's share [%]	2.9	4.2	4.8	5.9
Brazil's share [%]	0.9	1.2	1.2	1.3
OECD's share [%]	54.0	49.2	43.8	38.7

conflict in global politics in the 21st century would therefore divide the main drivers of climate change and the poor countries most affected by it, and that this line of conflict will not only divide North and South but will also have a South-South dimension. Based on a detailed analysis of the various greenhouse gas emissions, a more differentiated separation into 'main drivers' and 'affected' countries may be

appropriate among the industrialized countries too. However, in view of the development of North-South relations in the past and the ongoing prosperity gap, such considerations will be of secondary importance for the conflicts that can be anticipated.

There are many signs that the most powerful nations within the international system will come under very strong pressure to justify their actions

Table 8.3-2
Per capita greenhouse gas emissions for selected countries and groups of countries. Data under 'CO_2 emissions from fossil sources' also contain emissions from cement manufacturing. The total emissions include all the gases covered by the Kyoto Protocol including emissions from land use changes.
Source: CAIT WRI, 2007

	CO_2 emissions from fossil sources in 1990 [t CO_2/capita]	Total emissions in 1990 [t CO_2eq/capita]	CO_2 emissions from fossil sources in 2000 [t CO_2/capita]	Total emissions in 2000 [t CO_2eq/capita]
Global	4.0	7.3	4.0	7.0
Annex I countries	12.1	15.0	11.4	13.9
Non Annex I countries	1.8	5.1	2.2	5.1
USA	19.8	22.4	20.5	23.1
Brazil	1.5	17.6	1.8	12.8
Germany	12.5	15.4	10.3	12.5
Côte d'Ivoire	0.4	6.2	0.4	7.3
Zimbabwe	1.5	7.0	1.1	6.6
China	2.2	3.7	2.6	4.0
Ghana	0.4	2.2	0.4	2.6
India	0.7	1.5	1.1	1.8
Burkina Faso	<0.3	2.2	<0.3	1.8
Bangladesh	<0.3	0.7	<0.3	0.7

in the face of accelerating climate change. Instead of safeguarding stability, security, a balance of interests and multilateralism based on justice, the global and regional leading powers would be perceived as being the main drivers of climate change and therefore as drivers of international instability and global distributional conflicts. The opportunities to establish a functioning global governance architecture therefore decrease with rising global temperatures, bringing a global problem to light: While climate change can only be curbed effectively through international cooperation, the bases for constructive multilateralism will themselves be eroded as the world gets progressively hotter.

8.3.4
Climate change undermines human rights: Calling emitters to account

The divisions in global politics, outlined above, are likely to be amplified by a further combination of factors. The conflict constellations described in Chapter 6, which could be initiated and reinforced by climate change, ultimately undermine fundamental human rights: Food security and access to drinking water could be challenged by the impacts of climate change in affected countries and regions, destruction caused by rising sea levels and extreme weather conditions could put people's livelihoods at risk, and all this could trigger strong environmentally induced migration. Unabated climate change could threaten natural life-support systems, erode human security and thus contribute to the violation of human rights.

The main drivers of climate change are the CO_2-emitting industrialized countries, but increasingly the major newly industrializing countries as well (Section 8.3.3). Against the backdrop of rising global temperatures, growing awareness of the direct impacts of climate change on societies and inadequate mitigation efforts, these countries could increasingly be accused of knowingly causing human rights violations, or at least doing so in de facto terms. This could lead to a permanent shift in focus in the international human rights discourse. Whereas today, the democratic industrialized countries castigate the violation of human rights by unjust regimes, these countries themselves could – with good reason – find themselves in the dock in future, put there by the developing countries that are acutely affected by climate change and by international human rights organizations. Future human rights debates in the United Nations are likely to focus not only on the classic disputes about the violation of human dignity by authoritarian governments, but also on the threat to human rights resulting from climate impacts, especially as it is now recognised that human rights guarantees also entail certain active obligations. This includes, for example, the 'responsibility to protect', which makes it incumbent on states to take appropriate protective measures in the event of a threat to human rights. Unabated climate change could thus plunge the industrialized countries in particular into crises of legitimacy and limit their international scope for action. It is very possible that these crises of legitimacy vis-à-vis the rest of the world could also trigger internal crises of legitimacy in the democratic industrial societies themselves.

The correlations outlined above would also shift the parameters in international development policy and research. Over the last two decades, the view which has prevailed in development research is that the success and failure of development processes in developing countries have been influenced primarily by endogenous factors and actors. In the context of advancing climate change, however, this view may need to be reappraised. One factor which must be considered in this context is that the drivers of climate change are permanently impairing many developing countries' development prospects. The political responsibility for the development crises in the South would then no longer be mainly attributable to 'bad governance' in the poor countries. With rising temperatures, those countries which were once regarded as models for democracy, human rights protection and economic development would instead be regarded as being jointly responsible for global instability, insecurity and the destruction of economic potential. The G7 countries, which, as a value-based community of the world's leading democracies, claim to be controlling the fortunes of the world not only in their own but also in the global interest, would jeopardize their own credibility – the important basis of their international capacity to act.

8.3.5
Climate change triggers and intensifies migration

Migration is already a major and largely unresolved international policy challenge. Refugee flows in Africa are both the cause and the effect of spiralling violent conflicts. The European Union's impotence in dealing with African refugees and its failure to develop effective instruments to prevent migration are obvious. The US, meanwhile, is massively reinforcing its fortifications along its border with Mexico in order to limit the influx of illegal economic migrants. Migration and refugees from poor countries are politically sensitive and controversial issues in the industrialized countries, as the heated debates in Germany in the early 1990s, which led to the tightening up of asylum law, and the rise of xenophobic movements and parties in Europe illustrate very clearly.

There are many signs that the problem of migration will intensify worldwide as a result of climate change and its social impacts. Growing numbers of people will be affected, and the number of migration hotspots around the world will also increase. The associated conflict potential is considerable: 'Environmental migrants' are currently not provided for in international law, so people displaced as a result of climate change impacts have no formal rights. Conflicts can be expected between the countries which cause climate change and those affected by it, with key issues being the extent to which refugee and migration flows are genuinely triggered by rising global temperatures or by other environmental damage resulting from actions at national level. Disputes over compensation payments and the financing of systems to manage refugee crises will increase – and in line with the 'polluter pays' principle, the industrialized countries will have to face up to their responsibilities. The controversy over the development of international regimes to clarify which countries must admit climate refugees in future is likely to worsen political tensions. If global temperatures continue to rise unabated, migration could become one of the major fields of conflict in international politics in future.

8.3.6
Climate change overstretches classic security policy

Worldwide, climate change could lead to diffuse conflict structures, security threats, social destabilization processes and violence, therefore also posing a challenge to 'classic' security policy. Overstretched states, fragile subregions, migration flows, conflicts over access to water and food, or the failure of conflict management systems will be almost impossible to manage without support from police and military capacities. In this context, well-functioning, joined-up development and security policy will be crucial to restore stability and public order, as civilian conflict management and reconstruction assistance are reliant on a minimum level of security.

At the same time, experience with military operations which have aimed to stabilize weak states since the 1990s show that 'classic' security policy's options and capacities to act are limited. Highly equipped military contingents can occupy countries by force and topple governments, but have not proved particularly effective when it comes to stabilizing and bringing peace to societies and conflict situations in which it is difficult to distinguish between aggressor and defender, perpetrator and victim. A climate-

Box 8.3-2

Security threats in the 21st century: A comparison with strategic analyses from classic security policy

SECURITY-RELEVANT REGIONS

Military strategies focus very strongly on the Middle East region. The WBGU report, on the other hand, additionally highlights the security policy relevance of the African continent and Central Asia and pinpoints the Amazon region, the coastal regions of China and India, and Central America and the Caribbean as well.

COMPLEXITY OF SECURITY THREATS

The analyses by security policy think-tanks mainly highlight the problems of transnational terrorism, the proliferation of weapons of mass destruction, new resource conflicts and failing states as challenges to international security. The WBGU report shows that climate change could lead to new types of local, national and international conflict situations and insecurities which have rarely featured in most security policy scenarios to date. Unabated climate change could lead to four conflict constellations relating to freshwater, food production, storm and flood disasters and environmentally induced migration, which have the potential to destabilize societies and cause insecurity.

LONG-TERM NATURE

In their strategic analyses, military strategists consider the period to 2015/2020 – a much longer timeframe than those routinely deployed by policy-makers and planners in other policy fields. The WBGU report, however, makes it clear that in relation to climate change, an even longer perspective is required, i.e. to the mid 21st century and beyond, in order to identify the potential threats to societies posed by global environmental change and take appropriate preventive action wherever possible. The WBGU report thus urges a long-term perspective in policy-making.

Sources: EU, 2003; ISS, 2004; UN, 2004; Government of China, 2005; BMVg, 2006; White House, 2006

---

induced increase in the number of overstretched states or even the destabilization of entire subregions (Section 8.3.1) would thus limit the classic security policy options available to contain these conflictive dynamics and stabilize or even bring peace to such situations. Yet without adequate security guarantees, it is well-nigh impossible for civilian development policy measures to take effect. The result would be widening zones of instability and insecurity.

8.3.7
Summary: Overstretching the capacities of the global governance system

Unabated climate change is likely to become a major challenge for the international system in the coming decades. It heightens the interdependencies between all the world's societies and creates global risk potentials which can only be countered by policies that aim to manage global change. Every one of the six threats to international stability and security, outlined above, is itself hard to manage. The interaction between these threats intensifies the associated challenges and risks for international politics. It is almost inconceivable that in the coming years, a global governance system could emerge with the capacity to respond effectively to these global dynamics of conflict and instability from 2020 onwards. It is more likely that with ongoing climate change, distributional conflicts between the drivers and the victims will become increasingly urgent, as will the crises of legitimacy faced by the major CO_2 emitters. Growing tensions, conflicts and confrontations along the dividing line between the main drivers of climate change and the worst affected countries would steadily erode the prospects of establishing a global governance system based on cooperation, which is essential to master the world's problems. Against the backdrop of globalization, climate change is likely to further overstretch a still insufficient global governance system. The world could thus become a highly insecure place unless climate change can be controlled effectively.

The climate-induced challenges to stability and security in the global governance system, outlined above, differ fundamentally from the security problems which arose during the Cold War. The confrontation between the two ideological blocs in the second half of the 20th century focussed primarily on military superiority and deterrence, alliance-based security strategies to weaken the other side and peaceful coexistence to avoid military escalations. By contrast, the climate-induced security risks of the 21st century are almost impossible to mitigate through military spending and interventions (Box 8.3-2). Instead, an intelligent and well-crafted global governance strategy to mitigate these new security risks would initially consist of an effective climate policy, which would then evolve into a core element of preventive security policy in the coming decades. The more climate change advances, the more important adaptation strategies in the affected countries will become, and these must be supported by international development policy, e.g. strengthening states' governance capacities, agricultural policies, water management policies, food security programmes, reinforcement of disaster prevention mechanisms, etc. At international level, the focus will be on global diplomacy to contain climate-induced distributional conflicts, as well as the development of compensation mechanisms for the victims of climate change, global migration policy, and measures to stabilize the world economy. Climate change thus poses a challenge to international security, but classic, military-based security policy will be largely unable to make any major contributions to resolving the impending climate crises.

Research recommendations 9

9.1
Understanding the climate-security nexus – fundamentals

9.1.1
Climate research

A number of key statements on human-induced climate change have attained the status of certain knowledge in climate research today. This includes the fact that human activities greatly increase the concentration of greenhouse gases in the atmosphere, that this has already brought about marked global warming, and that, unless mitigation measures are put in place, it will lead to even greater warming in future. Many questions, notably regarding changes in precipitation distribution, extreme events and regional manifestations of climate change, remain unanswered, making it necessary to step up basic research on climate change.

A report such as this on 'Climate change as a security risk', however, cannot and should not recommend a programme of general climate research. WBGU therefore proposes recommendations for research only in areas where new knowledge can contribute to improving adaptation and conflict prevention, especially in relation to the conflict constellations analysed in this report.

SEA LEVEL
To enable successful and cost-effective adaptation in terms of coastal protection measures, there is a need to reduce uncertainties regarding future projections of sea-level rise. Inadequate understanding of the dynamics of continental ice sheets constitutes a major part of this uncertainty. Climate impact research should give greater priority to investigating the regional impacts of changes in sea level on coastal areas (e.g. stability of beaches and other types of coast, risk to valuable ecosystems such as mangroves and coral reefs, flood risk to infrastructure and human settlements, especially in densely populated urban agglomerations).

TROPICAL STORMS
One debate currently taking place within the research community relates to the possible future dynamics of storm intensity and the potential expansion of the geographical range of tropical storms. Research in this field is still in its infancy, because only a few studies (most of which are very recent) have hitherto focused on this topic. There are discrepancies between the theory (which predicts only a slight increase in the intensity of tropical storms as a result of climatic warming) and the considerable increase that has already been observed. The spatial resolution of current global climate models is generally too limited to be able to simulate tropical storms realistically. Reliable projections are therefore not available at the present time.

OTHER EXTREME WEATHER EVENTS
The incidence of droughts, intense precipitation events or severe storms is crucial in terms of the impact of climate change on human activities. At the same time, however, this is a topic associated with a particularly high level of uncertainty. The reason for this is that such extreme events are often highly localized and occur infrequently (which makes it difficult to identify trends in the observation data), and that extreme atmospheric conditions are particularly closely associated with non-linear physical mechanisms. A climate research programme focusing on extreme weather events is needed in order to improve estimations of potential risk.

MOUNTAIN GLACIERS
Predictions concerning the dynamics of glaciers continue to be dogged by major shortcomings. Reliable projections of the future development of glacier masses and the amount of discharge from glacial rivers are needed, especially for regions of the world where glaciers play an important role in the supply of water. A global overview should be compiled of the regions and towns that are at risk in this regard. Other risks arising as a result of changes in mountain glaciers (e.g. glacial lake outbursts) also need to be better observed and researched.

TIPPING POINTS
One issue in which considerable uncertainty is combined with great potential for harm is that of qualitative, non-linear system changes, or 'tipping points' in the climate system (Section 5.3). Enhanced research efforts are required in this regard to improve the assessment of risks.

9.1.2
Environmental and climate impact research

Great progress has been made in environmental and climate impact research over the last few decades. While excellent knowledge is available regarding certain processes and selected organisms or systems, major deficits exist in terms of global surveys and reliable extrapolations. Data of this sort are essential, however, in order to make sufficiently plausible predictions regarding the impacts of climatic changes on, for example, the quantity and quality of food production and availability of drinking water.

SOILS, DISTRIBUTION OF VEGETATION AND LAND USE
While the distribution of vegetation types (natural, semi-natural and anthropogenically influenced) at global level is well known, not least thanks to the contribution of remote sensing techniques, considerable gaps in knowledge remain with regard to the distribution of soil types and the current state of soils (e.g. degradation, compaction, salination, erosion) and to the corresponding land uses (e.g. types of farmland, irrigation agriculture). Moreover, in view of future climate change, projections of these parameters into the future will be particularly important in enabling scientists to make assertions with a high degree of spatial resolution regarding the trends and vulnerability of natural, semi-natural areas and areas with major anthropogenic influence. Integrated programmes involving institutions such as ESA, NASA, FAO or UNEP would make sense in this context.

REGIONAL VULNERABILITY STUDIES
Particularly in view of the considerable uncertainties regarding the precise regional pattern of impending climatic changes, a need exists for regional vulnerability studies aimed at identifying the areas, or sectors, in a given region that are highly vulnerable to climatic changes as well as special risks or opportunities. Vulnerability studies of this sort have so far been carried out only for a few regions and sectors (e.g. for Europe; Schröter et al., 2005). Very few studies exist as yet for especially vulnerable developing countries, where the need is greatest. There is great potential here for scientific cooperation between local experts and research institutes with an international focus.

SYSTEM KNOWLEDGE AND EXTREME EVENTS
While the impact of specific environmental or climatic factors (such as nitrogen deposition, temperature and CO_2 concentrations) on plants, soils and, to some extent, whole ecosystems is already well known, gaps in knowledge generally appear where several factors are at work simultaneously. If, for example, temperatures rise as a result of climate change, and if precipitation simultaneously becomes more sporadic, with the availability of water for irrigation also reduced, interactions may occur that have a negative impact on agricultural output or quality, to the extent that there may be total crop failure. In this context, the effects on soils and perennial vegetation such as forests or grazing systems remain poorly understood. Moreover, if major or rapid climatic changes occur, be it uniquely or repeatedly, in systems that are already weakened, it is not yet possible to estimate either the short-term effects – such as growth retardation – or the long-term effects with sufficient accuracy. They certainly cannot be estimated early enough for the purposes of agricultural management, because of a lack of long-term observational or experimental studies, notably for the particularly sensitive systems in the high latitudes and the tropics. Even in the temperate latitudes, however, current systemic knowledge relating to combined stress factors is too scant for reliable models or early warning systems. This situation could be corrected by means of long-term cooperative programmes (e.g. involving CGIAR, IGBP and IHDP, and especially the Global Land Project – GLP) with national partners in the localities affected.

FEEDBACKS BETWEEN BIOSPHERE AND CLIMATE
An important gap in knowledge that is currently being widely debated relates to the feedback mechanisms from the land surface, or biosphere, to the climate system. While the direct impact of the climate on the biosphere has been the focus of major research programmes for some time (e.g. IGBP), less research has been carried out on the feedback mechanisms resulting from these climatic changes in the biosphere. These include, for example, changes in distribution ranges and physiological adaptations in plants. The impacts of an altered biosphere on the energy balance of the atmosphere is still poorly understood. Targeted support for integrative projects at national or European level could close this knowledge gap.

Socio-economic consequences of climate change

Considerable variability still persists in the findings of economic analyses relating to the costs of climate change and the costs of mitigation policy. Model comparison exercises and sensitivity analyses of the anticipated mitigation costs along the lines of those carried out in the Integrated Modeling Comparison Project (Barker et al., 2006; Edenhofer et al., 2006) should therefore be pursued further. Research into the costs of adaptation measures also needs to be developed. Furthermore, the global cost estimates carried out hitherto should be given a stronger regional and sub-regional focus on which region-specific recommendations for action can be based. More research is also needed into the impacts of climatic change on individual economic sectors (IPCC, 2007b). Estimates carried out to date need to be developed further to include the sectoral and regional perspective; this is in order to enable relevant actors to prepare better for the structural and economic policy implications of climate change and mitigation policy. One of the biggest obstacles to producing differentiated estimates for the regional and sectoral levels is the lack of data sets. Efforts aimed at systematic data collection, especially in developing countries, need to be strengthened. Since the security risk in many regions will increase as a result of climate change, research should focus not only on the economic consequences of climate change in vulnerable regions, but also on the consequences of climate-induced destabilization of states and societies for global economic development.

9.1.3
Early warning systems

The importance of early warning systems in terms of saving human lives and crisis prevention is beyond question. Although there have already been numerous attempts to establish systems of this sort, a fully operational global early warning system for climate-related environmental changes and their potential risks is not yet on the horizon. WBGU supports the appeal made by the United Nations in 2006 to develop a comprehensive early warning system that, in addition to providing information on sudden-onset natural disasters, would also provide forecasts and impact assessments of gradual environmental change. Because social and political crises are increasingly interconnected with changes in the environment and natural disasters, it is also recommended that information on the status of political and humanitarian crises should be made an integral part of the warning system. In WBGU's view, a system of this sort is indispensable for creating a uniform knowledge base in all regions of the world concerning current and future hazards and risks. WBGU therefore recommends supporting international research activities in this field. In particular, the Global Survey of Early Warning Systems (UN, 2006a) should be considerably expanded to include the capacities and requirements of individual countries and regions with regard to early warning.

Of special significance in this regard – alongside basic research in the natural sciences and technology – is the question of how risks with a range of different spatial and temporal dimensions can be integrated into a system that meets requirements regarding information content and clarity. Research in this context should be directed primarily towards the needs of the relevant decision makers and guidelines drawn up for using forecasts of events that have a certain probability of occurrence.

An additional point to be borne in mind is that early warning does not end with the issuing of warnings. The main purpose of a global early warning system is to enhance the knowledge base at nation-state level, hence interfaces to national warning chains should be established and supported at an early stage. Most notably, ways must be found to overcome the 'last mile problem', in other words providing up-to-the-minute warning to people directly at risk, in varying socio-economic circumstances. An important role is played in this by institutional factors, the specific legal situation that exists, and the level of trust prevailing between the population and the local authorities. Furthermore, knowledge about local environmental changes that exists within the local population must be better utilized, as this is useful for a local early warning system.

Overall, global monitoring systems still fail to take adequate account of prevailing socio-economic circumstances. These issues should be made a new focal point of the Earth System Science Partnership, the joint coordination project of DIVERSITAS, IGBP, IHDP and WCRP.

9.1.4
Social destabilization through climate change

The complex impacts of global climate change on particular types of society have barely been investigated to date. The present report is intended to help highlight processes of social destabilization occurring as a result of climate change and to present them in a systematic manner. However, it also illustrates the fact that analysis of the causal links between climate change and social destabilization requires drawing

together different strands of theory that have hitherto coexisted largely in separation from each other.
- Findings from research into the underlying causes of conflict, violence and war, from research into environmental conflicts, from vulnerability research, from research on disaster management and on the reasons why governments and institutions fail (governance research) should be systematically interlinked, with a view to developing concepts and theoretical frameworks of reference that could provide a basis for adequately reconstructing the impacts of climate change on the stability of societies. There is great potential for synergies in the process of bringing together these different approaches, all of which address a similar set of questions from a variety of perspectives.
- The impacts of climate change on different types of society (such as democracies or autocracies) and different types of country (e.g. weak and fragile states) with differing levels of socio-economic development will vary enormously. Hence, alongside efforts to bring together the different strands of theory mentioned above, empirical studies that differentiate between types of society and levels of development are also important. Development research can make an important contribution in this regard.
- The social sciences and natural sciences need to cooperate more closely for the purpose of investigating the societal implications of climate change. Over the last few decades, the debate on environmental conflict has been conducted almost exclusively among social scientists. The elaboration of 'conflict constellations' in this report shows, however, that the mechanisms arising as a result of direct interaction between environmental change and social change can be understood only if there is cooperation between the social sciences and the natural sciences. It is exceedingly difficult, for example, for the social sciences to interpret the information that a global or regional temperature rise of X degrees is anticipated. The impacts of this temperature rise on land-use systems, water balance and other natural resources must first be captured and assessed by environmental and agricultural scientists. Declining agricultural potential, scarcity of water and resources, and environmental stress may lead to social instability and conflict, but they may also be attenuated by cooperation. The social sciences can examine the way in which climate change and resulting changes to biogeophysical conditions affect the relevant social, political and economic institutions and processes, and investigate the dynamics that favour cooperation or conflict, social stability or instability in this context. Close communication between the natural sciences and the social sciences is absolutely essential to enable robust research to be carried out into 'climate change – changes to biogeophysical conditions – relevance of the changes in biogeophysical conditions for societies and their stability' as a sequence, into the interdependencies between 'climate change', 'changes to biogeophysical conditions' and 'social stability/instability', and into the direct interactions between biogeophysical and social processes. Great efforts should be made by universities, non-university research establishments, research funding organizations and the various scientific disciplines to strengthen interdisciplinary cooperation across the high barriers that continue to divide the social sciences and the natural sciences. Research approaches such as those conceived by the German Research Ministry BMBF in its programme of socio-ecological research need to be pursued and developed further.

9.2
Policies to prevent and contain conflict

9.2.1
Research and policy focused on the long term

CONFLICTING OBJECTIVES OF ENVIRONMENTAL SCIENCE AND POLICY FORMULATION

One fundamental problem in environmental science is how to translate scientific knowledge into practical policy. Over the last two decades a sophisticated field of research has developed in this area, its aim being to examine critically the circumstances in which scientific policy advice is given, the potential it has to influence change, as well as questions concerning legitimacy and responsibility (Jasanoff, 1990, 2004; Haas, 2004; Farrell and Jäger, 2005). This is especially relevant both to the problem of assessing the consequences of technology and to the scientific underpinnings of global environmental policy (WBGU, 2000). The publication of international assessment reports, such as those from IPCC, and the Millennium Ecosystem Assessment has generated a great deal of public interest and exerted considerable political influence. Such reports serve to gather together existing knowledge and to make it publicly available in a form helpful for decision-making.

Nonetheless, it remains difficult to develop concrete policies based on research findings with a long-term focus that are inherently uncertain (Sprinz, forthcoming). There is a continuing need for clarity here with regard to what scientific policy advice based on solutions-focused research with a long-term perspective is capable of delivering and, not least, unresolved issues surrounding its legitimization.

Long-term focus as a research challenge

Global climate change occurs in the medium and long term, and its impacts on societies will unfold primarily over the long term. It is important, therefore, that (natural) scientific climate research and social scientific studies on climate impacts, along with research on mitigation, adaptation and stabilization strategies, are oriented towards the long term. However, the social sciences are clearly loath in general to say anything about long-term trends in societies, political systems or economies. This is understandable on the one hand, given the complexity of social processes; on the other, though, it is unsatisfactory given the challenges facing societies as a result of climate change. The behaviour of actors in the present (emitting CO_2, for example) is generating consequences for the future (impacts of climate change on social systems) for which societies would be much better able to prepare if they were informed about possible future scenarios.

The long-term processes of climate change make it necessary for social scientific research to also be focused on the long term if societies are to have a chance to prepare early on for the difficult challenges they are likely to encounter. For the social sciences to orient their efforts more on projections into the future, the main requirement is a more effective set of methods. In addition to applying more of a long-term focus using existing methodological approaches (scenario techniques, various methods of regression analysis and computer simulation models, for example), it will also be important to develop new methods. For example, it bears asking whether the approaches used in other scientific disciplines that have a longer-term orientation could, in certain circumstances, be applied to social scientific contexts as well.

Ways of dealing with uncertainty and irreversibility are crucial when it comes to focusing social scientific research on long-term factors. Today's decisions have to be made in a context of great uncertainty regarding the nature, extent and timing of the future consequences – some of which will be irreversible – of various options for action. Action taken 'too soon' may entail having to accept excessive costs of damage mitigation measures; action taken 'too late' may generate high costs as a result of irreversibility that could have otherwise been avoided. Social scientific research should outline aids to decision-making that enable both phenomena to be dealt with appropriately – that is, ones that are in accord with the preferences of a society.

The value of keeping practical options open when it comes to making decisions in conditions of uncertainty and when irreversibility is a given is described in economics as the 'option value'. An ability to determine accurately the option value of various climate policy alternatives would be extraordinarily helpful in the search for the best possible emissions reduction pathway (Arrow and Fisher, 1974; Ulph and Ulph, 1997; Gollier and Treich, 2003).

It will never be possible to predict with certainty the developments societies are likely to face in the long term; but it should be possible to make more robust statements about 'probabilities', 'conceivable futures' and 'possible pathways' than those available today.

Long-term focus for political, economic and societal actors

Long-term processes such as those foreseeable in the context of climate change are generally dealt with unsatisfactorily by political, economic and societal actors. Slowly encroaching phenomena and gradual processes (rising sea levels and demographic change, for example) are just as hard for politicians, economists and leading social figures to deal with as are events that, while having been predicted, occur suddenly (such as storm surges, droughts or conflicts). There are a number of plausible reasons for the failure of politics and the market in the face of questions and problems posed by the future:

- Climate change mitigation can be achieved only if all the large-scale emitters implement adequate policy measures; in other words, it depends on joint – collective – action. Since society as a whole benefits from individual actors' investment in mitigation measures but the individual enjoys only minimal benefits in proportion to the costs incurred, an incentive exists for all actors to not act themselves but rather to participate in the measures implemented by others at no cost to themselves. This results in a lower reduction in greenhouse gas emissions than would be economically efficient.
- There is often a powerful interplay at work between short-term and long-term actions. Decisions made today in conditions of uncertainty may constrain the options for action that are available in the future. This is also referred to as 'lock-in', or path dependence. For example, the decision to invest in gas-fired power stations today will determine a country's energy mix for the next 40 years.
- Today's interests may be opposed to future interests. The efforts of newly industrializing and developing countries to achieve rapid economic development pose a threat to the long-term goal of avoiding dangerous climate change if the development process is highly resource-intensive.

Due to climate change and processes of globalization, long-term dynamics (in the period covering approx.

2020–2100) are beginning to play an increasingly important role for societies. For this reason, research efforts are required in developing incentive systems, institutional innovations and norm and value systems that might help to strengthen societies' problem-solving capacity in the face of relevant future problems and to encourage a long-term focus in politics, industry and society. Insights from New Institutional Economics, governance research and social scientific theories of political control, along with studies of science and risk, are especially important in this regard.

9.2.2
Adaptation strategies in developing countries

The more powerful climate change turns out to be, the more far-reaching and complex the economic, political and societal impacts will be on developing countries. Climate change generates economic costs and devalues agricultural potential, established development strategies as well as sectoral patterns of specialization; it forces countries to establish new infrastructures. It also gives rise to health problems, triggers migratory movements, causes distribution conflicts and power shifts in societies and puts pressure on their governance capacity. In addition, the impacts of climate change are highly context-dependent, meaning that there is a need to develop adaptation strategies specific to particular regions and countries. Current development dynamics in Africa, for example, illustrate why there is an urgent need for action and research. At the present time many African governments and international development organizations are once again (for good reason) targeting their investments towards agriculture, which had been neglected during the 1990s. So far, though, the medium- to long-term impacts of climate change have barely been taken into account in this process. What might a sustainable form of agriculture look like in Africa in 2030, one that were capable of dealing with even greater aridity, extreme weather events and altered soil conditions? Which regions will be even agriculturally useful at all? Which crops would it make economic and ecological sense to plant? What agricultural strategies, technologies and infrastructures will be needed in the future? Adaptations to future changes in the natural environment need to begin sooner rather than later, as it takes time to build up and restructure institutions and infrastructures. Research initiatives are needed soon, given that the creation of certain knowledge about likely changes and conceivable adaptation strategies is also time consuming.

Research programmes focusing on adaptation strategies in developing countries are still in their early stages worldwide, however. The World Bank, FAO and other international organizations are beginning to undertake efforts to concentrate on a few sectors (including agriculture). At the end of 2006 – in the context of negotiations on climate protection – a special fund was set up to provide support for research, and development organizations are beginning to seek scientific expertise. The present report shows that the potential impacts of a global rise in temperature beyond the 2°C guard rail will be extremely serious for the developing countries. It also shows that large gaps in knowledge exist regarding these impacts in specific developing regions and countries. Given the complexity and extent of the task, it is important for an international – or at least European – division of labour to be established in this newly developing field of research.

9.2.3
Developing preventive strategies to stabilize fragile states

Taking the various definitions and criteria into account, about 30 states can currently be categorized as weak or fragile (Section 4.2.2). Apart from a few general recommendations, considerable confusion and uncertainty exists in the international community about how to deal with these countries, given that most traditional options for action do not work. On the one hand, a key characteristic of such states is that they have weak state structures, meaning that established forms of development cooperation are unable to find any suitable points of contact and their intended effects are barely able to come to fruition. On the other hand, the donor countries are generally reserved when it comes to organizing and channelling public support through non-state actors. Apart from issues of legitimacy under international law, the main concern here is that support of this kind may undermine the authority of existing public institutions and further weaken the sovereignty and capacity of the governments responsible. Military interventions are generally considered as an option only when existing conflicts have become so intense that they pose a serious threat to regional stability or the interests of other countries. Such interventions can prevent a situation from deteriorating further, for example by using measures designed to consolidate peace to prevent intrastate violence from flaring up again. But they have only a very limited capacity to contribute to the construction of a state system capable of functioning over the long term.

In view of the additional problems likely to exert pressure on governments as a result of climate change, the need to devise coherent strategies and effective instruments for stabilizing weak and fragile states appears more urgent than ever. Therefore, the task of social scientific research must be to evaluate existing preventive policy instruments on the basis of current and past case studies and to develop them further. The main issue to be explored here is what combination of individual measures is effective for the various instances and stages of state fragility and how these should be timed and coordinated internationally. The heterogeneity of the affected states needs to be taken into account much more than it has been up to now. This may mean providing support for governance structures that do not conform to the prevailing western-influenced, ideal-typical understanding of the state that exists on the donor side. To this end, it would be worthwhile considering how the legitimacy of public institutions in postcolonial states can be increased without relinquishing universal principles such as the separation of powers, the rule of law and the protection of human rights.

For this purpose WBGU recommends that support be provided for empirical field research in particular. The aim of this would be to generate robust, policy-relevant knowledge about the mode of functioning and legitimacy of public institutions and non-state actors in weak and fragile states and to be better able to understand perceptions of security at national and local level in the affected regions. Careful attention should always be paid in the process to ensure that analysis of the concrete security implications of climate change does not further any 'securitization' of the climate policy debate or displace the priority issues of climate change mitigation and adaptation.

9.2.4
International institutions in the context of global change and climate-induced conflicts

INSTITUTIONS AT THE INTERFACE OF
ENVIRONMENTAL SCIENCE AND POLICY
FORMULATION

Translating scientific findings into practical policies also presents challenges for research on environmental institutions. The principal issue here is to gather together existing knowledge and to make it publicly accessible in such a way that decisions about the problems reported can be made rapidly and rationally. It is also necessary to examine critically the circumstances in which scientific policy advice is given, the potential it has to influence change, as well as questions concerning legitimacy and responsibility (Haas, 2004; Farrell and Jäger, 2005). This is especially relevant both to the problem of assessing the consequences of technology and to the scientific underpinnings of global environmental policy (WBGU, 2000). The dissemination of international studies, such as the IPCC reports and the Millennium Ecosystem Assessment, has been extraordinarily widespread. The challenge is to gather together existing knowledge and make it publicly accessible in a way that facilitates decision-making about the problems reported.

INTERNATIONAL HUMAN DIMENSIONS
PROGRAMME ON GLOBAL ENVIRONMENTAL
CHANGE

The International Human Dimensions Programme on Global Environmental Change (IHDP) – supported by the German government, among others – and especially its sub-programmes Institutional Dimensions of Global Environmental Change (IDGEC) and Global Environmental Change and Human Security (GECHS, Section 3.1.1.4), have provided important new insights for the development of research on global environmental institutions in the past few years, rendering outstanding services in the sphere of interdisciplinary methods. This has made it possible to lend further precision to research questions concerning the characteristics of international institutions and the interrelationships between them, and to build on them steadily with the aim of creating a better synthesis between the findings from social scientific and natural scientific environmental research (Young, 2002). Follow-on research should address, among other things, issues relating to institutional architecture, taking explicit account of the potential capacity of non-state actors, the capacity of social systems to learn and adapt, and the legitimacy and responsibility of global decision-making structures; this also means addressing questions regarding the allocation of global goods and resources (Biermann, 2007).

In addition, the IHDP resolutely promotes scientific exchange between North and South and has proven its worth not least in ensuring that German research is integrated into international research networks. WBGU therefore recommends that the German government continue to support the research work being coordinated under the conceptual umbrella of the IHDP.

CAPACITIES OF INTERNATIONAL INSTITUTIONS AND
ORGANIZATIONS

The genesis, transformation and effectiveness of institutions in international environmental policy have represented a very dynamic field of research within International Relations since the early 1990s (Jakobeit, 1998; Zürn, 1998a; Sprinz, 2003). What has been largely neglected in this, however, is the specific role played by the international organizations crucially involved in shaping the institutional architecture of international environmental policy and their administrative agencies (Bauer and Biermann, 2007). The UNEP secretariat, for example, is entrusted with the task of monitoring a large number of multilateral environmental agreements. Given that the reform of the United Nations is proceeding somewhat sluggishly – not least in the environmental sphere (Section 10.3.1.2) – WBGU suggests a closer examination of the concrete practical options available to international organizations and how these might be improved in the service of better intergovernmental cooperation and a more effective implementation of multilateral environmental agreements.

Establishing an International Environmental Court
In view of the limited capacity of the International Court of Justice in The Hague to settle disputes in environmental conflicts, the idea of setting up an International Environmental Court capable of making binding legal judgments has been proposed in various quarters (Rest, 1998; Postiglione, 1999, 2002; MacCallion, 2000; Pauwelyn, 2002). Given the potential implications of environmentally induced conflicts, WBGU considers it worthwhile to explore the idea of such a legal body. A number of issues still require detailed scientific debate, however.

Since international legal treaties and customary international law contain few justiciable obligations directly related to the environment – that is, ones capable of being enforced in a court of law – it bears asking, first of all, what role an international tribunal for the environment would be capable of playing at the present time. One point to be considered is that questions relevant to the environment are also dealt with through existing arbitration systems in the context of WTO and the international law of the sea (International Tribunal for the Law of the Sea). The debate about setting up a specific International Environmental Court thus demands precise clarification with regard to which legal claims could be effectively asserted.

A tribunal of this sort could play a specific role if the way were also opened up to private individuals to lodge complaints. However, this too presupposes that adequate international legal guarantees are in place to provide legal protection for the environmentally related rights of the individual. Legal rights related directly to the environment can indeed arise out of the specific human rights recognized in international law, as the actions taken by the European Court of Human Rights in relation to the European Convention on Human Rights illustrate. A globally valid agreement providing similarly concrete and justiciable guarantees in relation to the rights of the individual does not exist at this time. It would thus be worthwhile examining whether guarantees aimed at protecting the rights of the individual in practice could be established as part of efforts to elaborate further specific environmental agreements and, if so, how this might be done.

WBGU proposes that the current debate be pursued further and – not least in connection with the reform efforts at the United Nations and its specialized agencies – that an analysis be conducted regarding the circumstances in which the idea of an International Environmental Court could be implemented and built upon, as well as what other developments could be considered for existing arbitration mechanisms. It would be important to clarify in detail who would be able to assert what rights against whom, and equally important to establish the scope of jurisdiction a legal body should have, along with its options for imposing sanctions, in order to improve the effectiveness of the institutions concerned and to avoid creating new conflicts over objectives among existing institutions or encouraging any further fragmentation of the multilateral system.

9.3
Conflict constellations and their prevention

Climate-induced conflict constellations are sets of causal linkages at the interface between climate impacts and society, the dynamics of which have the potential to bring about social destabilization or violence. WBGU has identified specific conflict constellations associated with degradation of freshwater resources, declining food production, increasing storm and flood disasters, and with environmentally induced migration. These, according to the current status of knowledge, are the most important conflict constellations. However, other constellations could conceivably emerge and gain greater prominence in the future. One example might be changes in the transmission patterns of infectious diseases. Such indirect consequences of climate change should be systematically analysed for their potential to trigger violence in society.

As yet, the study of causal links between climate impacts and violent conflict is in its earliest stages and still requires both empirical validation and theoretical underpinning. This represents a vast interdisciplinary field of basic research that must be taken forward by both climatologists and social scientists.

Climate change on the scale expected will give rise to major changes in many countries, particularly developing and newly industrializing countries. These may take the form of the conflict constellations described. The capacity of a country to cope with such adverse events in future depends critically on the availability of robust, region-specific assessments of the consequences of climate change and, based on these findings, anticipatory strategies for adapting to climate impacts that cannot be avoided. Developing countries in particular are found to have major deficits in this area. WBGU therefore recommends stepping up research cooperation with developing countries to address the areas of adaptation and disaster mitigation.

9.3.1
Degradation of freshwater resources

REGIONAL MONITORING AND MODELLING OF THE WATER BALANCE

The use of scenarios and ensemble modelling in devising water management strategies is at a very early stage and still requires a great deal of research. One aspect of this involves downscaling global scenarios to the regional level and taking account of the specific demands on water management that arise as a result. Models devised on this basis should take into account both the regional river basin area and the corresponding urban region. The impacts of climate change on groundwater resources should also be analysed more comprehensively. Another aspect involves developing practical ways of making findings available – such as the visualization of simulations of different scenarios (e.g. climate change scenarios) and various courses of action – in order to facilitate participatory approaches and to ensure that decision-making processes in water management are grounded in a solid source of information.

ENHANCING WATER PRODUCTIVITY IN THE AGRICULTURAL SECTOR

The efficiency of water use in agriculture needs to be improved. This applies mainly to regions with poorly functioning irrigation systems and physical water scarcity, which are especially vulnerable to poverty and food crises. Emphasis should be placed on researching and implementing integrated strategies appropriate to local circumstances which collectively optimize technologies, choice of varieties and production processes (e.g. integration of livestock farming, multiple usage of irrigation water), in order to achieve improved water productivity within a given catchment area.

DESALINATION USING RENEWABLE ENERGIES

Coastal regions and islands with low precipitation have the option of guaranteeing a reliable supply of drinking water by processing (decontaminating) and desalinating seawater or brackish water. Small desalination systems driven by renewable energies, such as evaporation systems or distillation processes involving solar thermal components, represent a sustainable solution, especially for remote regions. However, there is a need for further development of small facilities in particular, in order to increase reliability (of water quality) and reduce costs. There is still great potential, in terms of energy efficiency, for improving the already widespread reverse osmosis process. When combined with projects to electrify rural areas, e.g. using photovoltaic technology and wind-powered generators, the ability to store drinking water brings significant economic synergies. Research and development of such systems should be supported in the context of development cooperation, as should their introduction onto the market. Consideration should also be given to the environmental impacts of water extraction and the re-circulation of brine.

ANALYSING VIRTUAL WATER TRADE

To enable a comprehensive analysis of virtual water trade, further reliable data should be collected at all levels. Questions regarding suitable methods of calculation need to be answered, particularly in relation to appropriate ways of setting quantitative and spatial system boundaries when apportioning consumption to specific actors (particularly agricultural production). The use of virtual water trade as a strategy to limit regional water demand should be assessed in terms of the risks and opportunities it entails for the economic development of a country. The extent to which international virtual water trade requires regulation by an international institution, e.g. with regard to sustainability criteria, also merits exploration.

IMPLEMENTING INTEGRATED WATER RESOURCES MANAGEMENT

Due to the lack of institutional capacity in developing countries, there is a need for social science research to be carried out locally on the actual implementation of integrated water resources management systems. In view of past conflicts in some regions in the wake of water sector privatization, the question arises as to how planning and decision-making processes in water policy should ideally be structured in order to avoid conflict and ensure a good supply. One relevant issue is the compatibility of public and private forms of participation.

PREREQUISITES FOR TRANSBOUNDARY WATER COOPERATION

In spite of existing (meta-)studies of formal contractual agreements, there is still a need for a more in-depth analysis of local mechanisms and of the effectiveness of existing cooperation agreements. The role of climate change as a driver of future water shortage or of increased variability in the availability of water should be more fully integrated in social science research. This raises questions about suitable institutions for transboundary water cooperation in the face of advancing climate change as well as about appropriate guidelines for the assistance rendered by development cooperation institutions. The need for research on water cooperation is not limited to the shared use of rivers and lakes, but should increasingly include transboundary groundwater resources.

9.3.2
Decline in food production

CHALLENGES OF CLIMATE CHANGE AND ENVIRONMENTAL DEGRADATION FOR AGRICULTURAL RESEARCH AND RESEARCH INFRASTRUCTURE

Agricultural research should give greater priority to the foreseeable consequences of global climate change. In addition to exploring ways of ensuring regular and adequate crop yields, studies should also investigate the potential for increasing food production under conditions of environmental change. The prime concern should be to seek solutions that are appropriate to specific regions of the world. For example, the temporal distribution of rainfall in Africa acts as a limiting factor on a 'New Green Revolution', because extreme fluctuations in annual rainfall mean that it is not worth applying fertilizers every year to boost production. The main thrust of research in this regard must be towards achieving constant guaranteed yields in the face of fluctuating precipitation.

In addition, knowledge transfer in the areas of soil protection and crop improvement must be optimized, adapted and developed further. This implies research at various levels, from the molecular potential of important commercial crops to the prevention of unproductive water loss during irrigation. Another point is the identification of geographical-climatic risks by means of efficient monitoring systems and technology-supported climatic forecasts (using GPS/GIS). Studies should also examine whether plant water loss in relation to carbon gain can be optimized further. The development of appropriate farming regimes is also of considerable strategic importance in terms of safeguarding food production. In the context of its participation in the Consultative Group on International Agricultural Research (CGIAR), Germany should continue to press for research support that pays greater attention to these aspects. CGIAR's four current Challenge Programs point in the right direction in this regard. It is vital to maintain research cooperation with (African) developing countries and to strengthen their local research capacity.

LINKAGES BETWEEN GLOBAL WARMING, YIELD LOSSES AND RISKS OF VIOLENCE

Very little research has been carried out into the relationship between global warming and production losses in agriculture, and particularly into that between crises in farming or food production and risks of violence. Systematic empirical analyses could enable more precise conclusions to be drawn regarding the socio-economic and political factors involved in exacerbating conflict, making it possible to develop more precisely defined approaches to crisis prevention. This can only succeed, however, if scientific guidance for decision-makers is drafted with due consideration of the effects of climate change, particularly in scenarios and very long-range forecasts of crop yields and crop losses.

9.3.3
Increase in storm and flood disasters

PATTERNS OF BEHAVIOUR, COMMUNICATION AND DECISION-MAKING IN DISASTER SITUATIONS

Disaster prevention measures must always be adapted to the framework conditions of the society affected. Normally the implementation of such measures takes socio-economic conditions into account, but when emergency plans are drawn up, they are frequently based only on general assumptions about individual and social behaviour in disaster situations. These assumptions do not always correspond to the behaviour patterns that are likely arise in reality (Auf der Heide, 2004). Yet well-founded knowledge about individual, social and administrative patterns of behaviour, communication and decision-making in disaster situations is a fundamental prerequisite for effective early-warning systems, emergency plans and other disaster risk management measures (information campaigns, infrastructure planning, etc.). So far the relevant research has been rather fragmentary, and only in rare cases has it been worked up systematically. It is necessary to broaden this knowledge base and make the findings available to political decision-makers. In this connection, WBGU welcomes the German federal government's new research programme for civil security, which places a clear focus on pattern recognition, action strategies and organizational forms (BMBF, 2007). It points out, however, that to place the issues in a global perspective requires corresponding research on the socio-economic framework conditions in newly industrializing and developing countries.

ADAPTATION STRATEGIES FOR COASTAL CITIES AND DELTA REGIONS

Climate change poses the greatest challenge to storm-prone cities on coastal plains and in delta regions. Local land subsidence, changes in land use and settlement structure and in the local hydrological system often contribute further to an increased risk of disaster. In many cities and delta regions, the question of possible adaptation strategies still remains largely unresolved. While dams, dykes and tidal barrages are being constructed in many locations, comprehensive sea defence measures must also take account of the complexity of the urban infrastructure as well as the highly dynamic nature of social and economic processes. In some circumstances, the construction of physical defences may encourage a concentration of residential and industrial land uses in threatened areas, thus raising disaster risks to a new level. The points mentioned here could be considered in the context of sustainability research within the thematic area 'Megacities of Tomorrow' planned by the German Federal Ministry for Education and Research (BMBF).

Particularly in the cities and delta regions of newly industrializing and developing countries, ways must also be found of halting the concentration of people and material assets in threatened areas, and of harnessing existing regulations and incentive systems to encourage such adaptation measures. Due to the great complexity of urban spaces with their intermeshed residential and commercial/industrial uses, the development of such strategies calls for comprehensive interdisciplinary studies of the kind that generally exceed the competence of individual institutions. Research approaches should therefore pool the strengths of different scientific disciplines (hydrology, urban planning, geography, engineering sciences, sociology, etc.) and practice-oriented institutions (city administrations, the fire service, port authorities, etc.) for the purpose of devising adaptation strategies.

9.3.4
Environmentally induced migration

Environmentally induced migration – or 'environmental migration' for short – has so far received little attention from a scientific perspective. Studies have been confined to the field of conflict research; in the mid-1990s environmental migration was examined as a potential conflict factor and the issue was addressed by, among others, the Institute for Migration Research and Intercultural Studies of the University of Osnabrück, as part of its research into migration, multiculturalism, ethnicity and conflict (Bade, 1996). Despite this, patterns of cause and effect relating specifically to environmental migration remain largely unexamined. As part of a focus on migration within the 'Socio-economic Sciences and the Humanities' theme, the EU's 7th Research Framework Programme has set aside funding for a more far-reaching analysis of the causes and patterns of migration movements in general (EU-Kommission, 2007). Research on international migration and integration is already underway among research groups such as the International Migration, Integration and Social Cohesion Network, funded by the EU's 6th Research Framework Programme (IMISCOE, 2007). Nevertheless, these programmes have not yet dealt explicitly with the environmental and climate-related causes of migration. It is therefore important that German and European funding programmes and research networks should formulate a specific research programme relating to climate-induced and environmental migration. WBGU outlines below the central questions that future research needs to answer.

CAUSE-AND-EFFECT PATTERNS OF ENVIRONMENTAL MIGRATION

There is still widespread disagreement among those involved in the systematic study of the subject as to exactly how environmental migration should be defined. This lack of clarity stems primarily from the paucity of information available about the following questions: what kind of influence does environmental change, as one factor among others, have on an individual's decision to migrate? How does it relate to other factors that may trigger migration? What forms of migration – in terms of scope, direction and duration – are to be expected as a result of different kinds of environmental change? Finally, what are the implications of environmentally induced migration for the migrants' regions of origin, transit locations and ultimate destinations? The task of specifying this cause-and-effect mechanism more closely first requires further case studies analysing the role of environmental factors in causing migration and the role of environmentally induced migration in causing or exacerbating conflict. The next step would involve meta-analysis of this information in order to arrive at statements that can be generalized. The results of these meta-analyses would make it possible to distinguish environmental migrants more clearly from other types of migrant, to specify their number more precisely, to identify the regions most affected, to model early warning systems for environmental migration and to draw up appropriate prevention and response strategies. Exemplary in this regard is the Environmental Change and Forced Migration Scenarios project being undertaken by the Center on Migration, Citizenship and Development at the University of Bielefeld in cooperation with other European project partners as part of the EU's 6th Research Framework Programme. The project is due to run for 24 months (2007–2008), during which time data will be collected and analysed statistically in order to improve the state of information available on refugee movements caused by climatic and environmental factors (COMCAD, 2007).

Interdisciplinary research on environmental migration

In the specialist literature there has been little coordinated or coherent study up to now on the way environmental change may cause or exacerbate conflict by inducing migration; such work as has been done has been limited to isolated sub-disciplines. Interdisciplinary research would be beneficial here, enabling the findings from scientific analysis of climate change impacts, migration research and conflict research to be integrated; this in turn would contribute to a broader understanding of the cause-and-effect patterns involved.

Strategies promoting acceptance of migration policy measures

Long-term government resettlement programmes represent one possible response to severe environmental change that political decision-makers should consider. However, such programmes do bring with them an array of socio-economic and political problems. Experience has shown that where such resettlement has been carried out, those affected – despite careful planning based on principles of social sustainability – often show little acceptance of such large-scale state interventions in the lives of long-standing communities, thus giving rise to conflict in its various forms. Interdisciplinary research therefore needs to identify the most suitable strategies both for organizing the migration process itself and for promoting the acceptance of such measures among the affected populations in the regions of origin and places of destination.

International legal instruments for the protection of environmental migrants

At the legal level, states currently have no specific obligations under international law with regard to the treatment of environmental migrants, nor do any other legal mechanisms exist to protect affected individuals. In the interests of improving the legal status and protection of environmental migrants, consideration should be given to ways of remedying this omission in international law. It is particularly important to clarify at which legal level statutory provisions should apply. In the view of WBGU the preferred solution would involve a special multilateral umbrella convention on environmentally induced migration. The results of the studies on environmental migration proposed above should be drawn upon to help answer the question of which concrete measures – including legal obligations and norms designed to protect environmental migrants – should be covered by such a convention. It is particularly important that an equitable system be designed for distributing the burdens arising from environmentally induced migration. Since it can be assumed that environmentally induced migration is affected by a number of causal factors, it is necessary to define and analyse the essential criteria and to integrate them in a practicable and normatively appropriate way into the legal mechanisms to be developed.

Research is also needed into the bodies to be entrusted with implementing a regime for the protection of environmental migrants at international level. The question of whether UNHCR should assume particular responsibilities in this field needs to be addressed, as does, where relevant, the concrete forms of institutionalized cooperation with UNHCR.

A final point that needs to be clarified concerns the conditions under which the international legal principles governing state responsibility are capable of providing grounds for holding particular states responsible for 'causing' cross-border environmental migration. It is particularly important to analyse the potential scope of the duty of non-intervention along with issues of attribution and causality. Even if the principle of state responsibility can only apply in a subsidiary way – i.e. if concrete obligations under international law have actually been violated – rigorous enforcement of relevant rights could nevertheless provide an incentive for states to desist from behaviour that leads to environmental migration.

Recommendations for action 10

10.1
WBGU's key findings

In the past, conflicts triggered by environmental changes were localized and manageable. Advancing climate change, however, will radically alter the framework conditions, for every region of the world will be affected, albeit to a varying extent. If climate policy fails, climate change could become a major international security risk in the 21st century. Policy-makers must therefore lose no time in developing strategies based on the following findings.

A NARROW WINDOW OF OPPORTUNITY FOR CLIMATE POLICY
The first of WBGU's findings offers a major opportunity, but also presents a considerable challenge to policy-makers. Global climate change is a gradual process which, so far, has not posed any serious risks to human security. It is only likely to impact severely on international security in the medium to long term. If in the next two decades, the international community is successful, through effective climate policy, in setting a course towards an emissions pathway that makes compliance with the 2°C guard rail possible, such climate-induced conflicts can probably be avoided. However, any delay in setting this course towards a reversal of global emissions trends will mean that even higher emissions reduction rates will be required later on. There is also a risk that such a delay will entrench path dependencies in favour of emissions-intensive technologies, making compliance with the 2°C guard rail more expensive, more difficult and ultimately impossible (WBGU, 2007). In other words, there is a narrow window of opportunity for successful prevention, making swift action essential. An effective climate change mitigation policy, combined with preventive and adaptive strategies to curb the impacts of climate change and avoid security-relevant dynamics, must begin now, when climate-induced conflict risks are still in their infancy. Putting this type of conflict potential on the political agenda is difficult, however, for the international community is currently preoccupied with very different types of security risks (e.g. nuclear proliferation, terrorism, resource scarcity). A competition for political attention, financial resources and conceptual input is therefore emerging between rival security agendas, further complicated by the anticipated transition from a unipolar world order towards a multipolar power constellation. This poses a major challenge to a political system which is normally geared towards short-term perspectives and problems.

TYPES OF CONFLICT CONSTELLATION
The second finding of WBGU's analysis is that global climate change could lead to food crises, freshwater shortages, extreme weather events with massive destructive force, and increased migration in the various regions of the world (Chapter 6). The conflict constellations analysed by WBGU show that unabated climate change will increase human vulnerability, worsen poverty and thus heighten societies' susceptibility to crises and conflicts. The specific threats will depend on the dynamics of climate change, local environmental conditions and the affected societies' and actors' crisis management capacities. The present report identifies examples of regions that will be especially hard hit (Chapter 7) and outlines appropriate responses to the various conflict constellations.

SOCIETAL VULNERABILITY
The third finding is that overlapping and mutually reinforcing conflict constellations, described in the report, can trigger destabilization processes within societies whose impacts could extend far beyond the generally localized environmental conflicts of the past and present (Chapter 8). Societal vulnerability can lead to crises. The more drastic the climate change, the more severe the impacts of social destabilization and conflicts may be. Whereas the present debate is dominated by the challenge of implementing the Millennium Development Goals, unabated climate change would not only undo the current efforts to reduce poverty; it would also transform many developing regions into places of insecurity, characterized by state failure, a breakdown of law and order, conflicts, violence and human depriva-

tion. Climate change thus has the potential to reverse development efforts and undermine human security, and could therefore ultimately jeopardize national and regional security as well.

DESTABILIZATION OF STATES

The fourth finding is that climate change is unlikely to lead to classic inter-state wars. A more probable scenario is that it will create diffuse cross-border patterns and zones of destabilization which are difficult to contain and almost impossible to manage, with widening insecurity and an increase in the propensity for violence in regions especially at risk from climate change. Climate change will intensify existing conflicts: more frequent droughts and soil degradation, for example, will exacerbate land-use conflicts. As a consequence, large-scale deforestation and the associated displacement of communities will be fostered and could increase internal and cross-border refugee flows. States which are already fragile will thus be further weakened as a result of additional environmental stress. However, climate change will also create new conflict threats. Sea-level rise and storm and flood disasters could in future threaten cities and industrial regions along the east coasts of China and India, but could also confront New York with completely new challenges. A sea-level rise of just 50cm would put the habitats and livelihoods of millions of people in the Nile and Ganges Deltas at risk. The melting of the glaciers would jeopardize water supply in the Andean and Himalayan regions. Climate change could ultimately have quite incalculable consequences for states and societies if major ecosystems or regional climate systems were to change fundamentally as a result of unabated global warming (Section 5.3). For example, the collapse of the Amazon rainforest could occur, with far-reaching and as yet unpredictable consequences for the water and soil balance of the entire South American continent. The loss of the Asian monsoon would have equally severe impacts on agriculture, food security and the stability and adaptive capacities of the affected societies.

THREAT TO THE INTERNATIONAL SYSTEM

The fifth finding is that an escalation in the described patterns of social destabilization in various regions of the world has implications not only for the affected societies but for the international system as a whole (Chapter 8). Migration, for example, could become unmanageable. Distributional conflicts and crises of legitimacy between the main drivers of climate change (especially the industrialized countries and, in future, China and India as well) and those most affected (particularly the developing countries) would be likely to occur. Development policy and security policy, which are already largely unequal to the task of promoting stability in the world's circa 30 fragile states, would be completely overwhelmed by the ensuing challenges. The development of the world economy could be permanently impaired, and the international system as a whole could face a crisis of legitimacy, challenging its capacity to act. The globalized world would become a more insecure place. These major impacts of climate change are not considered in the USA's or the EU's current security strategies. There is therefore a need for political action. Furthermore, the analysis of the complex social impacts of climate change shows that many different political actors have a role to play in avoiding dangerous climate change and mitigating its impacts. 'Climate change as a security risk' poses a challenge for the entire political system in Germany and, indeed, the European Union.

ROAD MAP TO MITIGATE CLIMATE CONFLICTS

In light of this threat analysis, WBGU proposes a road map which takes account of these findings and prevents the emergence and escalation of climate conflicts. Between now and around 2020, the environmental impacts of climate change will probably continue to be manageable, at least to the extent that major climate-induced regional or global security threats are unlikely to occur. This offers a window of opportunity to avert dangerous climate change through a resolute climate policy. Assistance should be provided to the developing countries so that they can devise adaptation strategies to cushion the emerging impacts of climate change. Within this timeframe, the international community can still develop and implement an effective prevention strategy against massive climate-induced security risks. WBGU's policy recommendations therefore focus primarily on this window of opportunity.

If a comprehensive prevention strategy fails, the world must prepare itself for climate change impacts from around 2020 which are likely to trigger increasing destabilization processes, food and water crises, more frequent extreme weather events and increased migration. During this phase, the costs of mitigating the social and security impacts of climate change will rise considerably.

During the period from 2040 to around 2100, unabated climate change would result in environmental changes which are very likely to overstretch the capacities of societies, national and regional actors, the global economy and the international system. During the second half of the 21st century, climate-induced conflicts could then become a feature of the international order, and global crisis management would take the place of efforts to shape globalization constructively.

10.2
Scope for action on the part of the German government

Germany
The key starting point for addressing the complex issues associated with 'climate change as a security risk' is international climate policy, which can only be successful if it is global in scope. Viewed in terms of its problem-solving capacities, Germany is a minor and relatively insignificant actor compared with the major powers, i.e. the USA and China (and India, too, in future). The German federal government can, however, utilize its international role as a pioneer and leader of an intelligent climate policy in order to shape international policy pro-actively. To play this role effectively and bring appropriate influence to bear on international climate policy, however, it would have to take on a range of tasks which would not only involve climate policy actors such as the Federal Environment Ministry, but almost every federal ministry. The German government's strategies in the fields of foreign and security policy, development policy, economic, trade and energy policy, transport policy and agricultural and European policy all have a role to play here. In light of the complex challenges posed by climate change, the diversity of the relevant actors and the importance of interlinked initiatives involving several different ministries, the Federal Chancellery should coordinate this process.

Europe
In the context of the newly emerging power constellations, the European countries – all of which are relatively small – cannot make an effective contribution to a viable global climate policy without the European Union. The EU is the key global actor with an explicit and pro-active commitment to the concept of effective multilateralism. As major powers, the USA, but also China and India, tend to place their reliance primarily on their own power potential. So without major commitment from the EU, a multilateral follow-up agreement to the Kyoto Protocol is highly unlikely to succeed.

For the reasons stated, the EU is a key actor in global climate policy. Yet it is by no means the largest emitter of greenhouse gases. The USA is currently responsible for 22 per cent of global CO_2 emissions and China for 17 per cent, with around 15 per cent coming from the EU. The percentage of emissions from China, the USA and India will probably continue to rise substantially to 2030. Viewed solely in terms of the EU's own reduction potential, its room to influence global climate protection is therefore relatively limited. The EU should therefore utilize its pioneering role in climate protection at home primarily as a lever, enabling it to act as a credible catalyst for international climate policy.

So that Europe can exert the greatest possible influence in the interests of a progressive international climate policy, it is absolutely essential that national egoisms within the EU are set aside; coherent strategies must be developed instead and international agreements implemented consistently and resolutely. Besides the EU's undoubtedly important role as a negotiating party in the various international environmental processes, this applies especially to its own Common Foreign and Security Policy (CFSP). The EU should gear the CFSP towards a pro-active mitigation strategy aimed at curbing global warming (in line with the 2°C guard rail). The German government should therefore work vigorously for consistency and coherence in the EU's climate policy and encourage moves towards a climate-compatible CFSP.

The global level
The German government should step up its efforts to develop and consolidate a constructive multilateral global governance architecture. It can thus make a contribution to curbing the global political turbulence which is likely to arise from the forthcoming 'multipolarization of the world order', and prevent the reconfiguration of the international system from creating a highly unstable scenario in which rising and declining world powers (China, India and the USA) are locked in confrontation (Section 4.3). This is vital if effective solutions to global challenges are to be developed and implemented. The international community must recognize that anthropogenic climate change constitutes a global threat in every respect, and that solutions must be based on a just and comprehensive system of multilateralism.

In that sense, international climate policy could form the ideal arena for confidence-building measures between states, provided that global climate change is recognised as the 'common enemy' of humankind. Climate policy – even more than efforts to combat terrorism or the proliferation of weapons of mass destruction – is thus the 'high politics' of the future. In this context, the EU will only be able to exert substantive influence on international policy-making if it 'speaks with one voice' in its dealings with other more powerful actors.

10.3
The window of opportunity for climate security: 2007–2020

This section outlines nine policy initiatives which would have to be implemented by 2020 or thereabouts in order to avoid security-relevant climate impacts. If this window of opportunity is missed, climate-induced social crises and security problems are very likely to occur in the subsequent decades.

Table 10.3-1 depicts three types of initiatives. Firstly, efforts to stabilize or promote a cooperative international environment and strengthen the UN are important (Initiatives 1 and 2). A particular challenge is that from now to 2020 – the window of opportunity for the development and implementation of an effective global climate policy – turbulence and tensions may well arise in the international arena as a result of the emergence of China and, in future, India as key actors in global governance (Section 4.3). There will only be a viable prospect of achieving significant advances in international climate policy if this fundamental shift in global power relations takes place without major conflicts and geopolitical rivalries between the 'old' and the 'new' world powers.

Secondly, initiatives to abate climate change and avoid its dangerous impacts are required (Initiatives 3–5). Successful mitigation strategies will therefore act as preventive security policy measures as well. Effective climate policy can help avert international crises and conflicts in future. The UN report *Confronting Climate Change* (UN, 2007) refers in this context to strategies aimed at 'avoiding the unmanageable'.

Thirdly, adaptation strategies are crucial, especially in the developing countries, also in preparation for the anticipated increase in migration processes (Initiatives 6–9). These initiatives must also address the already unavoidable impacts of climate change, described in the UN report *Confronting Climate Change* as measures aimed at 'managing the unavoidable' (UN, 2007). They must include the establishment of global early warning systems.

This package of initiatives shows that climate change as a security risk poses a challenge not only to German, European and international environmental and security policy. Almost every department of government has a role to play in averting the national and international crises and security risks that will otherwise result from climate change.

10.3.1
Fostering a cooperative setting for a multipolar world

10.3.1.1
Initiative 1: Shaping global political change

The global economic and political power shifts which will result from the rise of China and India in particular are likely to trigger major changes in international relations. There are many signs at present that China and India have good prospects of establishing themselves as world powers alongside the USA. This is likely to weaken and perhaps even end the USA's dominance of the unipolar world order which has developed since the end of the Cold War. Under these circumstances, major turbulence can be expected in world politics. Although this may not necessary lead to military conflicts between major powers or their proxies, there is a risk that the present multilateral institutional architecture could become destabilized, making international cooperation more difficult and thus creating security risks.

In order to be prepared for this turbulence and shape it in a constructive and peaceful way, it is essential to recognise and respond to the global trends in time. However, insisting on the global political status quo is both unrealistic and counterproductive. If a re-emergence of confrontational rivalry between major powers is to be avoided and the many global problems facing humankind are to be resolved, cooperative structures are required, now more than ever, to manage these shifts in power relations. Germany and the European Union could establish themselves as key drivers for change here and, in the process, cushion and limit their own relative loss of power. In order to move in this direction, the following structural mechanisms are important and must be taken into account (Kupchan et al., 2001; Messner, 2006).

In dealings with the USA, it must be borne in mind that superpowers generally find it difficult to accept the need to move away from a strategy of 'global hegemony' towards a strategy of 'shared global leadership'. In relations with China and India, it is essential to consider that these countries are now starting to perceive their rapidly expanding global significance and are therefore likely to have little interest in challenging conventional notions of sovereignty, the nation state and power politics. The recognition that in view of the nation state's limited scope for action and today's complex global interdependencies, the delegation of sovereignty does not necessarily entail forfeiting the capacity to manage one's own affairs – an insight which, in Europe, has matured over several decades of European integration and in the course of the globalization debate – has yet to become widely accepted in China and India. Ultimately, the constructive multilateralism which is desired cannot, if it is to be effective, simply perpetuate the form of transatlantic multilateralism which has developed under US hegemony since 1945. In order to ensure the acceptance and, above all, the constructive participation of the rising world powers, a multilateral order is needed which is viewed as fair by all the world's countries,

Table 10.3-1
Overview of the nine initiatives proposed by WBGU for the mitigation of destabilization and conflict risks associated with climate change.
Source: WBGU

Relevant policy area	Initiatives
Section 10.3.1: Fostering a cooperative setting for a multipolar world	
Foreign policy	*Initiative 1 – Shaping global political change:* Constructively managing the emergence of China and India as global powers alongside the USA; need for a strong European foreign policy; possible option to convene a world conference on geopolitical change; recognising climate change as a common threat to humankind.
Foreign, environmental and development policy	*Initiative 2 – Reforming the United Nations:* Gearing the present UN system more strongly towards prevention and a coordinated approach; reflecting on the role and tasks of the UN Security Council; strengthening the United Nations' capacities in the field of environmental policy; establishing a Council on Global Development and Environment.
Section 10.3.2: Climate policy as security policy I: Preventing conflict by avoiding dangerous climate change	
Environmental and foreign policy	*Initiative 3 – Ambitiously pursuing international climate policy:* Stipulating the 2 °C guard rail as an international standard; further developing the Kyoto Protocol; adopting ambitious reduction targets for industrialized countries (including the USA), and integrating the newly industrializing and developing countries; conserving natural carbon stocks.
Environmental, energy, economic and research policy	*Initiative 4 – Transforming energy systems in the EU:* Strengthening the EU's leading role; improving and implementing the Energy Policy for Europe; triggering an efficiency revolution; expanding renewables.
Environmental, development, research and economic policy	*Initiative 5 – Developing mitigation strategies through partnerships:* Establishing climate protection as a cross-cutting theme in development cooperation; agreeing decarbonization partnerships with newly industrializing countries (especially China and India); agreeing an innovation pact within the framework of G8+5.
Section 10.3.3: Climate policy as security policy II: Preventing conflict by implementing adaptation strategies	
Development and research policy	*Initiative 6 – Supporting adaptation strategies for developing countries:* Industrialized countries must assist developing countries on adaptation and mitigation of climate impacts; priorities: devising specific strategies for developing regions particularly at risk (e.g. Africa); mitigating water crises; gearing the agricultural sector to climate change; strengthening disaster prevention.
Security and development policy	*Initiative 7 – Stabilizing fragile states and weak states that are additionally threatened by climate change:* Stabilization of weak and fragile states to be taken into account to a greater extent in the German Action Plan 'Civilian Crisis Prevention, Conflict Resolution and Post-Conflict Peace-Building'; supporting and implementing the OECD's working principles; expanding the 'Whole-of-Government' approach to encompass the environmental dimension; boosting the civil society potential of weak states in international forums and networks.
Foreign, domestic and development policy	*Initiative 8 – Managing migration through cooperation and further developing international law:* Developing comprehensive international strategies for migration; integrating migration policy into development cooperation; including environmentally induced migration in international cooperation; enshrining the protection of environmental migrants in international law; permitting no weakening of the existing protection regime; adopting measures supplementary to the existing refugee regime.
Development and research policy	*Initiative 9 – Expanding global information and early warning systems:* Actively supporting the development of a comprehensive global early warning system to provide information about all types of natural hazard, epidemics and technological risks, regional climate change and impacts, and environmental problems; improving the implementation of early warning information at national and local level.

Germany and the EU: Pioneers and models of best practice

Germany can act as a pioneer here by undertaking the necessary advocacy work and, via the EU, working pro-actively at international level for the adoption of confidence-building measures. One option, for example, is to initiate and institutionalize a process modelled on the Conference on Security and Co-operation in Europe (CSCE) and aimed at confidence-building worldwide. International climate policy and development policy appear to offer ideal thematic fields for a move in this direction. This requires at least the following three preconditions to be fulfilled:

1. The formulation of comprehensive policy objectives which must be ambitious in the sense that they can make an adequate contribution to problem-solving.
2. The adoption and implementation of concrete and sometimes painful measures that make a credible contribution to achieving the policy objectives and are internationally verifiable.
3. Negotiation between international partners on an equal footing, even if some partners are still less powerful than others at present. This applies as a general principle to relations with all the developing countries but especially to the 'anchor' countries which in many cases have already established regional predominance and are increasingly emerging as world powers. Major anchor countries such as China, India, Indonesia, Pakistan, Nigeria, South Africa, Brazil and Mexico are playing an increasingly important role in international politics, due to their economic weight and political influence in their own regions and increasingly at global level as well.

There is much to suggest that the EU will only be successful in persuading China and India to move towards an effective climate policy, and in developing strategic technology partnerships with these countries in the transition to a low-carbon economy if Europe itself pursues consistent and credible policies to solve global problems. The hectoring tone often adopted by the West towards the rapidly expanding newly industrializing countries diverts attention from the industrialized countries' own climate policy responsibility and must give way to a culture of common responsibility and cooperation.

If cooperative multilateralism is to prevail in tomorrow's world, a pioneering role for Germany and the EU is essential, too, because the USA, on the one hand, currently shows no sign of any political will to accept the relative ascendancy of rival powers, and on the other, the rising Asian powers China and India are likely to be preoccupied with their own affairs for some time to come. In particular, it is still unclear whether, and how, these two countries will handle their growing global responsibility.

Furthermore, the EU is at risk of being marginalized from global politics unless it makes substantial efforts to establish itself as a partner for China and India – a partner which is not only economically relevant but which is also still a significant global actor. This role is by no means guaranteed by 'history', despite European governments' periodic assumptions to that effect. Chinese foreign policy already regards Europe as a peripheral issue; its main frame of reference is the USA, followed by Japan and the South/South-East Asian countries that are relevant in an economic and regional policy context.

In order to avoid any further loss of significance vis-à-vis the Asian powers, Germany and the EU should invest to a far greater extent than before in a coherent, future-oriented Common Foreign and Security Policy (CFSP), which inevitably means setting aside national egoisms. The EU can no longer afford to present a divided image to the world as it has done in the past, notably in advance of the Iraq war in 2003. Instead, within the framework of the CFSP, it should openly address the issue of the anticipated shifts in the political world order and accept the challenge of presenting itself as a constructive partner for the Asian region, on the one hand, without permanently damaging its transatlantic relations, on the other, which are based on historical ties and remain on a sound footing (Biermann, 2005).

Against this background, one issue to be explored is whether convening a world conference, akin to the CSCE process, to consider, at the highest level, the possible implications of the anticipated power shifts could help foster a positive climate of cooperation. The title for the conference could be broadly as follows: 'A New World Order for the 21st Century – Power Shifts for Global Governance' (for a discussion of the real opportunities afforded by the world conferences, which are often regarded as ineffective, see Fues and Hamm, 2001; Haas, 2002). The concerns and diffuse uncertainty experienced by many states and societies in the face of geopolitical change could perhaps then be channelled constructively. The aim would be to generate a positive mood that is conducive to a fresh start, emphasizing and building on the opportunities afforded by the anticipated changes.

Recognising climate change as a common threat to humankind

Climate policy and energy policy offer ideal fields of action for Europe to play a pioneering role. Here, Germany and the EU could bring influence to bear to change the global discourse, and thus the world political agenda, in such a way that the 'war on terror' is replaced as the hegemonic discourse in international relations. The intention is certainly not to play down the significance of the problem of global terrorism, but it could be argued that global climate change and poverty pose a greater collective threat to human security than terrorism. Former US President Clinton has also emphasized that although climate change appears more remote than terrorism, it actually constitutes the greater threat (Economist, 2006).

Climate change can certainly be regarded as a collective threat to humankind which can only be countered effectively through joint efforts by the international community as a whole (or at least through a strong 'coalition of the willing' involving all the major actors). If the efforts to curb climate change fail, this will have a damaging effect on humankind as a whole – not just individual states or societies – in the long term. Moreover, more intensive efforts to achieve fair, resolute and targeted international cooperation in the field of climate protection or poverty reduction are likely to strengthen the multilateral architecture as a whole and thus make a contribution to peaceful development in the world.

10.3.1.2
Initiative 2: Reforming the United Nations

It has become commonplace in international politics to diagnose a substantial need for reform of the United Nations and its institutions, notably in the field of environmental and development policy. The publication of a series of expert reports under the auspices of former UN Secretary-General Kofi Annan (UN, 2004, 2006b), Annan's own reform proposals, entitled *In Larger Freedom* and unveiled at the special session of the General Assembly in 2005 (UNSG, 2005) and the expectations aroused by the appointment of the new Secretary-General Ban Ki-moon are currently generating a relatively high level of interest in the UN's reform agenda among the member states. Ultimately, it is their political will that will determine the substance and tangible outcomes of possible reforms.

Among the member states, it is the industrialized countries in particular which must demonstrate that they are serious about their reform rhetoric and are willing to provide the financial resources that may be necessary for the reform process. At the same time, the developing countries must move away from a position which combines empty demands for a leading role in the United Nations with obstinate adherence to an anachronistic status quo (Fues, 2006). Otherwise, this reform dynamic will also quickly wane. The report *Delivering as One* (UN, 2006b), produced by the High-Level Panel on System-wide Coherence towards the end of Kofi Annan's term of office and focussing on the areas of development, humanitarian assistance and the environment, is encouraging, for it provides a realistic appraisal of the current situation and sets out largely pragmatic demands.

As environmentally induced conflicts and the associated security issues are likely to increase in significance, the question which arises is which role the United Nations and its various institutions should play in managing these problems. On the one hand, the institutions and mechanisms already existing within the UN system could be utilized. On the other hand, not least in light of the reform efforts mentioned above, targeted institutional adjustments are both desirable and conceivable.

REFLECTING ON THE ROLE AND TASKS OF THE UN SECURITY COUNCIL
In recent years, in relation to 'humanitarian interventions', there has been intensive debate about the conditions under which the United Nations may intervene in domestic crisis situations, generally those in which there is a threat to international security as a result of massive human rights violations (Ipsen, 2004; Janse, 2006). There is general consensus that when the Security Council, in a resolution, determines the existence of a threat to the peace pursuant to Article 39 of the United Nations Charter due to human rights violations within a state, it may deploy the instruments of collective security in accordance with Chapter VII of the Charter (measures to maintain or restore international peace and security) and apply sanctions (Kimminich, 1997).

In WBGU's view, severe environmental degradation and environmentally induced conflicts can be regarded as a threat to international security and world peace. The impacts of climate change constitute a particularly high potential risk in this context. Presumably, therefore, the Security Council is also authorized to take action under Chapter VII of the United Nations Charter in cases of widespread destruction of natural environmental goods and grave violations of international environmental law. A further argument in favour of this position is that climate change impacts, not least, on human rights and therefore could be said to constitute a threat to the peace pursuant to Article 39 of the United Nations Charter.

This means that the Security Council is in principle authorized to apply appropriate sanctions against the states responsible (Elliott, 2002; Fassbender, 2005). At the first Security Council meeting with Secretary-General Ban Ki-moon on 8 January 2007, Belgium and Peru called for the Security Council's mandate to be extended accordingly. The United Kingdom – one of the countries holding a veto in the Security Council – then gave further clear political backing to this issue a few months later, in April 2007, with Foreign Secretary Margaret Beckett stressing the security dimension of climate change in a speech to the Security Council. However, the extremely restrained position of the USA and China, as veto-holding powers, suggests that realistically, a formal expansion of the Security Council's agenda is unlikely to occur in the near future, especially given that this particular aspect was absent from the relevant report on UN reform, *A More Secure World: Our Shared Responsibility* (UN, 2004). Nonetheless, the fact that climate change has been addressed before the highest body in the United Nations has raised its media profile and is a symbolic step towards ensuring that the importance of this issue is not downplayed in international diplomacy, and has also established a practical basis for further discussion in future. WBGU therefore urges further exploring the pros and cons of a broader mandate for the Security Council within the overall framework of the UN reform debate.

One option in this context is to invoke the concept of 'responsibility to protect', by means of which the United Nations claims high moral authority. The responsibility to protect was developed specifically in order to prevent future failures by the United Nations such as those which weigh on the international community's conscience as a result of its inaction in Rwanda and the former Yugoslavia (Fröhlich, 2006; Hilpold, 2006). In view of the conflict potential that could be unleashed by the impacts of climate change, this principle could provide a remedy and be modified and expanded accordingly. The Security Council could perhaps charge the newly established UN Peacebuilding Commission with addressing the specific tasks arising from this principle, although it must be ensured that this does not overstretch the Commission's mandate to an impermissible extent. The Commission's primary task is to strengthen the capacities of the United Nations to propose strategies for sustainable peacebuilding in countries torn apart by civil war and to ensure that no resurgence of violence occurs in post-conflict situations (Weinlich, 2006). However, in view of the additional pressure from the problems associated with the possible impacts of climate change, especially in fragile civil-war countries, the Peacebuilding Commission should also be able to address environmentally induced security risks appropriately as part of its own mandate.

In general, it is important to emphasize that authorizing and implementing military measures, also in response to environmentally induced threats to international peace and security, may only take place in strict compliance with the United Nations' monopoly of force. Pre-emptive measures by individual countries or groups of countries outside the scope of the powers granted under Chapter VII of the United Nations Charter are therefore clearly incompatible with international law, also in the context of threatened environmental conflicts.

STRENGTHENING THE UNITED NATIONS' ENVIRONMENTAL POLICY CAPACITIES
WBGU reaffirms its often-stated recommendation that the United Nations Environment Programme (UNEP) be strengthened and formally upgraded (WBGU, 2001, 2005). The French government has launched a diplomatic initiative specifically with a view to moving in this direction. In the academic debate, too, the differences in the positions for and against such a reform have noticeably narrowed in recent years (Biermann and Bauer, 2005; Rechkemmer, 2005). Nonetheless, in view of some states' political opposition to such a move, it is highly unlikely to take place in the near future, although member states' commitment to UNEP as the 'global authority and environmental policy pillar' of the United Nations is constantly reiterated, most recently at the Global Ministerial Environment Forum in Nairobi in February 2007 (UNEP, 2007).

In WBGU's view, a substantive strengthening of UNEP should take place without delay. For practical reasons, therefore, this should initially begin below the threshold for UNEP's upgrading to the status of a UN specialized agency, as also proposed in the report *Delivering as One* on coherence in the UN system and in the debate at the Global Ministerial Environment Forum. In that sense, WBGU takes the view that all measures resulting not only in additional budgetary resources for UNEP but also providing the requisite stability for UNEP to engage in more secure medium- to long-term financial planning should be supported. This requires, firstly, multiannual financial commitments from donors and, secondly, a reduction in the proportion of earmarked funds.

In this context, the efforts to improve cooperation between UNEP and the many UN agencies with an operational mandate should also be supported. This applies especially to cooperation with the United Nations Development Programme (UNDP) (Biermann and Bauer, 2004; Fues, 2006), and, in light of the growing significance of environmentally induced migration, the Office of the United Nations High Commissioner for Refugees (UNHCR) (Section 9.3.4) as well. The relevant recommendations of the High-Level Panel on System-wide Coherence are to be welcomed here, as are the encouraging signals sent out by the heads of the relevant UN agencies at the first Global Ministerial Environment Forum to be held since the appointment of Achim Steiner as UNEP's new Executive Director (IISD, 2007; Dervis, 2007; Steiner, 2007). Ultimately, however, it is the responsibility of the member states to coordinate the priorities of the individual organizations in a way which facilitates effective coordination. WBGU therefore recommends that the efforts to improve cooperation among the relevant UN agencies at political level be given resolute support and critically monitored.

In light of UNEP's core competences in the fields of monitoring, assessment and early warning, WBGU recommends that the longstanding demands for a strengthening of UNEP's role as the central knowledge manager on issues relating to the global environment be followed up with action. In particular, the further development and implementation of the UNEP Environmental Observing and Assessment Strategy, which was developed by the UNEP Secretariat in a complex international consultation process, should be supported. This strategy is based on the comparative strengths of the Environment Programme (e.g. the Global Environment Outlook reports) and, among other things, prioritizes international networking by UNEP and scientific capacity-building. In particular, it is important to ensure that the academic contribution to decision-making on international environmental policy is free from any direct influence of political office-holders. One option which could be considered is the creation of a high-profile and high-level office of 'Chief Scientist' within UNEP, supported by a staff of scientists of international repute, who would act as scientific adviser at levels up to and including the UN Secretary-General (Najam et al., 2006).

Establishing a Council on Global Development and Environment

WBGU reiterates its call for the establishment of a high-level Council on Global Development and Environment within the UN system, which could emerge from a reform of the largely ineffective Economic and Social Council (ECOSOC) and pool the competencies of various existing UN bodies (WBGU, 2005). However, any major reform of ECOSOC requires an amendment of the UN Charter, and given that such an amendment must be adopted by a vote of two-thirds of the members of the General Assembly, including all the permanent members of the Security Council, there is little prospect of achieving this in the near future (WBGU, 2005).

WBGU therefore recommends that policy be guided by the pragmatic proposals made by the High-Level Panel on System-wide Coherence. These proposals include the establishment of a UN Sustainable Development Board, reporting to ECOSOC. The Board should be granted substantial political authority at the level of the heads of state and government and exercise joint supervision of the UN programmes in the areas of development, humanitarian assistance and the environment. It would increase system-wide coherence, inter alia, by ensuring that goal conflicts between the various agencies would have to be negotiated by the member states at the highest level. The Board could take action, in particular, in relation to problems which specifically affect the environment-security nexus and, if appropriate, refer them to the Security Council for further deliberation.

This model takes account of the fact that most member states are opposed to any merger of various UN agencies. At the same time, joint supervision could improve programmatic coherence, although this would have to be reflected in budget planning. As the boards of the relevant programmes already hold joint meetings to some extent, the institutionalization of this cooperation via a Sustainable Development Board is a feasible political option.

In line with the proposals made by the High-Level Panel, the existing joint meetings of the boards of the United Nations Development Programme (UNDP), the United Nations Population Fund (UNFPA), the United Nations Children's Fund (UNICEF) and the World Food Programme (WFP) would be merged into this strategic oversight body. In WBGU's view, UNEP – until it can be upgraded to a UN specialized agency – and UN-HABITAT (United Nations Human Settlements Programme) should also be merged into this oversight body. A further option to be considered is whether and how participation by the United Nations Framework Convention on Climate Change (UNFCCC), the Convention on Biological Diversity (CBD) and the United Nations Convention to Combat Desertification (UNCCD) could also be beneficial.

In terms of its relationship with this comprehensive supervisory body, the mandate of the Commission on Sustainable Development (CSD) could be made more specific, so that it would act as the policy link between the programmes supervised by the Board and the relevant UN specialized agencies (e.g. FAO, WHO, UNIDO, UNWTO). In this instance, it is important to consider the extent to which parallel coordination structures, notably the UN Development Group and the Environmental Management Group, are still required.

In view of the environment-development-security nexus, the creation of an oversight body for the UN programmes mentioned above is both sensible and necessary, in WBGU's view. The Sustainable Development Board would guarantee greater coherence at operational level in the United Nations' environmental and development policy, enabling the UN to develop a more targeted preventive response, not least to the escalation of environmentally induced conflicts, than is possible with the current fragmented UN apparatus. WBGU therefore recommends that the Board be equipped with the authority to issue and enforce binding instructions in the event of a threat of severe environmental impairment. Naturally, this means that before the Board is established, its powers, its legal and political position within the UN system and, above all, its relationship with the Security Council and ECOSOC as principal bodies within the system must be clearly defined.

10.3.2
Climate policy as security policy I: Preventing conflict by avoiding dangerous climate change

WBGU has made recommendations on many occasions on the specific form that an effective climate protection policy should take; the reader's attention is drawn to various previous reports and policy papers in this context (WBGU, 2003, 2004, 2005, 2007). For that reason, the following initiatives merely briefly outline, in key words, the current and important fields of action for climate protection.

10.3.2.1
Initiative 3: Ambitiously pursuing international climate policy

MAKING THE 2 °C GUARD RAIL AN INTERNATIONAL STANDARD

Specific international targets with a long-term focus increase the prospects of implementing a successful climate policy that initiates the global technological revolution and the shift in attitudes that are necessary to achieve the ultimate objective of the United Nations Framework Convention on Climate Change (UNFCCC): to stabilize the concentration of greenhouse gases in the atmosphere at a level that would prevent dangerous anthropogenic interference with the climate system. At international level, a consensus must therefore be reached on quantifying this objective, as set out in Article 2 of the UNFCCC. To this end, WBGU recommends the adoption, as an international standard, of a global temperature guard rail limiting the rise in globally averaged near-surface air temperature to 2 °C relative to the pre-industrial value. If the concentration of greenhouse gases in the atmosphere is stabilized at 450ppm CO_2eq, there is a realistic prospect, according to current knowledge, of achieving this objective. This will require a 50 per cent reduction in global greenhouse gas emissions by 2050 compared with a 1990 baseline.

GEARING THE KYOTO PROTOCOL TOWARDS THE LONG TERM

For the Kyoto Protocol's second commitment period, it is not simply a matter of renewing the existing commitments with new targets. Instead, the mechanism established under Article 9 of the UNFCCC to review the Kyoto Protocol should form the basis for the ambitious further development of the Protocol and its compliance mechanisms.

In WBGU's view, equal per capita allocation of emission entitlements on a global basis is the allocation formula which should be aimed for in the long term. All countries must ultimately play a part in achieving this goal.

- For the second Kyoto commitment period, the industrialized countries should adopt ambitious goals in the order of a 30 per cent effective reduction in greenhouse gas emissions by 2020 against the 1990 baseline. Germany should serve as a role model here: in WBGU's view, a 40 per cent reduction target for Germany is appropriate. However, the global climate protection goals can only be achieved if the USA also substantially reduces its greenhouse gas emissions.
- In order to integrate newly industrializing and developing countries into mitigation efforts to a greater extent, WBGU recommends the adoption of a flexible approach to the setting of reduction commitments and clear differentiation within this country group. The major newly industrializing countries in particular, which are now playing an increasingly influential role in world affairs, have a responsibility to act as a role model for other developing countries and commit to flexible targets, with fixed targets being the medium-term objective.

CONSERVING NATURAL CARBON STOCKS

Preserving the natural carbon stocks of terrestrial ecosystems should be a key goal of future climate protection policy alongside the reduction of greenhouse gas emissions from the use of fossil fuels. Tropical forest conservation should be a particular priority in this context. The UNFCCC process to reduce deforestation in developing countries should be pursued further. The aim should be to establish arrangements which provide clear incentives for tropical forest protection but which do not lead to any slackening of the industrialized countries' own mitigation efforts. WBGU therefore takes the view that arrangements which are for the most part separated from the Kyoto carbon market would initially be beneficial.

10.3.2.2
Initiative 4: Transforming energy systems in the EU

STRENGTHENING THE EU'S LEADING ROLE
In order to be a credible negotiating partner within the climate process, the European Union should achieve its Kyoto commitments and then set more far-reaching and ambitious reduction targets. In WBGU's view, a 30 per cent reduction target for greenhouse gas emissions by 2020 compared with the 1990 baseline and an 80 per cent reduction target by 2050 are appropriate.

IMPROVING AND IMPLEMENTING THE ENERGY POLICY FOR EUROPE
In WBGU's view, the proposals for an Energy Policy for Europe, unveiled by the European Commission in January 2007, point in the right direction and their basic elements should be adopted and rigorously implemented by the member states. Binding targets, threshold values and timetables are essential to make the Energy Policy for Europe more specific. However, WBGU also sees a need for improvement in relation to the expansion targets and individual technological options. Overall, the proposals should be geared more strongly towards sustainability criteria such as those proposed by WBGU in its 2003 report on sustainable energy systems.

TRIGGERING AN EFFICIENCY REVOLUTION
The proposals set out in the Energy Efficiency Action Plan, as well as existing directives and regulations, provide a sound basis for the necessary improvements in energy efficiency. The potential energy savings of 20 per cent by 2020, cited in the Action Plan and endorsed by the European Council, should be increased substantially through binding European rules, ambitious national targets and the rigorous enforcement of existing legislation. This applies especially to buildings, cars and product standards. Here, dynamic standards should be set which progressively lead to a reduction in energy input and emissions and thus establish long-term objectives for technological development as well (WBGU, 2007).

EXPANDING RENEWABLES
WBGU proposes that in addition to the targets put forward in the European Commission's Energy Policy for Europe and reaffirmed by the European Council, a binding target of 40 per cent of renewables in electricity generation by 2020 be adopted, along with a figure of 25 per cent of renewables in primary energy production. However, renewables expansion should not take place at the expense of other dimensions of sustainability (this applies e.g. to the expansion of bioenergy or hydropower; WBGU, 2004). Key prerequisites for the efficient integration of renewables are unimpeded access to the (national) grids and high-capacity trans-European grids. These grids would also enable the EU to enter into an energy partnership with North Africa (WBGU, 2007).

10.3.2.3 Initiative 5: Developing mitigation strategies through partnerships

ESTABLISHING CLIMATE PROTECTION AS A CROSS-CUTTING THEME IN DEVELOPMENT COOPERATION
In development cooperation too, path dependencies in favour of emissions-intensive technologies should be avoided, and high priority should be granted to the promotion of sustainable structures, e.g. in the energy supply. To this end, climate protection must be integrated from the outset as a cross-cutting theme in poverty reduction strategies, such as the Poverty Reduction Strategy Papers (PRSPs), or in ongoing planning processes (e.g. national environmental action programmes). In this context, the promotion of sustainable energy systems in order to overcome energy poverty should be a key focus of attention. Another important field of action for climate protection in developing countries is the avoidance of emissions from land use changes, especially deforestation. The German government has been working for some time to 'join up' poverty reduction and climate protection objectives in multilateral development cooperation, especially within the GEF and in the PRSP process in the World Bank and IMF. Similar priorities should be adopted in bilateral development cooperation as well. In WBGU's view, poverty reduction and climate protection strategies should therefore be 'joined up' more systematically, and far more rigorously than before, within the German Ministry for Economic Cooperation and Development (BMZ) and the German implementing organizations (especially GTZ and KfW), and also within the framework of donor coordination in the European Union (WBGU, 2005).

AGREEING DECARBONIZATION PARTNERSHIPS WITH NEWLY INDUSTRIALIZING COUNTRIES
Germany and the EU should enter into strategic decarbonization partnerships with those newly industrializing countries that are likely to play an important role in the future world's energy sector. The aim would be to move energy systems and energy efficiency towards sustainability, thus providing innovative impetus and acting as a role model on a worldwide basis. China and India should be priority partners in this area (WBGU, 2007).

AGREEING AN INNOVATION PACT WITHIN THE
FRAMEWORK OF G8+5

The G8+5 forum should be utilized to develop joint targets for the promotion of climate-compatible technologies and products. This group, comprising the world's leading industrial nations and newly industrializing countries, represents the heavyweights in the global political arena and accounts for around two-thirds of global greenhouse gas emissions. On the basis of national Road Maps charting the transformation of national energy systems in the interests of climate protection, a strategic 'Road Atlas for the Decarbonization of Energy Systems' could then be produced. By adopting joint parameters for efficiency and CO_2 emissions standards, and promoting comprehensive technological cooperation, the G8+5 countries have the potential to become the driving force in a global transformation of energy systems (WBGU, 2007).

10.3.3
Climate policy as security policy II: Preventing conflict by implementing adaptation strategies

10.3.3.1
Initiative 6: Supporting adaptation strategies for developing countries

Developing countries' vulnerability to the impacts of climate change can be greatly reduced through the timely adoption of adaptation measures. An effective climate regime which is viewed as fair by the international community and is viable in the long term must therefore give adequate priority to adaptation. As well as implementing specific adaptation projects, the primary focus must be on strategies and measures to improve the adaptive capacity of those countries – generally the poorer ones – which will be hardest hit by climate change. Timely adaptation measures should therefore be an integral element of their national policies, alongside measures aimed at avoiding further global warming. However, the capacities of states and societies to prepare for global warming vary very widely. If poor adaptive capacities coincide with political and social instability (Section 3.3), climate impacts can trigger major security risks (Chapter 7). Besides specific adaptive capacities, factors which determine political and social stability are therefore always of relevance as well. At domestic level, maintaining state functions (Section 4.2.1), adequate rights of participation and the promotion of civil society institutions are vital here, while externally, the capacity to maintain conflict-free relations with other countries must also be developed. This is important in identifying cooperative solutions to issues relating to the management of transboundary water resources, for example.

Most developing countries lack the capacities and resources to implement effective adaptation measures. The industrialized countries, as the main drivers of climate change, have a special responsibility to provide assistance to the developing countries to enable them to deal with the impacts of climate change.

These aspects – especially the security policy relevance of climate change to the industrialized countries as well – have yet to be fully recognised in many German and European development institutions. An awareness of the development implications of climate change is proving slow to develop. This is understandable, in light of the focus on 'classic' development policy issues such as economic development, poverty reduction and, above all, the Millennium Development Goals. WBGU has already drawn attention to the urgent need to make progress here, as any overstretching of the developing countries' adaptive capacities would inevitably undermine the success prospects of many well-established development measures as well (WBGU, 2005). Nonetheless, adaptation programmes should on principle be 'joined up' with relevant sectoral policies and programmes, and possible climate impacts must be given due consideration, especially in relation to major infrastructural projects.

The integration of the anticipated regional impacts of climate change in the development and implementation of poverty reduction policies offers a good opportunity to significantly improve and safeguard these countries' long-term capacities for policy action. For example, the risk potential of the climate-induced conflict constellations discussed in this paper should be taken into account from the outset in the formulation of poverty reduction strategies. These conflict constellations make it clear that special efforts for adaptation to climate change are required in at least three priority areas: freshwater availability, food production and disaster mitigation.

ADAPTING WATER RESOURCES MANAGEMENT TO
CLIMATE CHANGE AND AVOIDING WATER CRISES
- *Mainstreaming the impacts of climate change on water management as an issue in development cooperation:* Adaptation to climate change in the field of water resources management should be integrated systematically into bi- and multilateral development cooperation in the water sector, especially for regions which already suffer from water scarcity. Developing countries often lack the capacities to manage local climate impacts or develop any awareness of the problem of climate change, or of the need and opportunities for adap-

tation, among decision-makers and the general public. Capacity-building is also required to enable or facilitate international networking between experts in the developing countries and their counterparts abroad and encourage collaboration on projects and databases. In future, these should be priority areas for development cooperation in the field of water resources management. Adaptation to climate change should be established as an ongoing process, and should not simply entail the implementation of one-off measures.

- *Promoting international cooperation on the provision of information:* In order to adapt water resources management to the impacts of climate change, it is essential, as a basis for planning, to draw not only on past data, such as mean precipitation rates or variability in precipitation and runoff, but also on the findings of regional models which take account of climate change. High priority should therefore be given to regular provision and evaluation of current scientific data on the regional impacts of climate change on water availability. International cooperation is vital here, especially for the developing countries. One issue which should be explored is whether a universally accessible database could be established and maintained by the international community, containing interpreted regional data as a basis for water resources management. The provision of information is also important for transboundary rivers and lakes: reliable data on water availability and use are a key prerequisite for the development of cooperative transboundary water resources management and thus the avoidance of water conflicts and crises.
- *Reorienting water management towards action under increased uncertainty:* For effective action to be taken, there is often no need to await the development of appropriate forecasting models. Measures which improve adaptation to existing climate variability can often be applied to adaptation to future climate impacts as well. This is especially true of measures to improve the efficiency of water management, local water storage capacity, systems for the distribution of stored water, and demand management. Integrated water resources management offers a suitable framework here (Box 6.2-1).

GEARING AGRICULTURE TO CLIMATE CHANGE

- *Strengthening the agricultural sector through development cooperation in light of climate change:* The anticipated pressure on agriculture, especially in many developing countries, should be taken into greater account in development cooperation, with higher priority being given to this sector. First, as a basis for effective policies to mitigate climate impacts and other global environmental changes, viable scenarios and forecasting are required for the agricultural sector which must take adequate account of these factors. At present, however, climate change is not being considered to an adequate extent as a factor in the FAO scenarios and forecasts on food production trends and food security, and this could result in the wrong conclusions being drawn in long-term analyses. The German federal government should therefore work pro-actively within the FAO Council for a greater focus on global environmental changes and especially climate change and provide more support for relevant research projects. At the same time, in view of the anticipated drop in agricultural yields, development cooperation should focus to a greater extent on the development of rural regions. However, it is not enough simply to invest more resources in strengthening the agricultural sector. Instead, a new qualitative focus is required in agricultural development strategies: only if climate-induced risks are anticipated in programme development and implementation, for example, can agricultural development strategies be successful in the long term. In particular, the growing tension between food security, nature conservation and the cultivation of energy crops must be considered. In WBGU's view, more intensive production of energy crops should certainly not take priority over nature conservation and food security (WBGU, 2004). The promotion of a robust agricultural sector must therefore always go hand in hand with the development of other sectors, for in view of the anticipated drop in agricultural yields, a diversified and therefore less crisis-prone economy is required.
- *Reforming world agricultural markets:* The reform of world agricultural markets should be pursued vigorously in order to improve opportunities for market access and create market incentives for production in the developing countries. Given the uncertain prospects of a successful outcome to the World Trade Organization's Doha Round, whose key elements include the dismantling of barriers to entry to the industrialized countries' agricultural markets and the removal of subsidies that are harmful to development, this is a particularly urgent message. However, liberalization is likely to lead to price increases which, in the short to medium term, could have an extremely adverse effect on Low-Income Food-Deficit Countries (LIFDCs), greatly impairing their long-term prospects of economic and social development. This would further reduce these countries' opportunities to adapt to climate change and mitigate

the security risks with which it is associated. For development and security policy reasons, therefore, it is particularly important to establish compensation mechanisms for these countries, akin to those already in place in the WTO and the Bretton Woods institutions. In order to respond adequately to the scale of the problem, however, these mechanisms need far more reliable funding. In WBGU's view, the German government should endeavour to ensure that such compensation mechanisms are adequately endowed.

- *Taking account of many developing countries' growing dependency on food imports:* The liberalization of the agricultural markets and short-term compensation payments to low-income countries will not solve the long-term supply and demand problems faced by many developing countries, however. A number of developing countries will experience major drops in agricultural yields and growing dependency on farm imports, not least as a result of climate change. This has implications for many different policy areas (e.g. food, poverty reduction, trade, balance of payments and monetary policy), so there is a need for international development cooperation, trade and monetary policy, and indeed other policy fields, to start developing strategic solutions to these problems without delay. Arguably, international climate policy should also focus to a greater extent on this issue. Assessments should be conducted for instruments for the payment of compensation to developing countries whose agricultural sector is severely impacted by climate change.

STRENGTHENING DISASTER PREVENTION

- *Developing cross-sectoral approaches in development cooperation:* In view of the destabilizing impacts of natural disasters and the expected increase in major storm and flood events, development cooperation should develop and implement strategies for the mitigation of disaster risks to a greater extent. In this context, preference should be given, as a matter of principle, to cross-sectoral approaches rather than one-off measures. This is based on the recognition that isolated infrastructural projects (e.g. dam-building) may encourage the concentration of settlements and property in at-risk areas and therefore further increase the risk of disasters. When devising appropriate packages of measures, there should be a particular focus on emergency planning, adaptation of land-use planning, establishment of clear decision-making structures at an early stage, and the inclusion of disaster mitigation in education programmes. In particular, early warning systems should be embedded in comprehensive development programmes; in this context, besides the provision of technology, high priority must be given to clarifying information and decision-making structures within the relevant administrations and to the participation of the affected population. In view of the threat posed to major coastal cities and delta regions by storm and flood disasters, development cooperation should also focus on protecting forests and mangroves and reducing land subsidence in urban areas. Both these topics must be coordinated with land-use and water supply issues.

- *Integrating disaster risks into development strategies to a greater extent:* In many places, weaknesses in existing development strategies are a contributory factor in increasing disaster risks (WBGU, 2005). WBGU therefore recommends, in particular, that disaster mitigation be taken into account in consultations regarding Poverty Reduction Strategy Papers and in the major poverty reduction programmes. The topic should also be utilized to a greater extent as a starting point for good governance.

- *Reviewing disaster prevention in industrialized countries:* The efficient functioning of modern industrial societies is heavily dependent on transport, energy and information networks. The high vulnerability of these networks due to power outages, the failure of the transport infrastructure due to weather events, or the overloading of telecommunications systems is becoming increasingly apparent. WBGU therefore recommends a review of disaster mitigation systems in the industrialized countries, especially in light of the challenges posed by ongoing climate change. The anticipated impacts of climate change must be taken into account when planning highly sensitive infrastructures.

10.3.3.2
Initiative 7: Stabilizing fragile states and weak states that are additionally threatened by climate change

As described in Section 4.2, it must be assumed that states which, based on current information, are classified as weak or fragile, are poorly equipped to protect their societies effectively from the impacts of climate change. The analysis of conflict constellations in Chapter 6 shows that political, social and community stability is highly influential in determining whether a regional environmental crisis escalates into violent conflict. It is also likely that the additional problems caused by climate change will impede the stabilization of weak and fragile states, and may even trigger further destabilization, jeopardizing the attainment of the Millennium Development Goals as well.

The findings of the present report show that it is the poorest developing countries in the tropical regions, above all, that will in future be affected by fragility at internal level and by strong climate stress from outside. It is essential that in their response to weak and fragile states, Germany and its partners assess the scale of the phenomenon correctly and consider the long-term timeframes in which there is any prospect of success. There is therefore also a need to secure comprehensive and long-term funding. The implications of climate change for the scale, longevity and financing of possible German contributions to the stabilization of weak and fragile states should therefore be taken into account to a greater extent in the Action Plan 'Civilian Crisis Prevention, Conflict Resolution and Post-Conflict Peace-Building'.

This debate should be conducted first and foremost within the European Union framework, firstly, in order to gain a more accurate picture of whether, and how, the comparative advantages and specific capacities of the national implementing organizations can be deployed on a targeted basis within the nascent European development cooperation and how these resources can be pooled in order to facilitate burden-sharing; and secondly, in order to ensure the greatest possible coherence with the Common Foreign and Security Policy (CFSP).

In this context, WBGU recommends, in particular, the operationalization of the Solana strategy in line with the Barcelona Report of the Study Group on Europe's Security Capabilities, which prioritizes crisis prevention with the aim of avoiding military intervention as far as possible. WBGU is therefore critical of the proposals made by the Institute for Security Studies in Paris, which call for greater military engagement by Europe and the expansion of its flexible deployment capabilities (Section 2.2).

IMPLEMENTING THE OECD'S WORKING PRINCIPLES

In its response to weak and fragile states, the German government should continue to play an active role in the Fragile States Group set up by the OECD's Development Assistance Committee (DAC). In particular, it should drive forward the implementation and further development of the Principles for Good International Engagement in Fragile States and Situations (Box 4.2-3). This process has high priority within the DAC and will form a key part of the DAC's work programme for the next two years.

WBGU recommends that within the DAC process, efforts be made to ensure that appropriate account is taken, in the implementation of the Principles, of the risks arising from climate change. Specifically, weak and fragile states' capacities to manage environmental risks must be maintained and reinforced, and if necessary re-established, even under difficult political and economic conditions. First and foremost, this means putting in place the necessary conditions for effective measures for adaptation to anticipated climate change. Although major security impacts resulting from climate change are only expected in the medium to long term, it must nonetheless be borne in mind that crisis prevention today costs far less than crisis management at a later stage. What is still lacking, however, are appropriate instruments to assess the impacts of climate change in various country contexts. That being the case, the DAC's Fragile States Group and the German Inter-Ministerial Working Group on Crisis Prevention should make efforts to promote further development of early warning and prevention mechanisms.

EXPANDING THE 'WHOLE-OF-GOVERNMENT' APPROACH TO ENCOMPASS THE ENVIRONMENTAL DIMENSION

With the adoption of the Principles for Good International Engagement in Fragile States and Situations, the DAC's policy response to weak and fragile states, which is actively supported by the German government through the BMZ, has entered a new phase. With a view to implementing the Principles, the German government intends to promote the 'Whole-of-Government' approach in order to ensure that civilian measures aimed at the establishment and stabilization of public institutions are not regarded purely as a peripheral issue and the sole domain of development workers. WBGU endorses this approach and recommends that the German government pursue this course resolutely and bring appropriate influence to bear on the EU's Common Foreign and Security Policy as well. So far, the discussions have centred primarily on issues concerning the relative significance and sequencing of military and development policy measures. Environmental risks, on the other hand, barely feature in this debate, despite the fact that climate change is becoming increasingly apparent and will require comprehensive adaptation measures in areas such as agriculture, the financial sector and infrastructure, i.e. beyond the purview of the 'classic' ministerial portfolio for the environment.

WBGU therefore recommends that in the ongoing debate about the conceptual development and practical implementation of the 'Whole-of-Government' approach, the security and development policy relevance of climate change and the associated environmental changes be emphasized, thus further sharpening the profile of the DAC process. An intervention policy which does not take account, in its planning, of the increasing pressure on weak and fragile states resulting from the problems of climate change will substantially reduce the prospects of stabilizing these states.

BOOSTING CIVIL SOCIETY POTENTIAL IN WEAK AND FRAGILE STATES

Despite the difficult circumstances, political, social or academic institutions often exist in weak and fragile states, as well as committed and in some cases highly qualified individuals. In many cases, however, they are unable to gain a hearing and are often passed over in the distribution of scarce resources at internal level or disregarded during the planning of external measures. As a result, their promising potential remains untapped. This applies especially at international level, and this effect may be reinforced because donor countries are increasingly focussing their attention on so-called 'anchor countries' such as Brazil, China, India and South Africa. WBGU therefore proposes that ways be identified of raising awareness of, and providing targeted support, for the capacities available in weak and fragile states, e.g. through international forums and academic/civil society networks. In particular, it is important to ensure that civil society actors in weak and fragile states are not excluded from international networking processes. The IPCC's working methods demonstrate both the importance and the difficulty of facilitating the participation of weak and fragile states in international processes.

10.3.3.3
Initiative 8: Managing migration through cooperation and further developing international law

Environmentally induced migration is already taking place and because of the impacts of climate change it is likely to increase markedly. It is to be expected that more and more countries will be affected by environmentally induced migration. As a result of unabated climate change, migration could take place on a scale that far exceeds present capacities for dealing with the problem. This means that, as described in Section 6.5, the risk of conflicts is likely to increase significantly. In order to prevent an increase in security risks of this type caused by environmentally induced migration, there is a need for both preventive measures – i.e. strategies directed at the causes of environmentally induced migration – and measures for managing unavoidable environmentally induced migration.

DEVELOPING COMPREHENSIVE INTERNATIONAL STRATEGIES FOR MIGRATION

If the problem of environmentally induced migration is to be dealt with in a permanent and preventive manner, a comprehensive migration policy strategy is required which takes account of the interests of all stakeholders. Its long-term objectives must be geared towards the interests of the destination, transit and home countries alike. In WBGU's view, an approach which focusses primarily on the industrialized countries' internal security is too one-sided, reactive and, at best, only effective in the short term. In addition, such an approach would leave the developing countries to shoulder most of the burden arising from environmentally induced migration; these countries would thus find themselves dealing with a large proportion of the effects of anthropogenic climate change.

The migration policy so far adopted by the industrialized countries is unsatisfactory in this regard. The focus of EU policy, which up to now has concentrated too one-sidedly on short-term procedural issues, is particularly problematic (Parkes, 2006). Operational cooperation between the EU member states with the aim of managing migration has so far focussed largely on border controls, the organization of refugee camps and international cooperation on repatriation agreements. Prevention strategies and efforts to deal with environmentally induced migration do not feature in the numerous bilateral readmission agreements that have been concluded between industrialized countries and countries of origin (most recently, for example, between Spain and Morocco).

INTEGRATING MIGRATION POLICY INTO DEVELOPMENT COOPERATION

In the least developed countries, unabated climate change would increase the risk of people being forced to abandon their home regions due to the collapse of their natural life-support systems. Development cooperation can help people living in absolute poverty to adapt to such changed environmental conditions and thus make it easier for them to remain in their homes. However, development strategies must in future also pay more heed to the sustainability of development plans in light of foreseeable climate impacts at local level. For example, the agricultural development of a region likely to be strongly affected by drought in the future should be re-evaluated; if there are no alternative means of generating an income that are less dependent on climate, people should be supported in seeking alternative areas in which to settle. It can also be assumed that environmentally induced migration within and between affected states will increase in future, opening up a new field of action in development cooperation. The importance of a comprehensive, pro-active and development-oriented migration policy is increasingly being recognized at political level as well. One example is the recent development in relations between the EU and African countries (the Rabat Action Plan). In WBGU's view, the creation and expansion of opportunities for legal migration also need to be considered.

INCLUDING ENVIRONMENTALLY INDUCED MIGRATION IN INTERNATIONAL COOPERATION

Because of the scope of the problem, environmentally induced migration must be a topic of the consultations of, and measures adopted by, future international migration forums. Consideration only of the issue of economically motivated migration (as for example in 2006 at the first Euro-African Ministerial Conference on Migration and Development in Rabat and the EU-Africa Ministerial Conference on Migration and Development in Tripoli) is not enough. WBGU recommends that Germany and the EU should commit themselves more fully in this regard. Convincing migration policy plans arising from cooperation between the EU and the important countries of origin could be propagated as models for other countries. In addition, constructive proposals need to be made for the further development of international law relating to migration and refugees.

ENSHRINING THE PROTECTION OF ENVIRONMENTAL MIGRANTS IN INTERNATIONAL LAW

Under international legal principles governing state responsibility (the International Law Commission's Draft articles on Responsibility of States for internationally wrongful acts), those who cause climate change cannot be held responsible for consequential damage arising from climate change – nor, therefore, for the costs of climate-induced environmentally induced migration. Liability could, however, arise if there were a clearly attributable infringement of the duty of non-intervention under customary international law ('causing refugees' as intervention in the internal affairs of another state; for a detailed account see Achermann, 1997). The country concerned could accordingly be found liable if the causes of environmental degradation that give rise to migration can be clearly proven. Where the conditions are met, this approach should be more rigorously utilized. However, it will in the foreseeable future be difficult to prove that a specific case of migration is caused by climate change. Since the issue of accountability cannot normally be adequately clarified, and given the current state of research relating to environmental migrants, this approach can only rarely be applied. Other options for using international law to regulate the problem of environmental migrants and enshrining the protection of environmental migrants in international law therefore need to be pursued.

There is likely to be a marked increase in environmentally induced migration on a global scale in future. Yet problems arise from the fact that environmental migrants (despite the commonly used term 'environmental refugees') do not fit into any of the standard categories of international refugee and migration law. From a legal perspective, 'flight' is defined as the cross-border movement of refugees. 'Refugees', according to Article 1 A (2) of the Geneva Refugee Convention, are people who 'owing to well-founded fear of being persecuted for reasons of race, religion, nationality, membership of a particular social group or political opinion', are outside the country of their nationality. Given the current definition of a refugee in international law, states have no specific obligations with regard to the treatment of environmental migrants and no other legal protection mechanisms exist that are of benefit to affected individuals. In the interests of improving the legal status of environmental migrants and the protection available to them, ways of remedying this omission in international law should be considered (Chemillier-Gendreau, 2006; Section 9.4.3).

WBGU recommends the establishment of an trans-sectoral, multilateral Convention to regulate the legal position of environmental migrants.

MAXIM: PERMIT NO WEAKENING OF THE EXISTING PROTECTION REGIME

However, considerable concern (Keane, 2004) attaches to any proposed enlargement of the definition of a refugee in international law – as laid down in the Geneva Refugee Convention – with the aim of extending the protection accorded under the existing international refugee regime to include environmental migrants. This might be done by, for example, adding an additional protocol to the Refugee Convention (Conisbee and Simms, 2003). There is a significant risk that this would lead to a weakening of the protection available to persecuted groups and individuals – those exposed to a concrete risk to life for political, ethnic, religious or similar reasons. In view of the particular risks faced by refugees in their country of origin, the legal privileges granted to them under the Geneva Refugee Convention are appropriate and necessary. If conditions for the granting of refugee status were formulated more broadly, there is a danger that the existing standard of protection might be lowered, perhaps significantly. It is worth noting in this connection that the Convention Governing the Specific Aspects of Refugee Problems in Africa (OAU Refugee Convention) drawn up by the Organization of African Unity (OAU) adopts a considerably broader definition of the refugee than the

Geneva Convention; in practice, however, the OAU Refugee Convention is not adhered to and the level of protection it affords is very low or in some cases non-existent (Keane, 2004; Rankin, 2005). It also needs to be borne in mind that there are particular problems associated with the legal status of environmental migrants and the obligations imposed on states in this connection; these problems do not arise with those who are refugees under the Geneva Convention. No attempt should therefore be made to modify the Refugee Convention through the inclusion of an additional protocol that would recognize 'environmental refugees' or environmental migrants as refugees under the Convention with corresponding rights to protection. It is therefore inadvisable to become involved in attempts to create a very broad definition of the term 'refugee' and the rights associated with the Geneva Refugee Convention by means of an additional protocol to that Convention.

ADOPTING SUPPLEMENTARY MEASURES OUTSIDE THE EXISTING REFUGEE REGIME

It should, however, be possible to apply the protection mechanisms of the existing international refugee regime in adapted form to the category of environmental migrants, and a link with existing refugee law should be created. The relevant provisions in international law should not, however, be enshrined in existing refugee law in such a way as to dangerously extend the existing definition of a refugee and the associated legal status.

The aim should be to draw up an interdisciplinary multilateral Convention to regulate the legal status of environmental migrants. This could be linked to the Geneva Refugee Convention. In view of the problems expected to arise from environmentally induced migration, such a protection regime should be pursued with vigour immediately. In the view of WBGU a future protection regime to benefit environmental migrants should include the following as a minimum:

- Recognition of environmental damage as a possible cause of migration.
- Specification of the conditions under which the protection regime would apply.
- Granting of at least temporary leave to remain as a form of protection for environmental migrants. By analogy with the thinking behind the principle of non-refoulement (which prevents persecuted people being moved to a country in which they are at risk of inhumane treatment), the states parties should undertake not to return environmental migrants to their country of origin if living conditions there are unacceptable.
- Protection measures to benefit internally displaced persons.
- Creation of a quantitative distribution formula which would involve the whole international community in admitting environmental migrants. In order to ensure that individual potential receiving countries are not overburdened, the hosting capacity of destination countries should be taken into account.
- Creation of an equitable formula for distributing the costs of receiving refugees among the whole international community. In order to achieve a fair and efficient distribution of costs, this distribution formula should be based on the international legal principle of common but differentiated responsibilities. According to this principle, the principal responsibility for bearing the costs lies with the countries that contribute most to causing global greenhouse gas emissions and that also have the greatest financial resources (Principle 7 of the Rio Declaration; Article 3 (1) and Article 4 (1) of the UNFCCC). At the same time, because environmentally induced migration is multi-causal, additional criteria should be defined in order to achieve a burden-sharing arrangement that is as fair as possible (Section 9.3.4).
- Establishment of an obligation on the part of industrialized and newly industrializing countries to compensate for the financial burdens associated with climate-induced environmental damage. Such payments could be viewed as compensation for actually causing damage. In accordance with the 'polluter pays' principle, these payments should be linked to a country's specific level of emissions. At the same time the 'ability-to-pay' principle could be taken into account by, for example, including GDP in the calculation.

WBGU furthermore recommends involving UNHCR as closely as possible in negotiations on the drawing up of relevant international treaties. It is also desirable that the terms of such treaties should make provision for cooperation between UNHCR and the organs of the conventions involved. The question of whether and to what extent UNHCR could assume particular responsibilities in relation to environmentally induced migration should also be considered. All work in this area should aim to ensure that existing standards for the protection of human rights laid down in the relevant conventions are implemented as effectively as possible. WBGU therefore firmly supports the recommendations made by the Global Commission on International Migration (GCIM, 2005).

WBGU's final recommendation is that efforts already being made by the United Nations to protect internally displaced people (in particular by appointing a Representative of the Secretary-General on Internally Displaced Persons) should be strengthened, with the aim of achieving fuller recognition of their human rights and ensuring that the financial costs involved are shared.

10.3.3.4
Initiative 9: Expanding global information and early warning systems

Both the gradual changes caused by climate change and the natural disasters which are expected to occur with increasing frequency and intensity could destabilize the affected regions and, in extreme cases, constitute a major risk factor for national and international security. Well-functioning global information and early warning systems can therefore do much to mitigate these adverse effects and make a major contribution to conflict and crisis prevention.

These systems should provide timely information and warning in advance of extreme events and crises. The United Nations recognised the importance of early warning systems long ago, as is apparent, for example, from its designation of the 1990s as the International Decade for Natural Disaster Reduction (IDNDR, 1990–1999), its adoption of the ensuing International Strategy for Disaster Reduction (ISDR), its staging of a series of UN conferences on early warning systems, and the launch of the new International Early Warning Programme. In WBGU's view, the recommendations made in the UN's *Global Survey of Early Warning Systems* (UN, 2006a) point in the right direction. The German government, which has been active in this area for many years, should continue to participate in the development of a global early warning system. The aim should be to establish a system which is not confined to individual risks but addresses threats to human security on a comprehensive basis. First and foremost, it should draw on existing early warning capacities and pool current knowledge in order to make it accessible as easy-to-understand situation reports and forecasts for users. The early warning system should provide information about all types of natural hazard, epidemics and technological risks, and also take account of ongoing environmental problems. Finally, WBGU recommends the establishment of early warning capacities for political crises and regional conflicts so that any escalation of violence can be avoided or conflicts contained.

The system must also provide processed data on expected regional climate impacts, especially for developing countries which lack adequate capacities of their own to model and evaluate these data. But in industrialized countries too, the systematic provision of information on regional climate impacts is still in its infancy and is in some cases obstructed by disputes over competencies or limited data access rights. Decision-makers in developing countries and development agencies alike need to integrate adaptation to climate change into their work as a cross-cutting theme and are therefore reliant on the provision of such data. The aim should be establish a database which would collate regional forecasts with all their uncertainties – ideally gleaned from ensemble modelling or from the evaluation of a variety of models – and make them accessible in an easy-to-understand format for users and affected communities. Support must also be provided for the development of partnerships for the analysis and interpretation of data in order to identify requisite actions and devise adaptation strategies. The five-year programme of work on impacts, vulnerability and adaptation to climate change, adopted by the 12th Conference of the Parties (COP 12) to the UNFCCC, offers a good starting point for this process.

In order to establish this type of global information and early warning system, the activities of existing UN institutions and conventions (e.g. WMO, FAO, UNDP, UNEP, UNFCCC) and other forums such as ISDR or IPCC must be properly coordinated. This broad-based early warning system should contribute to the development, in all regions of the world, of a comprehensive knowledge base on anticipated climate impacts and present and future threats and risks, and would be a valuable resource for the various national organizations as they develop adaptation and mitigation measures.

In the further development of early warning and information systems on natural disasters and crises at national and local level, a key task is to ensure that the information is passed promptly to potentially affected communities and to national and international emergency response agencies. It is also important to establish and trial information and decision-making structures within the relevant administrations. There are numerous examples which show that the conversion of early warning information into generally accessible and culturally adapted guidance on how to respond in the event of a disaster is often a weak point in early warning. The timely involvement of potentially affected communities is also often neglected, and long-term education and training programmes are required in at-risk areas. Finally, people and organizations in the affected areas should also be given information on who is responsible for carrying out specific measures in the event of a disaster warning, and how affected individuals and communities should respond.

10.3.4
Securing the financing of the initiatives

The mitigation of environmentally induced security risks not only requires resolute political action by the relevant national and international actors, but also adequate financial resources to implement the required measures. Funding is needed, first and foremost, for strategies to mitigate security risks arising in the context of ongoing climate change. In addition to climate-related strategies, funding must also be mobilized for the international community's general conflict prevention activities in order to ensure that Initiatives 1 and 2 can be implemented effectively on a sound financial basis (Table 10.3-2).

The wide range of fields of action identified and the various policies required make it clear that very substantial funding is needed in order to respond effectively to climate-induced security risks. Admittedly, there are no precise estimates of the funding required for the individual initiatives (further research is needed here; see Chapter 9), but it is plausible to assume that cumulatively and adjusted for inflation, several million millions (10^{12}) euros will be required to 2100. Existing funding mechanisms such as the GEF, which finances mitigation and adaptation strategies in developing countries, are quite inadequate, given the scale of the challenges. The gap between the funding requirement and the financial resources available now and in the near future must be closed at national and international level through transfer payments and the restructuring of existing budgets.

WBGU has made recommendations on existing and new mechanisms to fund investment in mitigation and adaptation in many of its previous reports. These mechanisms are summarized below, assigned to the various initiatives according to their practicability, and supplemented with appropriate instruments to fund international cooperation and conflict prevention activities.

10.3.4.1
Avoiding dangerous climate change

If a climate policy were agreed and implemented worldwide that limited the rise in globally averaged near-surface air temperature to 2 °C relative to the pre-industrial value, the associated mitigation costs would entail a reduction of annual global GDP by around 1.5 per cent against the reference case during the next one hundred years (WBGU's calculations based on various scenarios; WBGU, 2003). These costs are still far lower than the costs of climate damage resulting from inaction (WBGU, 2007). Climate protection is therefore worthwhile from a macroeconomic perspective, although the costs of climate-induced security risks are not yet included explicitly and in full in the damage assessments. As the costs of the relevant measures arise today but the benefits are not felt for several decades, there is no guarantee – despite the obvious advantages of mitigation strategies – that the necessary investments will be made promptly or, indeed, at all. International coordination is therefore required in order to ensure that adequate financial resources are available for the requisite investment in mitigation programmes.

TRANSFORMING ENERGY SYSTEMS WORLDWIDE
There are currently no sufficiently precise estimates which show how the required funding can be satisfactorily estimated. One thing is certain, however: the transformation of energy systems worldwide should be a key funding priority, as reflected in Initiatives 4 and 5. This transformation is especially relevant in the developing countries, where almost 1.6 thousand million people currently have no access to electricity (IEA, 2002). Sustainable energy systems would close this gap and would make a major contribution not only to mitigating climate change but also to fostering economic development and reducing poverty (WBGU, 2005). This could also help reduce vulnerability to the impacts of climate change and offer a preventive response to climate-induced security risks.

In order to initiate the development and/or transformation of energy systems in these relatively capital-poor countries, however, financial assistance from the international community is essential. Existing funds, such as the World Bank Carbon Finance Unit and the GEF (World Bank, 2005c; GEF, 2005), should be strengthened with the injection of better and more reliable funding. In the further development of the Clean Development Mechanism, too, there should be a greater focus on sustainable energy systems. Additional sources of income can be harnessed though new financing instruments such as the introduction of emissions-dependent user charges for international aviation and shipping (WBGU, 2002), unless these emissions are already covered by other regulatory schemes (e.g. a national levy on airline tickets; inclusion of aviation in the EU Emission Trading Scheme). In the longer term, a system of internationally tradable quotas for renewable energies can also generate revenue (WBGU, 2004). Financial resources can also be mobilized by restructuring existing budgets. For example, two-thirds of the energy subsidies in the EU-15 (approx. € 22 thousand million per annum) still go to support fossil fuel production and consumption (WBGU, 2007). Just one-sixth of the EU's energy subsidies currently goes to support renewable energies (EEA, 2004). This situa-

Table 10.3-2
Overview of the instruments proposed by WBGU to fund the initiatives.
Source: WBGU

Relevant areas of funding	Funding priorities and sources of finance
Section 10.3.4.1 Avoiding dangerous climate change	Funding of mitigation measures, especially the transformation of energy systems worldwide and the conservation of terrestrial carbon stocks through appropriate international funds (World Bank, GEF); user charges for shipping and aviation; revenue from auctions of emissions certificates or restructuring of subsidies.
Section 10.3.4.2 Adapting to unavoidable climate change	Funding of adaptation measures, especially through development cooperation (boosting funding for official development assistance (ODA); expanding the UNFCCC adaptation funds; strengthening microfinance); establishing an environmentally induced migration fund.
Section 10.3.4.3 International conflict prevention	Financing of international conflict prevention measures, especially through an integrated approach to the financing of crisis prevention, development cooperation and military spending; strengthening the financial institutions in the UN system (supporting the Central Emergency Response Fund, strengthening the UN Trust Fund for the Consolidation of Peace).

tion is incompatible with a sustainable energy policy, so subsidies for fossil fuels and nuclear energy should be gradually reduced and, if appropriate, diverted to promoting renewables.

To facilitate a successful transition away from fossil fuels towards renewables, a massive increase in research into renewables is required. In WBGU's view, a ten-fold increase in direct public spending on research and development in the energy sector by 2020 (against the mean value for 1990–1995) in the industrialized countries is entirely justified (WBGU, 2007). The ensuing advances in the field of renewables and efficiency of energy supply systems should be shared with developing countries as well through development and technical cooperation.

CONSERVING TERRESTRIAL CARBON STOCKS
Alongside the transformation of energy systems worldwide, the protection of terrestrial carbon stocks, especially forests, should be a further funding priority. A large proportion of this global forest stock is located in developing countries, but is under threat from non-sustainable use and forest clearance. The industrialized countries should actively promote the conservation of these forests. The UNFCCC process to reduce deforestation in developing countries offers a good starting point and should be pursued as a matter of urgency. In particular, the Annex I countries under the UNFCCC regime should provide incentives, in the form of financial compensation for loss of income from alternative land use, to encourage these countries to refrain from clearing their forests. As a possible mechanism for the payment of compensation, WBGU has proposed the concept of compensation for abstaining from using an environmental resource – in short: non-utilization obligation payments (WBGU, 2002; Kulessa and Ringel, 2003).

10.3.4.2
Adapting to unavoidable climate change

In order to mitigate the security risks triggered by climate change, there is an urgent need to develop a globally coordinated adaptation strategy based on precautionary considerations. This strategy should address the impacts of climate change on the basis of compliance with the 2°C guard rail, as this is a realistic and acceptable target for mitigation policy. The required funding implied by such a strategy is currently difficult to quantify precisely. The Stern Review (2006) highlights the problems associated with attempting such estimates, but also suggests that the funding requirement for adaptation measures has so far been underestimated. For example, Stern concludes that the investment costs of making new infrastructure and buildings resilient to climate change in OECD countries could increase baseline investment costs by as much as 10 per cent, i.e. up to US$150 thousand million annually in these sectors alone. In developing countries which will be hardest hit by the impacts of climate change, the cost of making new buildings and infrastructure more resilient to climate change through effective adaptation measures will entail additional costs of as much as 20 per cent (Stern, 2006). These estimates are predicated on global temperature rises of 3–4°C, i.e. levels at which the need for adaptation is higher than with the climate change postulated as unavoidable for the purpose of this report. On the other hand, these estimates are imprecise in that they do not take account of all the affected sectors. It can therefore be assumed that global warming of just 2°C is also likely to require thousand millions of dollars of additional investment in many countries of the world.

Many developing countries generally contribute very little to anthropogenic climate change, but they will still have to adopt comprehensive adaptation measures which they often cannot afford due to a lack of capital. In line with the principles of solidarity and ability-to-pay and also the 'polluter pays' principle, the international community has an obligation to provide at least some of the requisite financial resources to pay for adaptation measures. An international compensation and adaptation regime is required, whose primary task is to generate adequate funding to compensate for climate damage and to finance adaptation strategies in developing countries. In this context, the question as to what amount of money is required must also be answered. In WBGU's view, there is an urgent need for research, not only to answer this question but also to develop an appropriate strategy for adaptation to unavoidable climate change (Chapter 9).

FINANCING VIA DEVELOPMENT COOPERATION

- *Increasing funding for official development assistance (ODA):* For the financing of adaptation measures in developing countries, it is important not to lose sign of the real goal of development cooperation: economic and social development is still the best adaptation strategy. Development generally increases a country's adaptive capacities, reduces its vulnerability to the impacts of climate change (WBGU, 2005) and thus makes a contribution, not least, to mitigating environmentally induced security risks. WBGU estimates the international funding requirement for a coordinated strategy comprising poverty reduction and environmental policy and aimed at the ecologically sustainable attainment of the Millennium Development Goals as being US$200–400 thousand million per year (WBGU, 2005). An important contribution to safeguarding the requisite financial resources could be made by boosting official development assistance (including the resources for debt relief for the least developed countries). In 2006, ODA amounted to around US$104 thousand million, i.e. 0.3 per cent of the combined gross national income (GNI) of the industrialized countries belonging to the OECD-DAC, which means that the funding of official development assistance (ODA) is still failing to reach the target of 0.7 per cent of gross national income agreed by the United Nations (OECD, 2007b). In May 2005, the European Union's development ministers set a new intermediate target for development aid of 0.56 per cent of donor countries' gross national income by 2010, which would put Europe on course to reach the UN's 0.7 per cent target by 2015 (EU-Kommission, 2006). This timetable must be rigorously adhered to in order to signal a measure of financial security for recipient countries and thus provide incentives for their own investment efforts. In light of the major problems faced, WBGU therefore also proposes that a 1 per cent target be adopted in the long term (WBGU, 2005). The current efforts by donor countries and multilateral donor organizations, especially the EU and Germany, to implement the principles of the Paris Declaration on Aid Effectiveness in practice point in the right direction (BMZ, 2005). In the context of climate-induced adaptation requirements in particular, new funding priorities should also be set and additional financial resources generated. This must include a shift of resources away from emergency relief towards prevention, albeit without jeopardizing the reliable provision of emergency relief.

- *Expanding the UNFCCC adaptation funds:* The funds so far established under the UNFCCC and the Kyoto Protocol, which aim to provide direct or indirect support for adaptation to climate change in developing and newly industrializing countries, are inadequate to meet the challenges described above, both in terms of their volume and their institutional structures. Before a comprehensive adaptation strategy is launched, the financing of these funds must, at the least, be established on a sound footing. WBGU recommends that the financial contributions made by individual states should be based on their contribution to global warming, i.e. their greenhouse gas emissions, and their economic capacities. In assessing the financial contribution to be made on the basis of the 'polluter pays' principle, however, only emissions produced since 1990 should be taken into account, as it is only since this point in time, when the IPCC's First Assessment Report was published (IPCC, 1990), that the international community can be said to have developed an adequate awareness of the climate problem. In specific terms, WBGU recommends that more resources should be made available to the Least Developed Countries Fund, which supports developing countries in their endeavours to prepare and implement National Adaptation Programmes of Action, and the adaptation 'window' of the Special Climate Change Fund, which augments the financing of adaptation projects through other measures. The still unresolved institutional issues relating to the Adaptation Fund established under the Kyoto Protocol should be addressed at the 13th Conference of the Parties. Furthermore, the option for additional contributions to be paid to boost the Adaptation Fund, which is initially being funded by surcharges on CDM projects, should also be utilized. Boosting the UNFCCC Funds is part of the financing of development cooperation. The financial resources

provided in the Funds should be new and additional; in other words, they should not be generated through the redistribution of existing development funds.
- *Strengthening microfinance:* WBGU has on many previous occasions advocated the expansion of microfinancing institutions with resources from international development cooperation, and reiterates the importance of these institutions for the developing countries, also in relation to environmentally induced security risks (WBGU, 2005, 2006). Microfinancing instruments such as microcredits or microinsurance have the potential to make a major contribution to poverty reduction and thus to decreasing the vulnerability of populations in developing countries. Against the background of growing environmental risks resulting from climate change, microinsurance in particular could play an increasingly important role. For example, index-based microinsurance is promising as an instrument to guard against weather-related damage in agricultural production (CGAP, 2006; UNEP, 2006; Churchill, 2006). However, despite experts' great hopes of microinsurance as a financing instrument to guard against natural disasters, it is important to emphasize that microinsurance cannot replace – but at best can only supplement – multilateral financial assistance.

ESTABLISHING A FUND FOR ENVIRONMENTALLY INDUCED MIGRATION

Environmental migrants will become increasing relevant in security policy terms, as discussed in Section 6.5. The management of environmentally induced migration should therefore be a further funding priority (Initiative 8). The International Dialogue on Migration launched by the International Organization for Migration (IOM) in 2001 offers an appropriate platform for fair and efficient burden-sharing between countries affected by environmentally induced migration and those which are not (IOM, 2001). Burden-sharing should satisfy the 'polluter pays' principle, described above, and the 'ability-to-pay' principle by linking contributions to the Fund to the level of country-specific greenhouse gas emissions and other indicators such as GDP per capita.

10.3.4.3
International conflict prevention

In Section 8.3.6, WBGU argues that climate-induced social destabilization in various regions of the world will be almost impossible to resolve through classic, military-based security policy. Therefore as a supplement to strategies which focus on climate change mitigation and adaptation to unavoidable climate change, international conflict prevention strategies should constitute a third funding priority (Initiatives 1 and 2). Initiative 7 focusses on the importance of prevention strategies and the need to restructure, and if necessary increase, existing military budgets to meet new security policy challenges. Key areas of activity in international conflict prevention include measures aimed at averting violent conflicts, e.g. through civilian crisis prevention, peacebuilding, security sector reform or the stabilization of fragile states. International humanitarian assistance, notably the provision of essential goods to victims of violent conflicts and natural disasters and protection from further persecution, is also important (Reinhardt and Rolf, 2006).

There are overlaps between the strategies and measures deployed in these two fields of action. In both areas, civilian activities are to some extent reliant on military backing and are funded from official development assistance (ODA). Based on the findings of the previous chapters, it is likely that both fields of action will face increasing climate-induced security risks in future. Alongside the key agencies within the UN system, the World Bank and the International Monetary Fund, the donor community includes individual states which carry out projects within the framework of their bilateral development cooperation (OECD, 2001; Raffinot and Rosellini, 2006). In the context of humanitarian assistance in particular, civil society actors – such as the Red Cross/Green Crescent or NGOs – also play a role as donors (Development Initiatives, 2006).

As a prerequisite for the efficient financing of conflict prevention, adequate resources must be provided by the donor community. At the same time, it is essential to avoid funding gaps in security-relevant emergency action or in measures adopted in post-conflict processes, for example. Adequate funding should also be provided for institutions which help to ensure that the available financial resources are deployed efficiently in relief and peacebuilding processes.

ADOPTING AN INTEGRATED APPROACH TO THE FINANCING OF CRISIS PREVENTION, DEVELOPMENT COOPERATION AND MILITARY SPENDING

The funding that is required for civilian crisis prevention and international humanitarian assistance is almost impossible to quantify precisely. Due to the clear substantive overlaps between civilian crisis prevention and development cooperation, WBGU takes the view that there is no need for an additional funding target for crisis prevention. Instead, the political focus should be geared entirely towards achieving the 0.7 per cent ODA target. In moving towards this target, however, it is important to ensure that international spending on crisis prevention is not factored into official development assistance to an excessive extent. Instead, what is required is the expansion of the official definition of development cooperation provided by the OECD Development Assistance Committee (OECD-DAC) in order to apply a restrictive approach to security-relevant tasks of a military nature (Brzoska, 2006).

There is currently a major gap between the 0.7 per cent ODA target and the development assistance actually being provided, with the industrialized countries belonging to the OECD-DAC allocating only 0.3 per cent of their combined gross national income for development (OECD, 2007b). In 2006, ODA flows totalled US$107 thousand million, compared with global military spending of US$1,118 thousand million in 2005, of which the USA accounted for around US$500 thousand million and Germany around US$35 thousand million. What's more, the developing and newly industrializing countries spend more than US$160 thousand million on the military (SIPRI, 2006). In view of this discrepancy, there have long been calls for military budgets to be radically restructured in favour of development cooperation (Brandt, 1980; UNDP, 1994; Büttner and Krause, 1995). In this context, WBGU proposes that security spending be critically reviewed, especially as regards its effectiveness for international peacebuilding, and adjusted accordingly. The German government should drive forward the international debate and negotiation processes on this issue within the EU, NATO and in the wider arena. For example, the Human Security Doctrine for Europe produced by the Study Group on European Security Capabilities (2004) calls for the reallocation of existing procurement spending away from traditional military equipment such as heavy tanks and artillery towards the smart manpower and equipment needed by a Human Security Response Force which is better able to respond to new security threats.

As military spending is realigned towards a security policy based on crisis prevention, the need for funding in the 'classic' areas of military spending should be reduced. This applies to the industrialized and the developing countries alike. The resources freed up in this way should be channelled into preventive measures, including development projects. The financial burdens on civilian crisis prevention and development cooperation could also be eased to some extent through the non-military use of former military infrastructures. However, experience with conversion projects has shown that conversion entails significant costs. To ensure that conversion genuinely helps to alleviate financial burdens in the long term, efficient conversion management is essential (Brzoska et al., 1995).

STRENGTHENING THE FINANCIAL INSTITUTIONS IN THE UN SYSTEM

International crisis prevention and peacebuilding regimes are well-established at UN level. They coordinate and supplement or replace the activities of individual states. However, the mechanisms to finance these regimes, which could also take action in the event of climate-induced crises, are inadequate, in WBGU's view.

- *Supporting the Central Emergency Response Fund:* The Central Emergency Response Fund (CERF) was established within the United Nations Office for the Coordination of Humanitarian Affairs (OCHA) in order to provide quick initial funding for life-saving assistance at the onset of humanitarian crises such as food crises (Section 6.3) or floods (Section 6.4) and thus prevent social destabilization and resulting violence. The objective of the Fund is to provide the resources for early action and response during crises in order to save lives. UN agencies such as UNICEF, WHO and FAO as well as the International Organization for Migration are able to access the Fund. To date, the Fund's annual spending has fluctuated between US$5–60 million. In 2005, the UN General Assembly voted to increase the resources available to the Fund from the original figure of US$50 million to $500 million, to be derived mainly from voluntary contributions from the UN member states as well as from the private sector. By the end of 2006, a total of US$275 million had been paid into the enhanced Fund. Germany is contributing to the Fund for the first time in 2007, but compared with major donors such as the United Kingdom (US$83.2 million) or Norway (US$57 million), its contribution for this year is relatively small, i.e. US$6.6 million (OCHA, 2006). In order to do justice to its role as an active partner in the international climate and security debate, the German government should support the Central Emergency Response Fund with appropriate contributions and lobby for a binding schedule for the financing of this Fund.
- *Strengthening the UN Trust Fund for the Consolidation of Peace:* In 2005, the UN General Assembly voted to create a Peacebuilding Commission in order to assist countries making the transition from war to peace. The work of the Commission is to be supported by a UN Trust Fund for the Consolidation of Peace, funded from voluntary contributions. The Trust Fund will also provide resources as swiftly as possible to introduce peace consolidation processes and reconstruction (UN, 2005). In view of the importance of these tasks, the German government should fulfil its responsibility in the international community by taking an active

role in financing the UN Trust Fund for the Consolidation of Peace and lobby for the adoption of rules to ensure regular contributions to the Fund in future.

10.4
Window missed – mitigation failed: Strategies in the event of destabilization and conflict

If, by around 2020, political efforts to obey the 2 °C guard rail through effective emissions reductions fail, the international community must prepare itself to deal with climate-related conflicts such as those described in the 'conflict constellation' scenarios in Chapter 6. In any event, a pro-active climate protection policy must remain in place with the aim of keeping global warming as close to the 2 °C guard rail as possible. At the same time, however, strategies for adaptation to unavoidable climate change must be intensified and reoriented towards the type of future that can then be expected. The greater the delay in commencing efforts to mitigate climate change and adapt to its impacts, the more expensive such efforts will become. Development in line with the negative scenario resulting from missed opportunities to protect the climate would thus entail far higher costs than a reference scenario in which compliance with the 2 °C guard rail is achieved (Stern, 2006; WBGU, 2007). The challenges possibly associated with this type of negative scenario are outlined below, both for the medium term (2020–2040) and the long term (to the end of the century).

Future climate policy measures would need, as a matter of principle, to be stepped up in line with the recommendations for current action, although the conditions pertaining once the window of opportunity is missed will be far more difficult than they are today. For example, lock-in effects resulting from entrenched structures in conventional energy systems are likely to obstruct willingness to undertake reforms of energy and economic policy to an even greater extent than is the case today, driving up the costs of a transformation of energy systems. External trade policy would be forced to develop strategies to cope with the destabilization of the global economy resulting from climate change.

In the field of development policy, the need for preventive measures to mitigate water and food crises and storm and flood disasters would substantially increase. Moreover, the growing number of weak and fragile states would result in the absorption of substantial development capacities. In the long term, a significantly degraded natural environment, resulting from climate change, would also be likely to result in development cooperation focussing more and more on stabilization and adaptation strategies in order to prevent human development from dropping back, rather than advancing human development as is currently the case.

Linked with this, the challenges posed to global management of migration would increase, and would absorb considerable political, economic and legal capacities. This would apply to internal migration within the affected countries, to migration between the developing regions most affected by climate change, and to migration from developing to industrialized countries. Against this backdrop, efforts to devise a solution for environmental migrants in international law would, in practice, be essential.

Unless climate change can be abated within a reasonable period of time, major disruptions in international relations must be expected, not least in the North-South context. In order to prevent crisis management from descending into chaos and the breakdown of the international system, the world's crisis management capacities and multilateral institutional architecture would have to be strengthened; in particular, the crisis management potential of the world's leading powers – the USA, China, India and Europe – would need to be pooled.

Not least, in the event of climate policy failing, substantially increased financial resources would be required – e.g. within the framework of an international compensation and adaptation regime (WBGU, 2007) – to mitigate the adverse effects of climate change in developing countries and to prevent the escalation of conflicts between the main drivers of climate change and the countries hardest hit by it. In short, besides resolute mitigation policies, a need would arise during the period 2020–2040 for strategies to prepare for climate-induced challenges and critical escalations in various regions of the world and in the international system.

If these efforts also fail, it is likely that, from the mid 21st century onwards, crisis management on a global dimension will have to contend with the proliferation of local and regional conflicts and the destabilization of the international system, threatening global economic development and overwhelming global governance structures. The political scope for peaceful management and mitigation of conflicts would narrow steadily, while the costs of crisis response and adaptation to climate change would escalate. In order to avoid these dangerous developments, climate policy must be put on track now.

References

Abel, W (1974) *Massenarmut und Hungerkrisen im vorindustriellen Europa: Versuch einer Synopsis.* Parey, Hamburg, Berlin.

Achermann, A (1997) *Die völkerrechtliche Verantwortlichkeit fluchtverursachender Staaten: ein Beitrag zum Zusammenwirken von Flüchtlingsrecht, Menschenrechten, kollektiver Friedenssicherung und Staatenverantwortlichkeit.* Nomos, Baden-Baden.

ACIA (Arctic Climate Impact Assessment) (2005) *Arctic Climate Impact Assessment.* Cambridge University Press, Cambridge, New York.

ACT (Action by Churches Together International) (2000) 'Dateline ACT India, Orissa Cyclone 01/00: "When the Ocean emptied ...".' ReliefWeb website, http://www.reliefweb.int/rw/rwb.nsf/db900SID/ACOS-64D6ZB?OpenDocument&cc=ind&rc=3 (viewed 27. January 2006).

Adams, R M, Fleming, R A, Chang, C C, McCarl, B A and Rosenzweig, C (1995) 'A reassessment of the economic effects of global climate change in on U.S. agriculture'. *Climate Change* **30**: 147–67.

ADB (Asian Development Bank) (2005) 'Asia Water Watch 2015. Are Countries in Asia on Track to Meet Target 10 of the Millennium Development Goals?' Asian Development Bank website, http://www.adb.org/Documents/Books/Asia-Water-Watch/default.asp (viewed 21. March 2007).

ADB (Asian Development Bank) (2006a) *Regional Cooperation Strategy and Program South Asia 2006–2008.* Asian Development Bank, Manila.

ADB (Asian Development Bank) (2006b) 'Bangladesh – Key Indicators'. ADB website, http://www.adb.org/Bangladesh/default.asp (viewed 3. June 2006).

ADB (Asian Development Bank) (2007) 'Resettlement Plans'. ADB website, http://www.adb.org/ressetlement/plans.asp (viewed 27. March 2007).

Addison, T and Murshed, M (2003) 'Explaining violent conflict: going beyond greed versus grievance'. *Journal of International Development* **15**(4): 391–6.

Adler, E and Barnett, M (1998) 'A framework for the study of security communities'. In Adler, E and Barnett, M (eds) *Security Communities.* Cambridge University Press, Cambridge, MA, pp29–65.

AFP (Agence France Press) (2000) 'Floods in Eastern India Trigger Food Riots'. ReliefWeb website, http://www.reliefweb.int/rw/rwb.nsf/db900SID/ACOS-64CQ3X?OpenDocument&cc=ind&rc=3 (viewed 27. June 2006).

Ahmad, M and Wasiq, M (2004) *Water Resource Development in Northern Afghanistan and Its Implications for Amu Darya Basin.* World Bank Working Paper No. 36. World Bank, Washington, DC.

AKUF (Arbeitsgemeinschaft Kriegsursachenforschung) (2007) 'Aktuelle Kriege und bewaffnete Konflikte. Liste der Kriege und bewaffneten Konflikte 2005'. AKUF, University Hamburg website, http://www.sozialwiss.uni-hamburg.de/publish/Ipw/Akuf/kriege_aktuell.htm#Liste (viewed 7. May 2007).

Alcamo, J, Flörke, M and Märker, A (2007) 'Future long-term changes in global water resources driven by socio-economic and climatic changes'. *Hydrological Sciences Journal* **52**(2): 247–75.

Alden, C (2005) 'The new diplomacy of the South: South Africa, Brazil, India and trilateralism'. *Third World Quarterly* **26**(7): 1077–95.

Ali, A (1999) 'Climate change impacts and adaptation assessment in Bangladesh'. *Climate Research* **12**: 109–16.

Ali, S S (2006) 'Fencing the Porous Bangladesh Border'. Worldpress website, http://www.worldpress.org/print_article.cfm?article_id=2723&dont=yes (viewed 6. March 2007).

Alpermann, B (2004) 'Dimensionen sozialer Probleme in der VR China – regionale und sektorale Facetten'. In Kupfer, K (ed) *"Sprengstoff China?" Dimensionen sozialer Herausforderungen in der Volksrepublik. Volume 17.* Asienhaus, Essen, p18.

Altvater, E (2003) '"Menschliche Sicherheit" – ein friedenspolitischer Begriff'. University Kassel website, http://www.uni-kassel.de/fb5/frieden/themen/Theorie/altvater.html (viewed 12. January 2006).

AMCEN (African Ministerial Conference on the Environment) and UNEP (United Nations Environment Programme) (2006) *Africa Environment Outlook 2: Our Environment, Our Wealth.* UNEP, Nairobi.

Amineh, M P (2006) 'Die Politik der USA, der EU und Chinas in Zentralasien'. *Aus Politik und Zeitgeschichte* **4**: 11–8.

Amnesty International (2001) 'Jahresbericht 2001'. Amnesty International Deutschland website, http://www2.amnesty.de/internet/deall.nsf/WJahresberichtAll?OpenView&Start=1&Count=200&Expand=5#5 (viewed 11. May 2006).

Arnell, N W (2004) 'Climate change and global water resources: SRES emissions and socio-economic scenarios'. *Global Environmental Change* **14**: 31–52.

Arnell, N W (2006) 'Climate change and water resources: a global perspective'. In Schellnhuber, H J, Cramer, W, Nakicenovic, N, Wigley, T and Yohe, G (eds) *Avoiding Dangerous Climate Change.* Cambridge University Press, Cambridge, New York, p23.

Arrow, K J and Fisher, A C (1974) 'Environmental preservation, uncertainty, and irreversibility'. *Quarterly Journal of Economics* **88**: 312–9.

Aspinall, E (2005) *The Helsinki Agreement: A More Promising Basis for Peace in Aceh? Policy Studies 20.* East-West Center, Washington, DC.

Asshoff, R, Zotz, G and Körner, C (2006) 'Growth and phenology of mature temperate forest trees in elevated CO_2'. *Global Change Biology* **12**: 848–61.

Auf der Heide, E (2004) 'Common misconceptions about disasters: panic, the "Disaster Syndrome," and looting'. In

11 References

O'Leary, M (ed) *The First 72 Hours: A Community Approach to Disaster Preparedness*. Universe Publishing, Lincoln, pp340–80.

Avery, W P and Rapkin, D P (1986) 'World markets and political instability within less developed countries'. *Cooperation and Conflict* **21**(2): 99–117.

Bächler, G (1998) *Zivile Konfliktbearbeitung in Afrika. Grundelemente für die Friedensförderungspolitik der Schweiz*. Working Paper 27. Swiss Peace, Bern.

Bächler, G and Schiemann-Rittri, C (1994) 'Umweltflüchtlinge als Konfliktpotential'. In Bächler, G (ed) *Umweltflüchtlinge: das Konfliktpotential von morgen?* agenda-Verlag, Münster, pp7–35.

Bächler, G V, Böge, S, Kötzli, S, Libiszewski, S and R., S K (1996) *Kriegsursache Umweltzerstörung. Ökologische Konflikte in der Dritten Welt und Wege ihrer friedlichen Bearbeitung. Volume 1*. Rüegger, Chur, Zurich.

Bächler, G and Spillmann, K R (1996a) *Kriegsursache Umweltzerstörung. Länderstudien von externen Experten. Volume 3*. Rüegger, Chur, Zurich.

Bächler, G and Spillmann, K R (1996b) *Kriegsursache Umweltzerstörung. Regional- und Länderstudien von Projektmitarbeitern. Volume 2*. Rüegger, Chur, Zurich.

Bade, K J (1996) *Migration – Ethnizität – Konflikt: Systemfragen und Fallstudien*. Schriften des Instituts für Migrationsforschung und Interkulturelle Studien (IMIS) der Universität Osnabrück. Volume 1. Universitätsverlag Rasch, Osnabrück.

Bamberger, R L and Kumis, L (2005) *Oil and Gas: Supply Issues After Katarina. CRS Report for the Congress*. Congressional Research Service (CRS), Washington, DC.

Bangert, Y, Delius, U, Geesmann, J, Meyer, S, Reinke, S and Veigt, K (2006) *Die Arktis schmilzt und wird geplündert. Indigene Völker leiden unter Klimawandel und Rohstoffabbau*. Menschenrechtsreport No 44. Gesellschaft für Bedrohte Völker, Göttingen.

Barends, F B J, Brouwer, F J J and Schroder, F H (1995) *Proceedings of the Fifth International Symposium on Land Subsidence. The Hague, Netherlands, 16–20 October 1995*. International Association of Hydrological Sciences (IAHS), Balkema, Rotterdam.

Barker, T, Qureshi, M S and Köhler, J (2006) *The Costs of Greenhouse Gas Mitigation with Induced Technological Change: A Meta-Analysis of Estimates in the Literature*. Tyndall Centre for Climate Change Research, Cambridge, UK.

Barkmann, U B (2006) 'In der Mongolei ist ein wachsendes Misstrauen zu beobachten. Das Reich der Mitte profitiert von den Bodenschätzen'. *Das Parlament* **56**(30/31): 9–10.

Barnett, J (2000) 'Destabilizing the environment–conflict thesis'. *Review of International Studies* **26**(2): 271–88.

Barnett, T P, Adam, J C and Lettenmaier, D P (2005) 'Potential impacts of a warming climate on water availability in snow-dominated regions'. *Nature* **438**: 303.

Bauer, S (2006) 'Land and water scarcity as drivers of migration and conflicts?' *Entwicklung und ländlicher Raum* **4**: 5–6.

Bauer, S and Biermann, F (2007) 'Aktenzeichen Weltumweltpolitik: Die Rolle internationaler Verwaltungsapparate in der Bearbeitung grenzüberschreitender Umweltprobleme'. In Jacob, K, Biermann, F, Busch, P-O and Feindt, P H (eds) *Politik und Umwelt. PVS-Sonderheft 38*. VS, Wiebaden.

Bauernfeind, W and Woitek, U (1999) 'The Influence of climatic change on price fluctuations in Germany during the 16th century price revolution'. *Climatic Change* **43**: 303–21.

Behringer, W (1999) 'Climatic change and witch–hunting: the impact of the little ice age on mentalities'. *Climatic Change* **43**: 335–51.

Bender, D, Berg, H and Cassel, D (Hrsg.) (2003) *Vahlens Kompendium der Wirtschaftstheorie und Wirtschaftspolitik. Volume 1*. Vahlen, Munich.

Berdal, M (2003) 'The UN Security Council: ineffective but Indispensible'. *Survival* **45**(2): 7–30.

Berdal, M and Malone, D M (Hrsg.) (2000) *Greed and Grievance: Economic Agendas in Civil Wars*. Lynne Rienner, Boulder, CO.

Berenskoetter, F S (2005) 'Mapping the mind gap: a comparison of US and European security strategies'. *Security Dialogue* **36**(1): 71–92.

Berkner, A (2000) *Braunkohlenplanung und Umsiedlungsproblematik in der Raumordnungsplanung Brandenburgs, Nordrhein-Westfalens, Sachsens und Sachsen-Anhalts*. ARL-Arbeitsmaterial Volume 265. Akademie für Raumforschung und Landesplanung, Hannover.

Berkner, A (2001) 'Braunkohlenbergbau und Siedlungsentwicklung in Mitteldeutschland. Gratwanderung zwischen Aufschwung, Zerstörung und neuen Chancen'. In Dachverein Mitteldeutsche Straße der Braunkohle (ed) *Braunkohlenbergbau und Siedlungen. Protokollband der Fachtagung "Braunkohlenbergbau und Siedlung"*. Dachverein Mitteldeutsche Straße der Braunkohle, Leipzig, pp8–19.

Bertelsmann Stiftung (2003) 'Den Wandel gestalten – Strategien der Entwicklung und Transformation. Ländergutachten 2003. Peru'. Bertelsmann Stiftung website, http://bti2003.bertelsmann-transformation-index.de/fileadmin/pdf/laendergutachten/lateinamerika_karibik/Peru.pdf (viewed 15. February 2007).

Biermann, F (2001) 'Umweltflüchtlinge. Ursachen und Lösungsansätze'. *Aus Politik und Zeitgeschichte* **B12**: 24–9.

Biermann, F (2005) 'Between the USA and the South: strategic choices for European climate policy'. *Climate Policy* **5**: 273–90.

Biermann, F and Bauer, S (2004) 'UNEP and UNDP. Expertise for the WBGU Report "World in Transition: Fighting Poverty Through Environmental Policy"'. WBGU website, http://www.wbgu.de/wbgu_jg2004_ex02.pdf

Biermann, F and Bauer, S (2005) *A World Environment Organisation. Solution or Threat to Effective International Environmental Governance?* Global Environmental Governance Series. Ashgate, Aldershot, UK.

Biermann, F, Petschel-Held, G and Rohloff, C (1998) 'Umweltzerstörung als Konfliktursache? Theoretische Konzeptualisierung und empirische Analyse des Zusammenhangs von "Umwelt" und "Sicherheit"'. *Zeitschrift für Internationale Beziehungen* **5**(2): 273–308.

Bigagaza, J, Abong, C and Mukarubuga, C (2002) 'Land scarcity, distribution and conflict in Rwanda'. In Lind, J and Sturman, K (eds) *Scarcity and Surfeit – The Ecology of Africa's Conflicts*. Institute for Security Studies, Pretoria, pp51–84.

BIS (Bank for International Settlements) (2006) *BIS Quarterly Review – March 2006 – International Banking and Financial Market Developments*. BIS, Basel.

Black, R (2001) *Environmental Refugees: Myth or Reality? UNHCR Working Paper No. 34*. University of Sussex, Sussex, UK.

Black, R and Gent, S (2006) 'Sustainable return in post-conflict contexts'. *International Migration* **44**(3): 15–38.

BMBF (Bundesministerium für Bildung und Forschung) (2007) *Forschung für die zivile Sicherheit. Programm der Bundesregierung*. BMBF, Bonn, Berlin.

BMU (Bundesministerium für Umwelt, Naturschutz und Reaktorsicherheit) (2007) *Time to Adapt: Climate Change and the European Water Dimension. Conclusions From the International Symposium, 12–14 February 2007*. BMU, Berlin, Bonn.

BMVg (Bundesministerium für Verteidigung) (2006) *Weißbuch zur Sicherheitspolitik Deutschlands und zur Zukunft der Bundeswehr*. BMVg, Berlin.

BMZ (Bundesministerium für wirtschaftliche Zusammenarbeit und Entwicklung) (2005) *Mehr Wirkung erzielen: Die Ausrichtung der deutschen Entwicklungszusammenarbeit auf die Millenniums-Entwicklungsziele. Die Umsetzung der Paris Declaration on Aid Effectiveness*. BMZ-Spezial Nr. 130. BMZ, Berlin.

BMZ (Bundesministerium für wirtschaftliche Zusammenarbeit und Entwicklung) (2006a) *Grenzüberschreitende Wasserkooperation. Ein Positionspapier des BMZ*. BMZ Spezial 135. BMZ, Berlin.

BMZ (Bundesministerium für wirtschaftliche Zusammenarbeit und Entwicklung) (2006b) *Der Wassersektor in der deutschen Entwicklungszusammenarbeit*. BMZ-Materialien No. 154. BMZ, Berlin.

Booth, K and Vale, P (1997) 'Critical security studies and regional insecurity: the case of Southern Africa'. In Krause, K and Williams, M C (eds) *Critical Security Studies: Concepts and Cases*. UCL Press, London, pp329–58.

Botta, A and Foley, A (2002) 'Effects of climate variability and disturbances on the Amazonian terrestrial ecosystems dynamics'. *Global Biogeochemical Cycles* 16(1070): doi: 10.1029/2000GB001338.

Boulding, K E (1962) *Conflict and Defense: A General Theory*. Harper and Brothers, New York.

Brandt, W (1980) *North-South: A Program for Survival. Report of the Independent Commission on International Development Issues under the Chairmanship of Willy Brandt*. MIT Press, Cambridge, MA.

Brauch, H G (2005) *Threats, Challenges, Vulnerabilities and Risks in Environmental and Human Security*. United Nations University (UNU)-EHS, Bonn.

Brauch, H G (2006) 'Regionalexpertise: Destabilisierungs- und Konfliktpotential prognostizierter Umweltveränderungen in der Region Südeuropa und Nordafrika bis 2020/2050. Expertise for the WBGU Report "World in Transition: Climate Change as a Security Risk"'. WBGU website, http://www.wbgu.de/wbgu_jg2007_ex01.pdf

Brazil (2004) *Brazil's Initial National Communication to the UNFCCC*. Government of Brazil, Brazil.

Bremer, S (2000) 'Who fights whom, when, where and why'. In Vasquez, J A (ed) *What Do We Know About War*. University Press of America, Lanham, MD, pp23–36.

Brock, L (1992) 'Security through defending the environment: An Illusion?' In Boulding, E (ed) *New Agendas for Peace Research*. Lynne Rienner, Boulder, CO, pp79–102.

Brock, L (1997) *The Environment and Security: Conceptual and Theoretical Issues. Conflict and the Environment*. Kluwer, Dordrecht.

Brock, L (1998) 'Umwelt und Konflikt im internationalen Forschungskontext'. In Carius, A and Lietzmann, K M (eds) *Umwelt und Sicherheit. Herausforderungen für die internationale Politik*. Springer, Berlin, Heidelberg, New York, pp39–56.

Brock, L (2004) *Vom erweiterten Sicherheitsbegriff zur globalen Konfliktintervention. Eine Zwischenbilanz der neuen Sicherheitsdiskurse. Arbeitspapier*. Hessische Stiftung Friedens- und Konfliktforschung (HSFK), Frankfurt/M.

Brown, L (2006) *Global Warming Forcing US Postal Population to Move Inland. An Estimated 250,000 Katrina Evacuees are Now Climate Refugees*. Earth Policy Institute Washington, Washington, DC.

Brzezinski, Z (2004) *The Choice. Global Domination or Global Leadership*. Basic, New York.

Brzezinski, Z and Mearsheimer, J J (2005) 'Clash of the titans'. *Foreign Policy* **146**: 24–47.

Brzoska, M (2006) *Analysis of and Recommendations for Covering Security Relevant Expenditures Within and Outside of Official Development Assistance (ODA). Paper 53*. International Center for Conversion, Bonn.

Brzoska, M, Kingma, K and Wulf, H (1995) *Military Conversion for Social Development. Report on BICC Panel Discussion at the World Summit for Social Development. Copenhagen, 8 March 1995*. BICC Report 5. Bonn International Center for Conversion (BICC), Bonn.

Bucknall, J, Klytchnikova, I, Lampietti, J, Lundell, M, Scatasta, M and Thurman, M (2003) *Irrigation in Central Asia. Social, Economic and. Environmental Considerations*. World Bank, Washington, DC.

Buhaug, H and Gleditsch, K S (2005) *The Origin of Conflict Clusters: Contagion or Bad Neighborhoods? Paper prepared for the Third European Consortium for Political Research General Conference, Budapest 8–10 September 2005*. Centre for the Study of Civil War at the International Peace Research Institute, Oslo.

Burkett, V R, Zikoski, D B and Hart, D A (2005) 'Sea–Level Rise and Subsidence: Implications for Flooding in New Orleans, Louisiana'. USGS Geological Survey website, http://www.nwrc.usgs.gov/hurricane/Sea-Level-Rise.pdf (viewed 21. March 2006).

Büttner, V and Krause, J (Hrsg.) (1995) *Rüstung statt Entwicklung? Sicherheitspolitik, Militärausgaben und Rüstungskontrolle in der Dritten Welt*. Nomos, Baden-Baden.

CAIT WRI (Climate Analysis Indicator Tool – World Resources Institute) (2007) 'Total GHG Emissions in 2003 (excludes land use change)'. CAIT WRI website, http://www.cait.wri.org (viewed 8. February 2007).

Carius, A (2003) *Umweltpolitik als Instrument ziviler Krisenprävention. Hintergrundpapier zum Fachgespräch von Bundesumweltministerium und Adelphi Research*. Adelphi Research, Berlin.

Carius, A, Bächler, G, Pfahl, S and March, A (1999) *Umwelt und Sicherheit. Forschungserfordernisse und Forschungsprioritäten. Studie im Auftrag des BMBF*. Ecologic, Berlin.

Carius, A and Dabelko, G D (2004) 'Institutionalizing responses to environment, conflict, and cooperation'. In United Nations Environment Programme (UNEP) (ed) *Understanding Environment, Conflict, and Cooperation. Chapter Two*. United Nations Environment Programme (UNEP), New York, pp21–33.

Carius, A, Dabelko, G D and Wolf, A T (2004) 'Water, Conflict and Cooperation'. *ECSP Report* **10**: 60–6.

Carius, A and Lietzmann, K M (1998) *Umwelt und Sicherheit. Herausforderungen für die internationale Politik*. Springer, Berlin, Heidelberg, New York.

Carius, A, Petzold-Bradley, E and Pfahl, S (2001) 'Umweltpolitik und nachhaltige Friedenspolitik. Ein neues Thema auf der internationalen Agenda'. *Aus Politik und Zeitgeschichte* **12**: 6–13.

Carius, A, Tänzler, D and Winterstein, J (2006) 'Weltkarte von Umweltkonflikten: Ansätze zur Typologisierung. Expertise for the WBGU Report "World in Transition: Climate Change as a Security Risk"'. WBGU website, http://www.wbgu.de/wbgu_jg2007_ex02.pdf

Cassel-Gintz, M (2006) 'Karten zur Bodendegradation und Versalzung. GIS-II. Expertise for the WBGU Report "World in Transition: Climate Change as a Security Risk"'. WBGU website, http://www.wbgu.de/wbgu_jg2007_ex03.pdf

Castles, S (2002) *Environmental Change and Forced Migration: Making Sense Of The Debate. New Issues In Refugee Research Working Paper 70*. Refugees Studies Centre University of Oxford. Evaluation And Policy Analysis Unit, Oxford.

CGAP Working Group on Microinsurance (2006) 'Improving Risk Management for the Poor. Microinsurance 10'. CGAP Working Group on Microinsurance website, http://www.microfinance.lu (viewed 15. January 2007).

Chambers, B (2005) 'The Barriadas of Lima – slums of hope or despair? Problems and solutions?' *Geography* **90**(3): 200–24.

Chatel, T (2006) 'Wasserpolitik in Spanien – eine kritische Analyse'. *Geographische Rundschau* **58**(2): 20–8.

Chaudhuri, S and Ravallion, M (2007) 'Partially awakened giants: uneven growth in China and India'. In Winters, L A and Yusuf, S (eds) *Dancing with Giants: China, India, and the Global Economy*. World Bank, Institute for Policy Studies, Washington, DC und Singapur, pp157–88.

Chemillier-Gendreau, M (2006) 'Faut-il un statut international de réfugié écologique?' *Revue européenne de droit de l'environnement*: 446–52.

Choudhury, D (1994) *Constitutional Development in Bangladesh – Stresses and Strains*. Oxford University Press, Oxford, New York, Karachi.

CIA (Central Intelligence Agency) (2006) *The World Factbook 2006*. CIA, Washington, DC.

Clark, W A V (2006) 'Environmentally Induced Migration and Conflict. Expertise for the WBGU Report "World in Transition: Climate Change as a Security Risk"'. WBGU website, http://www.wbgu.de/wbgu_jg2007_ex04.pdf

CNA Corporation (2007) *National Security and the Threat of Climate Change*. CNA Corporation, Washington, DC.

Cody, E (2005a) *China grows more wary over rash of protests*. Washington Post, Washington, DC.

Cody, E (2005b) *China's party leaders draw bead on inequity*. Washington Post, Washington, DC.

Collier, P, Elliot, L, Hegre, H, Hoeffler, A, Reynal–Querol, M and Sambanis, N (2003) *Breaking the Conflict Trap: Civil War and Development Policy*. Oxford University Press, Oxford.

Collier, P and Hoeffler, A (2004) 'Greed and grievance in civil war'. *Oxford Economic Papers* **56**(4): 563–95.

COMCAD (Center on Migration Citizenship and Development) (2007) 'Research Project EACH-FOR – Environmental Change and Forced Migration Scenarios'. University Bielefeld COMCAD website, http://www.comcad-bielefeld.de/cgi-bin/pagemaker.pl?name=environmentalchange&aufb=2%A71,3%A72,undefined,28,3,11,,IMG_6,IMG_2 (viewed 26. March 2007).

Commission on Human Security (Hrsg.) (2003) *Human Security Now*. Commission on Human Security, New York.

Conca, K (2006) *Governing Water – Contentious Transnational Politics and Global Institution Building*. MIT Press, Cambridge, MA.

Conisbee, M and Simms, A (2003) *Environmental Refugees. The Case for Recognition*. New Economic Foundation, London.

Coplin, S C (1999) 'Houston–Galveston, Texas: managing coastal subsidence'. In Galloway, D L, Jones, D R and Ingebritsen, S E (eds) *Land Subsidence in the United States: U.S. Geological Survey Circular 1182*. U.S. Geological Survey (USGS), Washington, DC, pp35–48.

Cordesman, A H and Kleiber, M (2007) *Chinese Military Modernization*. Center for Strategic and International Studies (CSIS), Washington, DC.

Cosgrove, W J and Rijsberman, F R (2000) *World Water Vision: Making Water Everybody's Business*. Earthscan, London.

Coudrain, A, Francou, B and Kundzewicz, Z W (2005) 'Glacier shrinkage in the Andes and consequences for water resources – Editorial'. *Hydrological Sciences Journal* **50**(6): 925.

Cowling, S A and Shin, Y (2006) 'Simulated ecosystem threshold responses to co-varying temperature, precipitation and atmospheric CO_2 within a region of Amazonia'. *Global Ecology and Biogeography* **15**: 553–66.

Cox, P M, Betts, R A, Collins, M, Harris, P P, Huntingford, C and Jones, C D (2004) 'Amazonian forest dieback under climate-carbon cycle projections for the 21st century'. *Theoretical and Applied Climatology* **78**: 137.

Cox, P M, Betts, R A, Jones, C D, Spall, S A and Totterdell, I J (2000) 'Acceleration of global warming due to carbon-cycle feedbacks in a coupled climate model'. *Nature* **408**: 184–7.

CPT (Comissão Pastoral da Terra) (2007) 'Daten aus der Datenbank der brasilianischen NRO Comissão Pastoral da Terra (CPT)'. CPT website, http://www.cptnac.com.br (viewed 19. March 2007).

Cramer, C (2003) 'Does inequality cause conflict?' *Journal of International Development* **15**(4): 397–412.

CRED (WHO Collaborating Centre for Research on the Epidemiology of Disasters) (2006) 'Emergency Events Database (EM-DAT)'. CRED website, http://www.em-dat.net/ (viewed 11. May 2006).

Crips, J (2003) *No Solutions in Sight: The Problem of Protracted Refugee Situations in Africa*. Working Paper 75. New York, United Nations High Commissioner for Refugees (UNHCR).

CSIS (Center for Strategic and International Studies) and IIE (Institute for International Economics) (2006) *China: The Balance Sheet*. CSIS und IIE, Washington, DC.

Cure, J D and Acock, B (1986) 'Crop responses to carbon dioxide doubling: a literature survey'. *Agricultural and Forest Meteorology* **38**: 127–45.

Daalder, I V and Lindsay, J M (2003) *America Unbound*. Brookings Institution Press, Washington, DC.

Daase, C (1991) 'Der erweiterte Sicherheitsbegriff und die Diversifizierung amerikanischer Sicherheitsinteressen. Anmerkungen zu aktuellen Tendenzen in der sicherheitspolitischen Forschung'. *Politische Vierteljahresschrift* **32**: 425–51.

Daase, C (1992) 'Ökologische Sicherheit: Konzept oder Leerformel?' In Wellmann, C (ed) *Umweltzerstörung: Kriegsfolge und Kriegsursache. Friedensanalysen. Band 27*. Suhrkamp, Frankfurt/M, pp21–52.

Daase, C (1996) *Vom Ruinieren der Begriffe. Eine Welt oder Chaos. Friedensanalysen Band 25*. Suhrkamp, Frankfurt/M.

Daase, C (2002) 'Terrorismus: Der Wandel von einer reaktiven zu einer proaktiven Sicherheitspolitik der USA nach dem 11. September 2001'. In Daase, C, Feske, S and Peters, I (eds) *Internationale Risikopolitik. Der Umgang mit neuen Gefahren in den internationalen Beziehungen*. Nomos, Baden-Baden, pp113–42.

Dalby, S (1992) 'Security, modernity, ecology: the dilemmas of post-cold war security discourse'. *Alternatives* **17**: 95–134.

Dalby, S (2002) 'Environmental change and human security'. *Isuma - Canadian Journal of Policy Research* **3**(2): 71–9.

Dalhuisen, J, de Groot, H and Njkamp, P (1999) *The Economics of Water. A Survey of Issues. Research Memorandum 1999–*

36. University Amsterdam, Faculteit der Economischen Wetenschappen en Econometrie, Amsterdam.

Daughters, R and Harper, L (2006) *Fiscal and Political Decentralization.* Mimeo. Interamerican Development Bank, Washington, DC.

DBR (Deutsche Bank Research) (2006) *Indien – Auf dem Weg zur Weltmacht?* DBR, Frankfurt/M.

de Soysa, I (2000) 'The resource curse: are civil wars driven by rapacity or paucity?' In Berdal, M and Malone, D (eds) *Greed and Grievance: Economic Agendas in Civil Wars.* Westview Press, Boulder.

de Soysa, I (2002) 'Ecoviolence: shrinking pie, or honey pot?' *Global Environmental Politics* **2**(1): 1–35.

de Soysa, I, Gleditsch, N P, Gibson, M, Sollenberg, M and Westing, A H (1999) *To Cultivate Peace: Agriculture in a World of Conflict.* International Peace Research Institute (PRIO), Oslo.

Debiel, T, Klingebiel, S, Mehler, A and Schneckener, U (2005) *Zwischen Ignorieren und Intervenieren. Strategien und Dilemmata externer Akteure in fragilen Staaten.* SEF Policy Paper 23. Stiftung Entwicklung und Frieden (SEF), Bonn.

Dervis, K (2007) *Putting UN Reform into Action: The UNDP-UNEP Poverty-Environment Facility. Statement by Kemal Dervis, Administrator of the United Nations Development Programme on the occasion of the UNEP Governing Council Meeting.* Nairobi: United Nations Development Programme (UNDP).

Deudney, D (1990) 'The case against linking environmental degradation and national security'. *Millennium* **19**(3): 461–76.

Deudney, D (1991) 'Environment and security: muddeled thinking'. *The Bulletin of the Atomic Scientist* **47**(3): 23–8.

Development Initiatives (2006) 'Global Humanitarian Assistance 2006'. Development Initiatives website, http://www.devinit.org (viewed 3. April 2007).

Diamond, J (2005) *Kollaps – Warum Gesellschaften überleben oder untergehen.* Fischer, Frankfurt/M.

Diehl, P F and Gleditsch, N P (Hrsg.) (2000) *Environmental Conflict: An Anthology.* Westview Press, Boulder, CO.

Dikich, A N and Hagg, W (2004) 'Climate driven changes of glacier runoff in the Issyk-Kul basin, Kyrgyzstan'. *Zeitschrift für Gletscherkunde und Glazialgeologie* **39**: 75–86.

Dimaranan, B, Ianchovichina, E and Martin, W (2007) 'Competing with giants: who wins, who loses?' In Winters, L A and Yusuf, S (eds) *Dancing with Giants: China, India, and the Global Economy.* World Bank, Institute for Policy Studies, Washington, DC und Singapur, pp57–80.

Dingwerth, K and Pattberg, P (2006) 'Was ist Global Governance?' *Leviathan Berliner Zeitschrift für Sozialwissenschaft* **34**(3): 377–99.

Donahue, J D and Nye, J S (Hrsg.) (2000) *Governance in a Globalizing World.* Brookings Institution Press, Washington, DC.

Douglas, I (2005) 'The urban environment in Southeast Asia'. In Gupta, A (ed) *The Physical Geography of Southeast Asia.* Oxford University Press, Oxford, New York, pp314–35.

Drury, A C and Olson, R S (1998) 'Disasters and political unrest: an empirical investigation'. *Journal of Contingencies and Crisis Management* **6**: 153–61.

du Castel, V (2005) *La mer de Barents: vers un nouveau "grand jeu"?* Institut Francais des Relations Internationales (IFRI), Paris.

Ebert, C (2006) 'Nasty Social Behavior Common after a Disaster. Press Release University at Buffalo'. University at Buffalo website, http://www.buffalo.edu/news/fast-execute.cgi/article-page.html?article=74630009 (viewed 11. May 2006).

ECLAC (Economic Commission for Latin America and the Carribean) (2006) *Social Panorama of Latin American.* ECLAC, Santiago de Chile.

Economist (2006a) 'A spectre haunting India: India's Naxalites'. *The Economist* (19–25 August): 44–6.

Economist (2006b) 'The politics of meltdown – Could climate change become a hot topic for the primaries?' *Economist* **379**(8479): 46.

Economy, C E (2004) 'The river runs black. The environmental challenge to China's future'. Ithaka: Cornell University Press.

Edenhofer, O, Lessmann, K, Kemfert, C, Grubb, M and Köhler, J (2006) 'Induced technological change: exploring its implications for the economics of atmospheric stabilization: Synthesis Report From the Innovation Modeling Comparison Project'. *The Energy Journal* (Special Issue): 57–107.

EEA (European Environmental Agency) (2004) *Impacts of Europe's Changing Climate.* EEA Report No. 2. EEA, Copenhagen.

Effendi, A, Nasution, A, Djarwoto, A, Murdohardono, D, Kertapati, E, Hidajat, R, Sutawidjaja, I S, Jäger, S, Manhart, A, Ranke, U, Rehmann, T, Dalmin, R and Weiland, L (2005) *Mitigation of Geohazards in Indonesia. A Contribution to the World Conference on Disaster Reduction Kobe, Hyogo, Japan.* Georisk Project, Badan Koordinasi Nasional (BAKORNAS), Department Dalam Negeri, Bandung, Indonesia.

Elbadawi, I and Sambanis, N (2000) 'How Much War Will We See? Estimating the Incidence of Civil War in 161 Countries'. World Bank website, http://www.worldbank.org/research/conflict/papers/incidencev2.pdf (viewed 7. May 2007).

Elliott, L (2002) 'Expanding the Mandate of the UN Security Council to Account for Environmental Issues'. United Nations University (UNU) website, http://www.unu.edu/inter-linkages/docs/IEG/Elliot.pdf (viewed 21. April 2006).

El-Hinnawi, E (1985) *Environmental Refugees.* United Nations Environment Program (UNEP), New York.

Emanuel, K (2005) 'Increasing destructiveness of tropical cyclones over the past 30 years'. *Nature* **436**: 686–8.

Energy and Environmental Analysis (2005) *Hurricane Damage to Natural Gas Infrastructure and Ist Effects on the U.S. Natural Gas Market.* Energy and Environmental Analysis Inc, Arlington, VA.

EPA (United States Environmental Protection Agency) (2006) *Global Anthropogenic Non-CO_2 Greenhouse Gas Emissions: 1990–2020.* EPA, Washington, DC.

Epstein, G S and Gang, I N (2004) *The Influence of Others on Migration Plans.* IZA Discussion Paper Series 1244. Forschungsinstitut zur Zukunft der Arbeit (IZA), Bonn.

EU (European Union) (2003) *Ein sicheres Europa in einer besseren Welt: Europäische Sicherheitsstrategie.* EU, Brussels.

EU Commission (2005) *EU Strategy for Africa: Towards a Euro-African Pact to Accelerate Africa's Development, {SEC(2005)1255}. Communication From the Commission to the Council, the European Parliament and the European Economic and Social Committee.* EU Commission, Brussels.

EU Commission (2006) *Annual Report 2006 on the European Community's Development Policy and the Implementation of External Assistance in 2005.* EuropeAid Co-operation Office, Brussels.

EU Commission (2007) *7th Framework Programme, FPp7.* EU, Brussels.

Faist, T (2005) *The Migration-Securits Nexus: International Migration and Security Before and After 9/11*. Working Paper 9. University Bielefeld, Center on Migration, Citizenship and Development (COMCAD), Bielefeld.

Fajnzylber, P and López, J P (2006) *Close to Home. The Development Impact of Remittances in Latin America*. World Bank, Washingon, DC.

Falleti, T G (2005) 'A sequential theory of decentralization: Latin American cases in comparative perspective'. *American Political Science Review* **99**: 327–46.

FAO (Food and Agriculture Organisation of the United Nations) (1997) *Irrigation Potential in Africa: A Basin Approach. FAO Land and Water Bulletin 4*. FAO Land and Water Development Division, Rome.

FAO (Food and Agriculture Organisation of the United Nations) (2002) *The State of Food and Agriculture 2002. Agriculture and Global Public Goods ten Years After the Earth Summit*. FAO, Rome.

FAO (Food and Agriculture Organisation of the United Nations) (2003) *World Agriculture: Towards 2015/2030*. FAO, Rome.

FAO (Food and Agriculture Organisation of the United Nations) (2005a) *Global Forest Assessment 2005. Progress Towards Sustainable Forest Management*. FAO Forestry Paper 147. FAO, Rome.

FAO (Food and Agriculture Organisation of the United Nations) (2005b) 'Global Forest Resources Assessment 2005'. FAO website, http://www.fao.org/forestry/site/32086/en (viewed 15. August 2006).

FAO (Food and Agriculture Organisation of the United Nations) (2005c) *Impact of Climate Change, Pests and Diseases on Food Security and Poverty Reduction*. Background Document, 31st Session of the Committee on World Food Security, May 2005. FAO, Rome.

FAO (Food and Agriculture Organisation of the United Nations) (2006a) *Crop Prospects and Food Situation*. No. 1 April 2006. FAO, Rome.

FAO (Food and Agriculture Organisation of the United Nations) (2006b) *Food Outlook June 2006*. FAO, Rome.

Faris, S (2007) 'The real roots of Darfur'. *The Atlantic Monthly* **299**(3): 2.

Farrell, A E and Jäger, J (2005) *Assessments of Regional and Global Environmental Risks: Designing Processes for the Effective Use of Science in Decisionmaking*. Resources for the Future, Washington, DC.

Fassbender, B (2005) *UN–Reform und kollektive Sicherheit Der Bericht des "UN High–level Panel on Threats, Challenges and Change" vom Dezember 2004 und die Empfehlungen des UN–Generalsekretärs vom März 2005*. Global Issue Papers No. 17. Heinrich Böll Foundation, Berlin.

Faust, J (2006) 'The political economy of decentralization in Latin America'. *Iberoamericana* **22**: 164–9.

Faust, J and Harbers, I (2007) *Political Parties and the Politics of Administrative Decentralization. Exploring Subnational Variance in Ecuador*. Mimeo. Deutsches Institut für Entwicklungspolitik (DIE), Bonn.

Faust, J and Messner, D (2004) *Europe's New Security Strategy. Challenges for Development Policy*. DIE Discussion Paper 3. Deutsches Institut für Entwicklungspolitik (DIE), Bonn.

Fearon, J and Laitin, D D (2003) 'Ethnicity, insurgency, and civil war'. *American Political Science Review* **97**(1): 75–90.

Fischer, G (2005) 'Feeding China in 2030'. *IIASA options* (Autumn): 12–6.

Fischer, G, Shah, M, Tubiello, F N and van Velthuizen, H (2005) 'Socio-economic and climate change impacts on agriculture: an integrated assessment, 1990–2080'. *Philosophical Transactions of the Royal Society B Biological Sciences* **360**: 2067–73.

Fischer, G, Shah, M and van Velthuizen, H (2002) *Climate Change and Agricultural Vulnerability. Contribution to the UN World Summit on Sustainable Development, Johannesburg 2002*. IIASA, Laxenburg.

Flintan, F and Tamrat, I (2006) 'Spilling blood over water? The case of Ethopia'. In Lind, J and Sturman, K (eds) *Scarcity and Surfeit. The Ecology of Africa's Conflicts*. Institute for Security Studies (ISS), Pretoria, Cape Town, pp243–319.

Foreign Policy (2006) 'The Failed State Index (May/June 2006)'. Foreign Policy and the Fund for Peace website, http://www.foreignpolicy.com/story/cms.php?story_id=3420 (viewed 7. May 2007).

Fravel, M T (1996) 'China's attitude toward U.N. peacekeeping operations since 1989'. *Asian Survey* **36**(11): 1102–21.

Freedom House (2006) 'Combined Average Ratings: Independent Countries 2006'. Freedom House website, http://www.freedomhouse.org/template.cfm?page=267&year=2006 (viewed 7. May 2007).

Friedlingstein, P, Cox, P, Betts, R, Bopp, L, von Bloh, W, Brovkin, V, Cadule, P, Doney, S, Eby, M, Fung, I, Bala, G, John, J, Jones, C, Joos, F, Kato, T, Kawamiya, M, Knorr, W, Lindsay, K, Matthews, H D, Raddatz, T, Rayner, P, Reick, C, Roeckner, E, Schnitzler, K G, Schnur, R, Strassmann, K, Weaver, A J, Yoshikawa, C and Zeng, N (2006) 'Climate-carbon cycle feedback analysis: Results From the (CMIP)-M-4 model intercomparison'. *Journal of Climate* **19**: 3337–53.

Fröhlich, M (2006) '"Responsibility to Protect" – Zur Herausbildung einer neuen Norm der Friedenssicherung'. In Varwick, J and Zimmermann, A (eds) *Die Reform der Vereinten Nationen*. Duncker & Humblot, Berlin, pp167–86.

Fuentes, V E (2003) *The Political Effects of Disasters and Foreign Aid: National and Subnational Governance in Honduras after Hurricane Mitch*. University of Florida, Gainesville, FL.

Fues, T (2006) 'Die gemischten Empfehlungen des High-Level Panels: Mehr Kohärenz im UN-Entwicklungsbereich?' *Informationsbrief Weltwirtschaft & Entwicklung. W&E-Hintergrund* **11**.

Fues, T and Hamm, B I (Hrsg.) (2001) *Die Weltkonferenzen der 90er Jahre: Baustellen für Global Governance. Band 12 "Eine Welt"*. Dietz, Bonn.

Fukuyama, F (1992) *The End of History and the Last Man*. Hamish Hamilton, New York.

Galloway, D L, Jones, D R and Ingebritsen, S E (1999) *Land Subsidence in the United States*. U.S. Geological Survey Circular 1182. Washington, DC, U.S. Geological Survey (USGS).

Gao, X J, Zhao, Z C and Giorgi, F (2002) 'Changes of extreme events in regional climate simulations over east Asia'. *Advances in Atmospheric Sciences* **19**: 927–42.

Gartzke, E, Li, Q and Boehmer, C (2001) 'Investing in the peace: economic interdependence and international conflict'. *International Organization* **55**(2): 391–438.

Gasana, J K (2002) 'Natural resource scarcity and violence in Rwanda'. In The World Conservation Union (IUCN) (ed) *Conserving the Peace: Resources, Livelihoods and Security*. IUCN, Gland, pp199–245.

Gates, S (2002) *Empirically Assessing the Causes of War*. Paper presented to the 43rd Annual Convention of the International Studies Association, New Orleans, LA, 24–27 March. Centre for the Study of Civil War, Oslo.

GCIM (Global Commisson on International Migration) (2005) *Migration in an Interconnected World: New Directions for Action*. Global Commission on International Migration (GCIM), Geneva.

GEF (Global Environment Facility) (2005) *GEF Global Action on Climate Change. GEF Support for Adaptation to Climate Change*. GEF, Washington, DC.

Geißler, N (1999) *Der völkerrechtliche Schutz der Internally Displaced Persons. Eine Analyse des normativen und institutionellen Schutzes der Internally Displaced Persons im Rahmen innerer Unruhen und nicht-internationaler Konflikte.* Duncker & Humblot, Berlin.

Geller, D S and Singer, D J (1998) *Nations at War: A Scientific Study of International Conflict*. Cambridge University Press, Cambridge, MA.

Giese, E (1997) 'Die ökologische Krise der Aralseeregion'. *Geographische Rundschau* **5**: 293–9.

Giese, E and Moßig, I (2004) *Klimawandel in Zentralasien*. Discussion Paper Nr. 17. University Gießen, Zentrum für internationale Entwicklung- und Umweltforschung (ZEU), Gießen.

Giese, E and Sehring, J (2006) 'Regionalexpertise: Destabilisierungs- und Konfliktpotential prognostizierter Umweltveränderungen in der Region Zentralasien bis 2020/2050. Expertise for the WBGU Report "World in Transition: Climate Change as a Security Risk"'. WBGU website, http://www.wbgu.de/wbgu_jg2007_ex05.pdf

Giese, E, Sehring, J and Trouchine, A (2004a) 'Zwischenstaatliche Wassernutzungskonflikte in Mittelasien'. *Geographische Rundschau* **56**(10): 10–6.

Giese, E, Sehring, J and Trouchine, A (2004b) *Zwischenstaatliche Wassernutzungskonflikte in Zentralasien*. Discussion Paper Nr. 18. University Gießen, Zentrum für internationale Entwicklungs- und Umweltforschung (ZEU), Gießen.

Gill, B (2007) *Rising Star: China's New Security Diplomacy*. Brookings Institution Press, Washington, DC.

Giorgi, F and Bi, X (2005) 'Updated regional precipitation and temperature changes for the 21st century from ensembles of recent AOGCM simulations'. *Geophysical Research Letters* **32**(L21715): doi:10.1029/2005GL024288.

Gleditsch, N P (1995) 'Geography, democracy, and peace'. *International Interactions* **20**(4): 297–323.

Gleditsch, N P (1998) 'Armed Conflict and The Environment: A Critique of the Literature'. *Journal of Peace Research* **35**(3): 381–400.

Gleick, P (1993) 'Water and conflict: fresh water resources and international security'. *International Security* **18**(1): 79–112.

Gleick, P (2003) 'Global freshwater resources: soft-path solutions for the 21st century'. *Science* **302**: 1524–8.

Gleick, P (2006) *The World's Water 2006–2007: The Biennial Report on Freshwater Resources*. Island Press, Washington, DC.

Gogolin, I and Pries, L (2003) 'Transmigration und Bildung. Beitrag für ZfE 1/2004, überarbeitete Fassung'. University Bochum website, http://www.ruhr-uni-bochum.de/soaps/download/publ-2004_lp_transmigrationundbildung.pdf (viewed 27. October 2006).

Golda-Pongratz, K (2004) 'The Barriadas of Lima: utopian city of self-organisation?' *Architectural Design* **74**(4): 38–45.

Goldman Sachs (2003) *Dreaming With BRICs: The Path to 2050. Global Economics Paper No 99*. Goldman Sachs Global Research Centres, New York.

Goldstein, A, Pinaud, N, Reisen, H and Chen, X (2006) *The Rise of China and India. What's in it for Africa?* Development Centre Studies. Organisation for Economic Development and Co-operation (OECD), Paris.

Goldstone, J (2002) 'Population and security: how demographic change can lead to violent conflict'. *Journal of International Affairs* **56**(1): 3–21.

Gollier, C and Treich, N (2003) 'Decision-making under scientific uncertainty: the economics of the precautionary principle'. *The Journal of Risk and Uncertainty* **27**: 77–103.

Government of China (2005) 'White Paper on China's Peaceful Development Road'. State Council Information Office website, http://www.china.org.cn/cnglish/2005/Dec/152669.htm (viewed 12. May 2006).

Grävingholt, J (2007) *Staatlichkeit und Governance: Herausforderungen in Zentralasien und im Südkaukasus*. DIE Analysen und Stellungnahmen 2. Deutsches Institut für Entwicklungspolitik (DIE), Bonn.

Gregory, J M, Huybrechts, P and Raper, S C B (2004) 'Threatened loss of the Greenland ice-sheet'. *Nature* **428**: 616.

Grimm, S and Klingebiel, S (2007) 'Governance-Herausforderungen in Afrika südlich der Sahara'. *DIE – Afrika Agenda 2007: Ansatzpunkte für den deutschen G8-Vorsitz und die EU-Ratspräsidentschaft. Discussion Paper* **18**: 41–55.

Gritti, E S, Smith, B and Sykes, M T (2006) 'Vulnerability of Mediterranean basin ecosystems to climate change and invasion by exotic plant species'. *Journal of Biogeography* **33**: 145–57.

Große Koalition (2005) *Gemeinsam für Deutschland – Mit Mut und Menschlichkeit. Koalitionsvertrag zwischen CDU, CSU und SPD, 11. November 2005*. Große Koalition, Berlin.

Grossmann, M (2006) 'Cooperation on Africa's international waterbodies: information needs and the role of information-sharing'. In Scheumann, W and Neubert, S (eds) *Transboundary Water Management in Africa. DIE Studies 21*. Deutsches Institut für Entwicklungspolitik (DIE), Bonn, pp173–236.

Grote, U, Engel, S and Schraven, B (2006) *Migration Due to the Tsunami in Sri Lanka – Analyzing Vulnerability and Migration at the Household Level*. Discussion Papers on Development Policy Nr. 106. Zentrum für Entwicklungsforschung (ZEF), Bonn.

GTZ (Deutsche Gesellschaft für Technische Zusammenarbeit) (2003) *Klimaschutz-Programm für Entwicklungsländer (CaPP). Anpassung an den Klimawandel: Ergebnisse eines Screenings der deutschen TZ-Projekte*. GTZ, Eschborn.

Gupta, A K, Anderson, D M and Overpeck, J T (2003) 'Abrupt changes in the Asian southwest monsoon during the Holocene and their links to the North Atlantic Ocean'. *Nature* **421**(23.01.): 324–5.

Gurr, T R (2000) *Peoples versus States: Minorities at Risk in the New Century*. USIP Press, Washington, DC.

GWP (Global Water Partnership) (2000) 'Integrated Water Resources Management. Technical Advisory Committee Background Papers No. 4'. GWP website, http://www.gwpforum.org/gwp/library/TACNO4.PDF (viewed 11. April 2007).

Haas, P M (2002) 'UN conferences and constructivist governance of the environment'. *Global Governance* **8**: 73–91.

Haas, P M (2004) 'When does power listen to truth? A constructivist approach to the policy process'. *Journal of European Public Policy* **11**: 569–92.

Haass, R (2005) *The Opportunity*. PublicAffairs, New York.

Hadley Centre for Prediction and Research (o. J.) 'Climate Change Projections. Electronical Data'. Hadley Centre for Prediction and Research website, http://www.metoffice.com/research/hadleycentre/models/modeldata.html (viewed 27. July 2007).

Hagmann, T (2005) 'Confronting the concept of environmentally induced conflict'. *Peace, Conflict, and Development* **6**(January): 1–22.

Hagopian, F and Mainwaring, S P (2006) *The Third Wave of Democratization in Latin America: Advances and Setbacks.* Cambridge University Press, Cambridge, New York.

Halbach, U (2002) *Stabilitätspolitik in Zentralasien und Kaukasien im Rahmen der "Anti-Terror-Allianz". SWP-Diskussionspapier der Forschungsgruppe Russland.* Stiftung Wissenschaft und Politik (SWP), Berlin.

Hamilton, J, Maddison, D and Tol, R (2005) 'Climate change and international tourism. A simulation study'. *Global Environmental Change* **15**(3): 253–66.

Hansen, G (2006) 'Displacement and Return'. Concillation Ressources (CR) website, http://www.c-r.org/accord/georab/accord7/displace.shtml (viewed 11. May 2006).

Hansen, J E (2005) 'A slippery slope: How much global warming constitutes "dangerous anthropogenic interference"?' *Climatic Change* **68**: 269–79.

Hansen, S A (2003) 'Neuer Ärger beim chinesischen Riesenstaudamm'. taz, Berlin.

Hare, B (2006) 'Relationship between increases in global mean temperature and impacts on ecosystems, food production, water and socio–economic systems'. In Schellnhuber, H J, Cramer, W, Nakicenovic, N, Wigley, T and Yohe, G (eds) *Avoiding Dangerous Climate Change.* Cambridge University Press, Cambrigde, New York, pp177–85.

Hauge, W and Ellingsen, T (1998) 'Beyond environmental scarcity: causal pathways to conflict'. *Journal of Peace Research* **35**(3): 299–317.

Hauswedell, C (2006) 'Erweiterte Sicherheit und militärische Entgrenzung'. *Blätter für deutsche und internationale Politik* **51**(6): 723–32.

Heberer, T and Senz, A-D (2006a) *Die Rolle Chinas in der internationalen Politik. Innen- und außenpolitische Entwicklungen und Handlungspotenziale.* DIE Discussion Paper 3. Deutsches Institut für Entwicklungspolitik (DIE), Bonn.

Heberer, T and Senz, A-D (2006b) 'Regionalexpertise: Destabilisierungs- und Konfliktpotential prognostizierter Umweltveränderungen in China bis 2020/2050. Expertise for the WBGU Report "World in Transition: Climate Change as a Security Risk"'. WBGU website, http://www.wbgu.de/wbgu_jg2007_ex06.pdf

Hecht, M (2004) *Nahrungsmangel und Protest: Teurungsunruhen in Frankreich und Preußen in den Jahren 1846/47. Studien zur Landesgeschichte.* Mitteldeutscher Verlag, Halle.

Hegre, H, Ellingsen, T, Gleditsch, N P and Gates, S (2001) 'Towards a democratic civil peace?' *American Political Science Review* **95**(1): 33–48.

Heilig, G (1999) 'Can China Feed Itself? A System for Evaluation of Policy Options'. IIASA website, http://www.iiasa.ac.at/Research/LUC/ChinaFood/index_h.htm (viewed 1. July 2006).

Heitzman, J and Worden, R L (1989) *Bangladesh: A Country Study.* Federal Research Division. Library of Congress, Washington, DC.

Held, D, McGrew, A, Goldblatt, D and Perraton, J (1999) *Global Transformations. Politics, Economics and Culture.* Stanford University Press, Stanford.

Held, M, Delworth, T L, Lu, J, Findell, K L and Knutson, T R (2005) 'Simulation of Sahel drought in the 20th and 21st centuries'. *PNAS* **102**(50): 17891–6.

Hély, C, Bremond, L, Alleaume, S, Smith, B, Sykes, M T and Guiot, J (2006) 'Sensitivity of African biomes to changes in the precipitation regime'. *Global Ecology and Biogeography* **15**: 258–70.

Hensell, S (2002) *Modernisierung und Gewalt in Mazedonien. Zur politischen Ökonomie eines Balkankrieges.* Working Paper University Hamburg – IPW. IPW Forschungsstelle Kriege, Rüstung und Entwicklung, Hamburg.

Herbst, J (1996) 'Responding to state failure in Africa'. *International Security* **21**(3): 120–44.

Herrfahrdt, E (2004) *Landwirtschaftliche Transformation, Desertifikation und nachhaltige Ressourcennutzung. Fallbeispiel Usbekistan.* DIE Studies 2/2004. Deutsches Institut für Entwicklungspolitik (DIE), Bonn.

Herrmann, G (2006) 'Gereiztes Klima nach dem Tauwetter. Die Erderwärmung lässt das Arktis-Eis schmelzen – nun streiten Anrainerstaaten um freiwerdende Bodenschätze wie Öl und Gas'. Süddeutsche Zeitung, Munich.

HIIK (Heidelberg Institute for International Conflict Research) (2005) *Conflict Barometer 2005. Crisis – Wars – Coups d'État – Negotiations – Mediations – Peace Settlements. 14th Annual Conflict Analysis.* HIIK, Heidelberg.

HIIK (Heidelberg Institute for International Conflict Research) (2006) *Conflict Barometer 2006. Crisis – Wars – Coups d'État – Negotiations – Mediations – Peace Settlements. 15th Annual Conflict Analysis.* : HIIK, Heidelberg.

Hillmann, F (1996) *Jenseits der Kontinente. Migrationsstrategien von Frauen nach Europa.* Centaurus, Pfaffenweiler.

Hilpold, P (2006) 'The duty to protect and the reform of the United Nations: a new step in the development of international law?' *Max Planck Yearbook of United Nations Law* **10**: 35–69.

Hippler, J (2003) 'USA und Europa: unterschiedliche Sicherheitspolitiken'. In Hauchler, I, Messner, D and Nuscheler, F (eds) *Globale Trends 2004/2005.* Fischer, Frankfurt/M., pp292–307.

Hitz, S and Smith, J B (2004) 'Estimating global impacts from climate change'. *Global Environmental Change* **14**: 201–18.

Hofmeier, R (2004) 'Regionale Kooperation und Integration'. In Ferdowsi, M A (ed) *Afrika – ein verlorener Kontinent.* Bayerische Landeszentrale für politische Bildungsarbeit, Munich, pp189–224.

Homer-Dixon, T F, Boutwell, J and Rathjens, G (1993) 'Environmental change and violent conflict'. *Scientific American* **268**: 8.

Homer–Dixon, T F (1990) *Environmental Change and Violence Conflict.* Canadian Environment and Sustainable Development Program. Institute for Research on Public Policy, Ontario, Canada.

Homer–Dixon, T F (1991) 'On the threshold. Environmental changes as causes of acute conflict'. *International Security* **16**(2): 76–116.

Homer–Dixon, T F (1994) 'Environmental scarcities and violent conflict. Evidence from cases'. *International Security* **19**(1): 5–40.

Homer–Dixon, T F (1999) *Environment, Scarcity, and Violence.* Princeton University Press, Princeton.

Hörig, R (2006) 'Der Hunger nach Energie – Der Subkontinent ist berüchtigt für ineffiziente Kohlekraftwerke'. *Das Parlament* (7./14. August): 12.

Horlemann, L and Neubert, S (2007) *Virtual Water Trade: A Realistic Concept for Resolving the Water Crisis?* Deutsches Institut für Entwicklungspolitik (DIE), Bonn.

Houdret, A and Tänzler, D (2006) 'Umwelt und Konflikte'. In Stiftung Entwicklung und Frieden (SEF) (ed) *Globale Trends 2007. Frieden, Entwicklung, Umwelt.* Fischer, Frankfurt/M., pp359–78.

Houscht, M P (2003) 'Bangladesch 2003: Hypotheken und Herausforderungen'. Friedrich-Ebert-Stiftung website, http://

library.fes.de/pdf-files/stabsabteilung/01741.pdf (viewed 9. February 2007).

Hoyos, C D, Agudelo, P A, Webster, P J and Curry, J A (2006) 'Deconvolution of the factors contributing to the increase in global hurricane intensity'. *Science* **312**: 94–7.

Hu, R L, Yue, Z Q, Wang, L C and Wang, S J (2004) 'Review on current status and challenging issues of land subsidence in China'. *Engineering Geology* **76**: 65–77.

Hubacek, K and Sun, L (1999) *Land Use Change in China: A Scenario Analysis Based on Input-Output Modeling.* IIASA Interim Report IR-99-073. IIASA, Laxenburg.

Hubacek, K and Vazquez, J (2002) *The Economics of Land Use Change.* IIASA Interim Report IR-02-015. IIASA, Laxenburg.

Hugo, G (1996) 'Environmental concerns and international migration'. *International Migration Review* **30**(1): 105–31.

Hummel, H (2006) 'Bedeutungswandel des Multilateralismus'. In Debiel, T, Messner, D and Nuscheler, F (eds) *Globale Trends 2007. Frieden, Entwicklung, Umwelt.* Fischer, Frankfurt/M., pp61–80.

Humphrey, J and Messner, D (2006a) 'China and India as emerging global governance actors'. *IDS Bulletin* **37**(1): 107–14.

Humphrey, J and Messner, D (2006b) 'Eigensinnige Riesen in multipolarer Welt'. *E+Z Entwicklung und Zusammenarbeit* **47**(5): 192–6.

Humphrey, J and Messner, D (2006c) *Instabile Multipolarität: Indien und China verändern die Weltpolitik. Analysen und Stellungnahmen: Band 1.* Deutsches Institut für Entwicklungspolitik (DIE), Bonn.

Huntington, S P (1993) 'The clash of civilizations'. *Foreign Affairs* **72**(3): 22–49.

Huntington, S P (1996) *The Clash of Civilizations and the Remaking of World Order.* Simon & Schuster, New York.

Hutchinson, C F, Herrmann, S M, Maukonen, T and Weber, J (2005) 'Introduction: the "greening" of the Sahel'. *Journal of Arid Environments* **63**: 535–7.

Hutyra, L R, Munger, J W, Nobre, C A, Saleska, S R, Vieira, S A and Wofsy, S C (2005) 'Climatic variability and vegetation vulnerability in Amazonia'. *Geophysical Research Letters* **32**: L24712.

IBGE (Instituto Brasileiro de Geografia e Estatística) (2007) 'Daten aus der elektronischen Datenbank des brasilianischen Bundesamts für Statistik'. IBGE website, http://www.ibge.gov.br (viewed 16. March 2007).

IDA (International Development Association) (2006) 'Rebuilding Post-Hurricane Mitch'. IDA website, http://web.worldbank.org (viewed 4. December 2006).

IEA (International Energy Agency) (2002) *World Energy Outlook 2002.* IEA, Paris.

IEA (International Energy Agency) (2006a) *Renewables in Global Energy Supply 2006. An IEA Fact Sheet.* IEA, Paris.

IEA (International Energy Agency) (2006b) *EA Bioenergy. Annual Report 2005.* IEA, Paris.

IEA (International Energy Agency) (2006c) *World Energy Outlook 2006.* IEA, Paris.

IFRC (International Federation of Red Cross) (1999) *World Disasters Report 1999.* IFRC, Geneva.

IISD (International Institute for Sustainable Development) (2007) 'Summary of the 24th Session of the UNEP Governing Council/Global Ministerial Environment Forum: 5–9 February 2007'. *Earth Negotiations Bulletin* **12**.

Imbusch, P and Zoll, R (2005) *Friedens- und Konfliktforschung. Eine Einführung.* VS-Verlag für Sozialwissenschaften, Wiesbaden.

IMF (International Monetary Fund) (2006) *World Economic Outlook Database (Spring 2006), International Financial Statistics (IFS).* IMF, Washington, DC.

IMF (International Monetary Fund) (2007) *World Economic Outlook Database, April 2007.* IMF, Washington, DC.

Imhasly, B (2006) 'Demokratie als Basis für Stabilität – und Unruhe'. *Das Parlament* (7./14. August): 2.

IMISCOE (International Migration Integration and Social Cohesion) (2007) 'IMISCOE Joint Programme of Research'. University Amsterdam IMISCOE website, http://www.imiscoe.org/research/programme/documents/IMISCOEJointResearchProgramme.pdf (viewed 26. March 2007).

INPE (Instituto Nacional de Pesquisas Espaciais) (2007) 'Data from the website of the Brazilian National Institute for Space Research INPE'. INPE website, http://www.inpe.br (viewed 16. March 2007).

International Crisis Group (2006) *Bangladesh Today. Asia Report 121.* International Crisis Group, Manila.

IOM (International Organization for Migration) (1996) *Environmentally-Induced Population Displacements and Environmental Impacts Resulting from Mass Migration. International Symposium, Geneva, 21–24 April 1996.* IOM, Geneva.

IOM (International Organization for Migration) (2001) *IOM 50th Anniversary Council. An International Dialogue on Migration, 27–29 November 2001.* IOM, Geneva.

IOM (International Organization for Migration) (2006) 'About Migration. Facts & Figures. Global Estimates and Trends'. IOM website, http://www.iom.int/jahia/page254.html#9 (viewed 27. October 2006).

IPCC (Intergovernmental Panel on Climate Change) (1990) *Climate Change. The IPCC Response Strategies.* IPCC, Geneva.

IPCC (Intergovernmental Panel on Climate Change) (2000) *Special Report on Emissions Scenarios.* IPCC, Geneva.

IPCC (Intergovernmental Panel on Climate Change) (2001) *Climate Change 2001: Impacts, Adaptation and Vulnerability. Contribution of Working Group II to the Third Assessment Report of the IPCC.* Cambridge University Press, Cambridge, New York.

IPCC (Intergovernmental Panel on Climate Change) (2007a) *Climate Change 2007: The Physical Science Basis. Contribution of Working Group I to the Fourth Assessment Report of the IPCC.* IPCC, Geneva.

IPCC (Intergovernmental Panel on Climate Change) (2007b) *Climate Change 2007: Impacts, Adaptation and Vulnerability. Contribution of Working Group II to the Fourth Assessment Report of the IPCC.* IPCC, Geneva.

IPCC (Intergovernmental Panel on Climate Change) (2007c) *Climate Change 2007: Mitigation of Climate Change. Contribution of Working Group III to the Fourth Assessment Report of the IPCC.* IPCC, Geneva.

Ipsen, K (2004) *Völkerrecht.* Beck, Munich.

ISS (Institute for Security Studies) (2004) European Defence. A Proposal for a White Paper. Report *of an Independent Task Force.* ISS, Paris.

IWMI (International Water Management Institute) (2007) *Water for Food, Water for Life. A Comprehensive Assessment of Water Management in Agriculture.* Earthscan, London.

Jachtenfuchs, M (2003) 'Regieren jenseits der Staatlichkeit'. In Hellmann, G, Wolf, K D and Zürn, M (eds) *Die neuen*

Internationalen Beziehungen. Forschungsstand und Perspektiven in Deutschland. Nomos, Baden-Baden, p495.

Jackson, R (1990) *Quasi–States: Sovereignty, International Relations and the Third World.* Cambridge University Press, Cambridge, UK.

Jacobs, A and Mattes, H (Hrsg.) (2005) *Un-politische Partnerschaft. Eine Bilanz politischer Reformen in Nordafrika/Nahost nach zehn Jahren Barcelonaprozess.* Konrad-Adenauer-Stiftung, Bonn.

Jacobsen, K (2002) 'Livelihoods in conflict. The pursuit of livelihoods by refugees and the impact on the human security of the host communities'. *International Migration* **40**(5): 95–123.

Jahn, E (1997) 'Migration movements'. In Bernhardt, R (ed) *Encyclopedia of Public International Law. Volume III.* Elsevier, Amsterdam, pp369–71.

Jahn, E (2000) 'Refugees'. In Bernhardt, R (ed) *Encyclopedia of Public International Law. Volume IV.* Elsevier, Amsterdam, pp72–6.

Jakobeit, C (1998) 'Wirksamkeit in der internationalen Umweltpolitik'. *Zeitschrift für Internationale Beziehungen* **5**: 345–66.

Janse, R (2006) 'The legitimacy of humanitarian interventions'. *Leiden Journal of International Law* **19**: 669–92.

Jasanoff, S (1990) *The Fifth Branch: Science Advisers as Policy-Makers.* Harvard University Press, Cambridge, MA.

Jasanoff, S (2004) *States of Knowledge. The Co-production of Science and Social Order.* Routledge, London.

Jenkins, R (2005) 'Chinas Gewicht'. *Entwicklung und Zusammenarbeit (E+Z)* **46**(10): 376–9.

Johnston, A I (2003) 'China and International Institutions'. In Wang, Y (ed) *Construction Within Contradiction. Multiple Perspectives on the Relationship between China and International Organizations.* China Development Publishing House, Peking.

Jones, A (2002) 'Case Study: Genocide in Bangladesh, 1971'. Gendercide website, http://www.gendercide.org/case_bangladesh.html (viewed 21. March 2006).

Jones, S (2005) *Katrina's Impact on World Energy Markets.* Purvin Gertz Inc., Houston.

Kagan, R (2002) 'Power and weakness'. *Policy Review* **113**: 3–28.

Kagan, R (2003) *Of Paradise and Power. America and Europe in the New World Order.* Alfred A. Knopf, New York.

Kaiser, J (2001) 'Ecological restoration: NRC Panel pokes holes in Everglades scheme'. *Science* **291**(5503): 959–61.

Kaldor, M (Hrsg.) (1999) *New and Old Wars. Organized Violence in a Global Era.* Stanford University Press, Palo Alto, CA.

Kalter, F (2000) 'Theorien der Migration'. In Müller, U, Nauck, B and Diekmann, A (eds) *Handbuch der Demographie 1: Modelle und Methoden. Band 1.* Springer, Berlin, Heidelberg, New York, pp438–57.

Kaplan, R D (1994) 'The coming anarchy'. *Atlantic Monthly* **273**(2): 44–76.

Kaplan, S L (1985) 'The Paris Bread Riot of 1725'. *French Historical Studies* **14**(1): 23–56.

Kaplinsky, A (2005) 'The Impact of Asian Drivers on the Developing World. Asian Drivers Programme – Background Papers'. Institute of Development Studies (IDS) at the University of Sussex website, http://www.ids.ac.uk/ids/global/Asiandriversbackgroundpapers.html (viewed 8. February 2007).

Kaplinsky, R (2006) 'China and the global terms of trade'. *IDS Bulletin (Special Issue)* **37**(1): 43–53.

Karaev, Z (2005) 'Water diplomacy in Central Asia'. *The Middle East Review of International Affairs* **9**(1): 63–9.

Karim, H and Gnisci, D (2004) *2004–2006 Initial Work Programme.* OECD Sahel and West Africa Club Secretariat. The Governance, Conflict Dynamics, Peace and Security Unit, Paris.

Katzenstein, P J (2005) *A World of Regions. Asia and Europe in the American Imperium.* Cornell University Press, Ithaca, NY.

Kaufmann, D, Kraay, A and Mastruzzi, M (2006) *Governance Matters V: Governance Indicators for 1996–2005.* World Bank, Washington, DC.

Keane, D (2004) 'The environmental causes and consequences of migration: a search for the meaning of "environmental refugees"'. *Georgetown International Environmental Law Review* **16**: 209–23.

Kemfert, C and Schumacher, K (2005) *Costs of Inaction and Costs of Action in Climate Protection – Assessment of Costs of Inaction or Delayed Action of Climate Protection and Climate Change.* DIW Politikberatung kompakt. Deutsches Institut für Wirtschaftsforschung (DIW), Berlin.

Kennedy, P (1988) *The Rise and Fall of the Great Powers: Economic Change and Military Conflict from 1500–2000.* Harper Collins, New York.

Kennedy, P and Connelly, M (1994) 'Must it be the rest against the west?' *Atlantic Monthly* **12**: 61–91.

Kennedy, P, Messner, D and Nuscheler, F (Hrsg.) (2002) *Global Trends and Global Governance.* Pluto Press, London.

Keohane, R O and Nye, J S (1977) *Power and Interdependence. World Politics in Transition.* Little Brown, Boston, MA.

Keohane, R O and Nye, J S (1987) 'Power and interdependence revisited'. *International Organization* **41**(4): 725–53.

Keohane, R O and Nye, J S (2000) 'Globalization: what's new? What's not? (and so what?)'. *Foreign Policy* **118**: 104–19.

KfW (Kreditanstalt für Wiederaufbau) (2002) *Transportinfrastruktur – Investitionen in die Transportinfrastruktur als Beitrag zur Armutsbekämpfung.* Diskussionsbeitrag Nr. 30/2002. KfW, Frankfurt/M.

Kibreab, G (1994) 'Migration, environment and refugeehood'. In Zaba, B and J., C (eds) *Environment and Population Change.* International Union for the Scientific Study of Population. Derouaux Ordina Editions, Liège, Belgium, pp115–29.

Kienast, F, Wildi, O and Brzeziecki, B (1998) 'Potential impacts of climate change on species richness in mountain forests – an ecological risk assessment'. *Biological Conservation* **83**: 291–305.

Kimminich, O (1997) *Einführung in das Völkerrecht.* Francke, Tübingen, Basel.

Klaphake, A and Scheumann, W (2001) 'Politische Antworten auf die globale Wasserkrise: Trends und Konflikte'. *Aus Politik und Zeitgeschichte* **48–49**: 3–13.

Klaphake, A and Voils, O (2006) 'Cooperation on international rivers from an economic perspective: current state and experiences'. In Scheumann, W and Neubert, S (eds) *Transboundary Water Management in Africa. DIE Studies 21.* Deutsches Institut für Entwicklungspolitik (DIE), Bonn, pp103–72.

Klare, M and Volman, D (2006) 'America, China & the scramble for Africa's oil'. *Review of African Political Economy* **108**: 227–309.

Klein, R J T, Nicholls, R J and Thomalla, F (2002) 'The resilience of coastal megacities to weather related hazards

– a review'. In Kreimer, A, Arnold, M and Carlin, A (eds) *Proceedings of the Future Disaster Risk: Building Safer Cities.* World Bank, Washington, DC, pp111–37.

Klenke, C (2006) 'Ein gewaltiges Krisenpotential. Militärische Aspekte des Taiwan-Konfliktes'. *Das Parlament* **56**(30731): 3.

Klingebiel, S and Roehder, K (2004) *Entwicklungspolitisch–militärische Schnittstellen. Neue Herausforderungen in Krisen und Post–Konfliktsituationen.* Deutsches Institut für Entwicklungspolitik (DIE), Bonn.

Klingebiel, S and Roehder, K (2005) 'Entwicklungs- und Sicherheitspolitik. Neue Schnittstellen in Krisen- und Post-Konflikt-Situationen'. In Messner, D and Scholz, I (eds) *Zukunftsfragen der Entwicklungspolitik.* Nomos, Baden-Baden, pp391–403.

Korf, B (2005) 'Rethinking the greed-grievance nexus: property rights and the political economy of war in Sri Lanka'. *Journal of Peace Research* **42**(2): 201–17.

Körner, C (2006) 'Plant CO_2 responses: an issue of definition, time and resource supply'. *New Phytologist* **172**(3): 393–411.

Kramer, J (1989) *Kein Deich – kein Land – kein Leben: Geschichte des Küstenschutzes an der Nordsee.* Rautenberg, Leer.

Kreutzmann, H (2004) 'Mittelasien – politische Entwicklung, Grenzkonflikte und Ausbau der Verkehrsinfrastruktur'. *Geographische Rundschau* **56**(10): 3–9.

Kröhnert, S (2003) 'Theorien der Migration. Berlin-Institut für Weltbevölkerung und globale Entwicklung – Online Handbuch'. Berlin-Institut für Weltbevölkerung und globale Entwicklung website, http://www.berlin-institut.org/pdfs/Kroehnert_Theorien_der_Migration.pdf (viewed 27. October 2006).

Kulessa, M and Ringel, M (2003) 'Kompensationen als innovatives Instrument globaler Umweltschutzpolitik: Möglichkeiten und Grenzen einer Weiterentwicklung des Konzepts am Beispiel der biologischen Vielfalt'. *Zeitschrift für Umweltpolitik & Umweltrecht* **26**(3): 263–85.

Kupchan, C A (2003) *The End of the American Era. U.S. Foreign Policy and the Geopolitics of the Twenty-first Century.* Alfred A. Knopf, New York.

Kupchan, C A, Adler, E, Coicaud, J-M and Khong, Y F (2001) *Power in Transition: The Peaceful Change of International Order.* UNU Press, Tokyo.

Kurlantzick, J (2006) 'China's Charm: Implications of Chinese Soft Power'. Washington, DC: Carnegie Endowment for International Peace.

Lacina, B (2004) 'From side show to centre state: civil conflict after the cold war'. *Security Dialogue* **35**(2): 191–205.

Lailufar, Y (2004) 'Bangladesh-India tussles'. *South Asian Jounal* website, http://www.southasianmedia.net/magazine/journal/bangladeshindia_tussels.htm (viewed 27. October 2006).

Lal, M, Nozawa, T, Emori, S, Harasawa, H, Takahashi, K, Kimoto, M, Abe-Ouchi, A, Nakajima, T, Takemura, T and Numaguti, A (2001) 'Future climate change: implications for Indian summer monsoon and its variability'. *Current Science* **81**(9): 1196.

Lamb, H H (1995) *Climate, History and the Modern World.* Routledge, London.

Lambach, D (2005) *Schwäche und Zerfall von Staaten. Operationalisierung eines schwierigen Konzepts. Paper gelesen anlässlich "Krieg, Gewalt und prekärer Frieden": Nachwuchstagung der Arbeitsgemeinschaft Friedens– und Konfliktforschung, 14.–16. January, Bocholt.* Arbeitsgemeinschaft Friedens- und Konfliktforschung, Bocholt.

Lan, X, Simonis, U E and J., D (2006) *Environmental Governance in China.* Wissenschaftszentrum Berlin für Sozialforschung (WZB), Berlin.

Landsteiner, E (1999) 'The crisis of wine production in late sixteenth–century Central Europe: climatic causes and economic consequences'. *Climatic Change* **43**: 323–34.

Langhammer, R (2005) 'China and the G-21: a new north-south divide in the WTO after Cancún?' *Journal of the Asia Pacific Economy* **10**(3): 339–58.

Le Billon, P (2001) 'The political ecology of war: natural reources and armed conflicts'. *Political Geography* **20**: 561–84.

Le Billon, P (2002) 'Risiko Ressourcenreichtum: Ursachen und Wirkung der "Neuen Kriege"'. In Jung, A (ed) *medico Report 24: Ungeheuer ist nur das Normale: Zur Ökonomie der "neuen" Kriege.* Mabuse-Verlag, Frankfurt/M, pp28–49.

Le Monde Diplomatique (2007) *Atlas der Globalisierung.* taz Verlag, Berlin.

Lee, E S (1972) 'Eine Theoerie der Wanderung'. In Széll, G (ed) *Regionale Mobilität.* Nymphenburger Verlagshandlung, Munich, pp115–29.

Leemans, R and Eickhout, B (2004) 'Another reason for concern: regional and global impacts on ecosystems for different levels of climate change'. *Global Environmental Change* **14**: 219–28.

Lefebvre, G (1932) *La Grande Peur de 1789.* Libraire Armand Colin, Paris.

Leighton, M (2006) 'Desertification and migration'. In Johnson, P M, Mayrand, K and Paquin, M (eds) *Governing Global Desertification. Linking Environmental Degradation, Poverty and Participation.* Ashgate, Aldershot, pp43–59.

Lenton, T M, Held, H, Kriegler, E, Hall, J, Lucht, W, Rahmstorf, S and Schellnhuber, H J (2007) 'Tipping elements in the Earth system'. *PNAS* (submitted).

Létolle, R and Mainguet, M (1996) *Der Aralsee. Eine ökologische Katastrophe.* Springer, Berlin, Heidelberg, New York.

Levermann, A, Griesel, A, Hofmann, M, Montoya, M and Rahmstorf, S (2005) 'Dynamic sea level changes following changes in the thermohaline circulation'. *Climate Dynamics* **24**: 347–54.

Levy, M A (1995) 'Is the environment a national security issue?' *International Security* **20**(2): 35–62.

Lieberthal, K and Lampton, D (1992) *Bureaucracy, Politics and Decision-Making in Post-Mao China.* University of California Press, Berkley, LA, Oxford.

Liedtke, H and Marcinek, J (2002) *Physische Geographie Deutschlands.* Klett, Gotha.

Lindemann, S (2006) 'Success and failure in international river basin management – the case of Southern Africa'. In Jänicke, M and Jacob, K (eds) *Environmental Governance in Global Perspective: New Aproaches to Ecological and Political Modernisation.* University Berlin, Environmental Policy Research Centre, Berlin.

Liniger, H, Weingarten, R and Grosjean, M (1998) *Mountains of the World: Water Towers for the 21st Century.* University Bern. Mountain Agenda. Center for Development and Environment, Bern.

Lipschutz, R D (1995) *On Security. New Directions in World Politics.* Columbia University Press, New York.

Lobina, E (2000) 'Cochabamba – water war'. *Focus (PSI Journal)* **7**(2).

Lohrmann, R (2000) 'Migrants, refugees and insecurity. Current threats to peace?' *International Migration* **38**(4): 3–22.

Lonergan, S (1998) 'The role of environmental degradation in population displacement'. *Environmental Change and Security Project Report* **4**: 5–15.

Long, S P, Ainsworth, E A, Leakey, A D B, Nosberger, J and Ort, D R (2006) 'Food for thought: lower-than-expected crop yield simulation with rising CO_2 concentrations'. *Science* **312**: 1918–21.

Lüders, M (2003) 'Macht und Glauben in Zentralasien'. *Aus Politik und Zeitgeschichte* **B 37**: 49–54.

Lummel, P (2002) *Von der Hungersnot zum Beginn modernen Massenkonsums – Berlins nimmersatter Riesenbauch – Die Lebensmittelversorgung einer werdenden Weltstadt. Zeitschrift "Nahrungskultur: Essen und Trinken im Wandel"*. Landeszentrale für politische Bildung Baden-Württemberg, Stuttgart.

MA (Millennium Ecosystem Assessment) (Hrsg.) (2005a) *Millennium Ecosystem Assessment 2005. Ecosystems and Human Well-Being: Synthesis Report*. Island Press, Washington, DC.

MA (Millennium Ecosystem Assessment) (2005b) *Millennium Ecosystem Assessment 2005. Ecosystems and Human Well-Being: Biodiversity Synthesis*. World Resources Institute (WRI), Washington, DC.

MA (Millennium Ecosystem Assessment) (2006) *Millennium Ecosystem Assessment. Ecosystems and Human Well-Being: Current State and Trends. Volume 1*. Island Press, Washington, DC.

MacCallion, K F (2000) 'Environmental justice without borders: the need for an International Court of the Environment to protect fundamental environmental rights'. *George Washington Journal of International Law and Economics* **32**: 351–65.

Mahar, D (1988) Government Policies and Deforestation in Brazil's Amazon Region. World Bank, Washington, DC.

Mainwaring, S, Bejarano, A M and Pizarro, E (2006) *The Crisis of Representation in the Andes*. Stanford University Press, Stanford.

Mark, B G, Seltzer, G O and Geoffrey, O (2003) 'Tropical glacier melt water contribution to stream discharge: a case study in the Cordillera Blanca, Peru'. *Journal of Glaciology* **49**: 271–81.

Martin, A (2005) 'Environmental conflict between refugee and host communities'. *Journal of Peace Research* **42**(3): 329–46.

Marugg, M (1990) *Völkerrechtliche Definitionen des Ausdrucks "Flüchtling". Ein Beitrag zur Geschichte unter besonderer Berücksichtigung sogenannter de-facto-Flüchtlinge*. Helbing & Lichtenhahn, Basel, Frankfurt/M.

Massey, D S, Arango, J, Hugo, G, Kouaouci, A, Pellegrino, A and Taylor, J E (1998) *Worlds in Motion. Understanding International Migration at the End of the Millennium*. Clarendon Press Oxford, Oxford, NY.

Matthew, R, Brklacich, M and Mcdonald, B (2004) 'Analyzing environment, conflict and cooperation'. In United Nations Environment Programme (UNEP) (ed) *Understanding Environment, Conflict and Cooperation*. pp5–16.

Matthew, R and McDonald, B (2004) 'Networks of threats and vulnerability: lessons from environmental security research'. *ESCP Report* **10**: 36–42.

Matthew, R A and Fraser, L (2002) 'Global Environmental Change and Human Security: Conceptual and Theoretical Issues'. Irvine, CA: University of California.

McGregor, J (1993) 'Refugees and the environment'. In Black, R and Robinson, V (eds) *Geography and Refugees: Patterns and Processes of Change*. Belhaven, London, pp157–70.

Mearsheimer, J J (1990) 'Back to the future. Instability in Europe after the cold war'. *International Security* **15**(1): 5–56.

Medeiros, E S and Fravel, M T (2003) 'Victim no more. China's new diplomacy'. *Foreign Affairs* **82**(6): 22–35.

Mehler, A (2002) 'Structural stability: meaning, scope and use in an African context'. *afrika spectrum* **37**(1): 5–23.

Meinshausen, M (2006) 'What does a 2°C target mean for greenhouse gas concentrations? A brief analysis based on multi-gas emission pathways and several climate sensitivity uncertainty estimates'. In Schellnhuber, H-J, Cramer, W, Nakicenovic, N, Wigley, T M L and Yohe, G (eds) *Avoiding Dangerous Climate Change*. Cambridge University Press, Cambridge, New York, pp265–79.

Menzel, A, Sparks, T H, Estrella, N, Koch, E, Aasa, A, Aha, R, Alm-Kubler, K, Bissolli, P, Braslavska, O, Briede, A, Chmielewski, F M, Crepinsek, Z, Curnel, Y, Dahl, A, Defila, C, Donnelly, A, Filella, Y, Jatcza, K, Mage, F, Mestre, A, Nordli, O, Penuelas, J, Pirinen, P, Remisova, V, Scheifinger, H, Striz, M, Susnik, A, van Vliet, A J H, Wielgolaski, F E, Zach, S and Zust, A (2006) 'European phenological response to climate change matches the warming pattern'. *Global Change Biology* **12**: 1969–76.

Menzel, U (2003) *Die neue Hegemonie der USA und die Krise des Multilateralismus. Forschungsbericht*. Institut für Sozialwissenschaften der Technischen Universität Braunschweig, Braunschweig.

Messner, D (Hrsg.) (1998) *Die Zukunft des Staates und der Politik*. Dietz, Bonn.

Messner, D (2006) 'Machtverschiebungen im internationalen System: Global Governance im Schatten des Aufstieges von China und Indien'. In Debiel, T, Messner, D and Nuscheler, F (eds) *Globale Trends 2007. Frieden, Entwicklung, Umwelt*. Fischer, Frankfurt/M., pp45–60.

Messner, D, Schade, J and Weller, C (2003) 'Weltpolitik zwischen Staatenanarchie und Global Governance'. In Hauchler, I, Messner, D and Nuscheler, F (eds) *Globale Trends 2004/2005*. Fischer, Frankfurt/M., pp235–51.

Metz, C M (2001) *Dominican Republic and Haiti: Country Studies*. Federal Research Division. Library of Congress, Washington, DC.

Miles, L, Grainger, A and Phillips, O L (2004) 'The impact of global climate change on tropical forest biodiversity in Amazonia'. *Global Ecology and Biogeography* **13**: 553–65.

Milliken, J and Krause, K (2003) 'State failure, state collapse and state reconstruction: concepts, lessons, and strategies'. In Miliken, J (ed) *State Failure, Collapse and Reconstruction*. Blackwell, London, pp753–74.

Mitchell, T D, Hulme, M and New, M (2002) 'Climate data for political areas'. *Area* **34**: 109.

Mitra, S K (2006) 'Die Hindu-Bombe. Auf dem Weg in den Club der Atommächte'. *Das Parlament* (7./14. August): 5.

Miyan, A (2003) *Dynamics of Labor Migration – Bangladesh Context*. Mimeo. International University of Business Agriculture and Technology (IUBAT), Bangladesh.

Molle, F and Berkoff, J (2006) *Cities Versus Agriculture: Revisiting Intersectoral Water Transfers, Potential Gains and Conflicts. Comprehensive Assessment Research Report 10*. Comprehensive Assessment Secretariat, Colombo, Sri Lanka.

Montalvo, J and Reynal–Querol, M (2003) 'Ethnic polarization, potential conflict, and civil wars'. *American Economic Review* **95**(3): 796–816.

Morton, D C, DeFries, R S, Shimabukuro, Y E, Anderson, L O, Arai, E, del Bon Espirito-Santo, F, Freitas, R and Morisette, J (2006) 'Cropland expansion changes deforestation dynamics in the Southern Brazilian Amazon'. *Proceedings of the National Academy of Sciences of the USA* **103**: 14637–41.

Mouillot, F, Rambal, S and Joffre, R (2002) 'Simulating climate change impacts on fire-frequency and vegetation dynamics in a Mediterrannean-type ecosystem'. *Global Change Biology* **8**: 423–37.

MPI (Max Planck Institute for Meteorology) (2006) *Climate Projections for the 21st Century*. MPI, Hamburg.

Muller, E N and Weede, E (1990) 'Cross national variation in political violence: A rational action approach'. *Journal of Conflict Resolution* **34**(4): 624–51.

Münchener Rück (2004) *Welt der Naturgefahren. CD-Rom*. Münchener Rück, Munich.

Münchener Rück (2006) *Hurrikane – stärker, häufiger, teurer. Assekuranz im Änderungsrisiko*. Münchener Rück, Munich.

Münkler, H (2002) *Die neuen Kriege*. Rowohlt, Hamburg.

Münkler, H (2005) *Imperien. Die Logik der Weltherrschaft – Vom Alten Rome bis zu den Vereinigten Staaten*. Rowohlt, Berlin.

Myers, N (1993) 'Environmental refugees in a globally warmed world'. *Bioscience* **43**: 752–61.

Myers, N (2002) 'Environmental refugees: a growing phenomenon of the 21st century'. *Philosophical Transactions of the Royal Society of London Series B-Biological Sciences* **1420**: 609–13.

Najam, A, Papa, M and Taiyab, N (2006) *Global Environmental Governance. A Reform Agenda*. International Institute for Sustainable Development (IISD), Winnipeg, Manitoba.

Narain, S (2006) '"Die Reichen sollen bezahlen" – Die indische Umweltschützerin Sunita Narain über die Kosten des Klimawandels für ärmere Länder und die Verantwortung der Industriestaaten (Interview)'. *Die Zeit* (10. August): 19.

Neilson, R P, Pitelka, L F, Solomon, A M, Nathan, R, Midgley, G F, Fragoso, J M V, Lischke, H and Thompson, K (2005) 'Forecasting regional to global plant migration in response to climate change'. *BioScience* **55**: 749–59.

Nemani, R R, Keeling, C D, Hashimoto, H, Jolly, W M, Piper, S C, Tucker, C J, Myneni, R B and Running, S W (2003) 'Climate-driven increases in global terrestrial net primary production from 1982 to 1999'. *Science* **300**: 1560–3.

Nepstad, D C, Moutinho, P, Dias, M B, Davidson, E, Cardinot, G, Markewitz, D, Figueiredo, R, Vianna, N, Chambers, J, Ray, D, Guerreiros, J B, Lefebvre, P, Sternberg, L, Moreira, M, Barros, L, Ishida, F Y, Tohlver, I, Belk, E, Kalif, K and Schwalbe, K (2002) 'The effects of partial throughfall exclusion on canopy processes, aboveground production, and biogeochemistry of an Amazon forest'. *Journal of Geophysical Research – Atmospheres* **107**: Art. 8085.

Nepstad, D C, Verissimo, A, Alencar, A, Nobre, C, Lima, E, Lefebvre, P, Schlesinger, P, Potterk, C, Moutinho, P, Mendoza, E, Cochrane, M and Brooksk, V (1999) 'Large-scale impoverishment of Amazonian forests by logging and fire'. *Nature* **398**: 505–8.

Neubert, S (2002) *Wege zur Überwindung regionaler Wasserarmut. Politischer Wille und angepasste Managementstrategien entscheiden über zukünftige Verfügbarkeit der Ressource*. Analysen und Stellungnahmen (4/2002). Deutsches Institut für Entwicklungspolitik (DIE), Bonn.

Neubert, S and Herrfahrdt, E (2005) 'Integriertes Wasserressourcen-Management: ein realistisches Konzept für Entwicklungs- und Transformationsländer?' In Messner, D and Scholz, I (eds) *Zukunftsfragen der Entwicklungspolitik*. Nomos Verlagsgesellschaft, Baden-Baden.

New, M, Hewitson, B, Stephenson, D B, Tsiga, A, Kruger, A, Manhique, A, Gomez, B, Coelho, C A S, Masisi, D N, Kululanga, E, Mbambalala, E, Adesina, F, Saleh, H, Kanyanga, J, Adosi, H, Bulane, L, Fortunata, L, Mdoka, M L and Lajoie, R (2006) 'Evidence of trends in daily climate extremes over southern and west Africa'. *Journal of Geophysical Research* **111**(D14102): doi:10.1029/2005JD006289.

Nicholls, R J (1995) 'Coastal megacities and climate change'. *GeoJournal* **37.3**: 369–79.

Norby, R J, DeLucia, E H, Gielen, B, Calfapietra, C, Giardina, C, P, King, J S, Ledford, J, McCarthy, H R, Moore, D J P, Ceulemans, R, de Angelis, P, Finzi, A C, Karnosky, D F, Kubiske, M E, Lukac, M, Pregitzer, K S, Sacrascia-Mugnozza, G E, Schlesinger, W H and Oren, R (2005) 'Forest response to elevated CO_2 is conserved across a broad range of productivity'. *Proceedings of the National Academy of Sciences of the United States of America* **102**: 18052–6.

Nordås, R and Gleditsch, N P (2005) *Climate Conflicts: Common Sense or Nonsense?* Paper presented at the 13th Annual National Political Science Conference, Hurdalsjøen, Norway, 5–7 January 2005. Centre for the Study of Civil War (CSCW) at the International Peace Research Institute (PRIO), Oslo.

Nuscheler, F (2004) *Internationale Migration. Flucht und Asyl*. VS Verlag für Sozialwissenschaften, Wiesbaden.

Nussbaumer, J (2003) *Gewalt. Macht. Hunger – Schwere Hungerkatastrophen seit 1845*. Studienverlag, Innsbruck.

Nye, J S (2002) *The Paradox of American Power. Why the World's Only Superpower Can't Go It Alone*. Oxford University Press, Oxford, New York.

O'Leary, M (2004) *The First 72 Hours: A Community Approach to Disaster Preparedness*. iUniverse Inc. Book Publisher, Lincoln, NE.

O'Neill, K (2005) *Decentralizing the State: Elections, Parties, and Local Power in the Andes*. Cornell University Press, New York.

OCHA (Office for the Coordination of Humanitarian Affairs) (2006) 'Central Emergency Response Fund'. United Nations (UN) website, http://ochaonline2.un.org/Default.aspx?tabid=8770 (viewed 15. February 2007).

ODI (Overseas Development Institute) (2005) *Conflict in the Great Lakes Region. How is it linked with Land and Migration?* Natural Resource Perspectives 96. ODI, London.

OECD (Organisation for Economic Development and Co-operation) (2001) *The DAC Guidelines – Helping Prevent Violent Conflict*. OECD, Paris.

OECD (Organisation for Economic Development and Co-operation) (2007) *Fragile States: Policy Commitment and Principles for Good International Engagement in Fragile States and Situations. Final*. Document DCD/DAC(2007)29. OECD, Paris.

OECD DAC – Organisation for Economic Co-operation and Development - Development Assistance Committee (2003) 'Principles and Good Practice of Humanitarian Donorship'. OECD DAC website, http://www.reliefweb.int/ghd/outreach.html (viewed 25. April 2007).

OECD-FAO (Organisation for Economic Development and Co-operation - Food and Agriculture Organisation of the United Nations) (2005) 'Agricultural Outlook: 2005–2014'. OECD, FAO website, http://www.oecd.org/document/5/0,2340,en_2649_201185_35015941_1_1_1_1,00.html (viewed 27. October 2006).

OECD-FAO (Organisation for Economic Development and Co-operation - Food and Agriculture Organisation of the United Nations) (2006) 'Agricultural Outlook: 2006–2015'. OECD, FAO website, http://www.oecd.org/document/16/

0,2340,en_2649_201185_37032958_1_1_1_1,00.html (viewed 27. October 2006).

OHCHR (Office of the United Nations High Commissioner for Human Rights) (2006) 'Questions and Answers about IDPs'. OHCHR website, http://www.ohchr.org (viewed 11. May 2006).

Oki, T and Kanae, S (2006) 'Global hydrological cycles and world water resources'. *Science* **313**: 1068–72.

Oldeman, L R (1992) *Global Extent of Soil Degradation. ISRIC Bi-Annual Report 1991–1992.* International Soil Reference and Information Centre (ISRIC), Wageningen.

Oldeman, L R, Hakkeling, R T A and Sombroek, W G (1991) *World Map of the Status of Human-Induced Soil Degradation – An Explanatory Note. Global Assessment of Soil Degradation GLASOD.* International Soil Reference and Information Centre (ISRIC), Wageningen.

Opp, C (2004) 'Desertifikation in Usbekistan'. *Geographische Rundschau* **56**(10): 44–51.

Ottaway, M and Mair, S (2004) *States at Risk and Failed States: Putting Security First (Policy Outlook of the Democracy & Rule of Law Project).* Carnegie Endowment for International Peace and German Institute for International and Security Affairs (SWP), Washington, DC.

Oxfam (2006) *Kenyan Food Crisis Compounded by 'Serious Flaws' in Distribution System.* Press Release January 17. Oxfam, Oxford.

Parkes, R (2006) *Gemeinsame Patrouillen an Europas Südflanke. Zur Frage der Kontrolle der afrikanischen Einwanderung. SWP-Aktuell 44.* Stiftung Wissenschaft und Politik (SWP), Berlin.

Parmesan, C and Yohe, G (2003) 'A globally coherent fingerprint of climate change impacts across natural systems'. *Nature* **421**: 37–42.

Parry, M, Rosenzweig, C, Iglesias, A, Fischer, G and Livermore, M (1999) 'Climate change and world food security: a new assessment'. *Global Environmental Change* **9**: S51–S67.

Parry, M L, Rosenzweig, C, Iglesias, A, Livermore, M and Fischer, G (2004) 'Effects of climate change on global food production under SRES emissions and socio-economic scenarios'. *Global Environmental Change* **14**: 53–67.

Pathania, J (2003) 'India & Bangladesh – Migration Matrix. South Asia Analysis Group Paper No. 632'. South Asia Analysis Group (SSAG) website, http://www.saag.org/papers7/paper632.html (viewed 17. July 2006).

Paul, B K (2005) 'Evidence against disaster-induced migration: the 2004 tornado in northcentral Bangladesh'. *Disasters* **29**: 370–85.

Pauwelyn, J (2002) *A World Environment Court. International Environmental Governance (Gaps and Weaknesses/Proposals for Reform). Working Paper.* United Nations University (UNU), Tokyo.

Penuelas, J and Filella, I (2001) 'Responses to a warming world'. *Science* **294**: 793–5.

People's Daily Online (2007) 'Chinese Government Will Protect Migrant Workers Legal Rights'. People's Daily Online website, http://english.people.com.cn/200703/01/eng20070301_353438.html (viewed 6. March 2007).

Percival, V and Homer–Dixon, T F (1995) 'Environmmental scarcity and violent conflict: the case of Rwanda'. *The Journal of Environment and Development* **5**(3): 270.

Peru Cambio Climatico (2001) *Comunicación Nacional del Perú a la Convención de Naciones Unidas sobre Cambio Climático.* Cambio Climatico Peru, Lima, Peru.

Peterson, B J, Holmes, R M, McClelland, J W, Vörösmarty, C J, Lammers, R B, Shiklomanov, A I, Shiklomanov, I A and Rahmstorf, S (2002) 'Increasing river discharge to the Arctic Ocean'. *Science* **298**: 2171–3.

Peterson, L C and Haug, G H (2006) 'Variability in the mean latitude of the Atlantic Intertropical Convergence Zone as recorded by riverine input of sediments to the Cariaco Basin (Venezuela)'. *Palaeogeography Palaeoclimatology Palaeoecology* **234**: 97–113.

Phienwej, N and Nutalaya, P (2005) 'Subsidence and flooding in Bangkok'. In Gupta, A (ed) *The Physical Geography of Southeast Asia.* Oxford University Press, Oxford, New York, pp358–78.

Philips, D, Daoudy, M, McCaffrey, S, Öjendal, J and Turton, A R (2006) *Trans-boundary Water Coooperation as a Tool for Conflict Prevention and for Broader Benefit-Sharing.* Global Development Studies No. 4. Swedish Ministry of Foreign Affairs, Stockholm.

Phuong, C (2004) *The International Protection of Internally Displaced Persons.* Cambridge University Press, Cambridge, New York.

Pilardeaux, B (2004) 'Erschwerte Bedingungen – Globale Umweltveränderungen und Ernährungssicherung'. *Politische Ökologie* **90**: 19–22.

Postiglione, A (1999) *The Global Demand for an International Court of the Environment.* International Court of the Environment Foundation (ICEF), Napels.

Postiglione, A (2002) 'La necessità di una corte internazionale dell'ambiente'. *Rivista giuridica dell'ambiente* **17**: 389–94.

Potter, C, Klooster, S, Reis de Carvalho, C, Brooks Genovese, V, Torregrosa, A, Dungan, J, Bobo, M and Coughlan, J (2001) 'Modeling seasonal and interannual variability in ecosystem carbon cycling for the Brazilian Amazon region'. *Journal of Geophysical Research* **106**: 10423–46.

Potter, C, Tan, P-N, Kumar, V, Kucharik, C, Klooster, S, Genovese, V, Cohen, W and Healey, S (2005) 'Recent history of large-scale ecosystem disturbances in North America derived from the AVHRR satellite record'. *Ecosystems* **8**: 808–24.

Prugh, T, Flavin, C and Sawin, J L (2005) 'Changing the oil economy. Chapter 6'. In Worldwatch Institute (ed) *State of the World 2005: Redefining Global Security.* Worldwatch Institute, Washington, DC, pp100–20.

Rabobank (2005) *Country Report Peru.* Economic Research Department. Country Risk Research, Lima, Peru.

Raffinot, M and Rosellini, C (2006) *Out of Financing Trap? Financing Post-Conflict Countries and LICUSs.* Working Paper 29. Agence Française de Developpement, Paris.

Rahmstorf, S (2002) 'Ocean circulation and climate during the past 120,000 years'. *Nature* **419**: 207–14.

Rahmstorf, S (2007) 'A semi-empirical approach to projecting future sea-level rise'. *Science* **315**: 368–70.

Rahmstorf, S, Cazenave, A, Church, J A, Hansen, J E, Keeling, R F, Parker, D E and Somerville, R C J (2007) 'Recent climate observations compared to projections'. *Science*: DOI: 10.1126/science.1136843.

Raja, M C (2006) 'India and the balance of power'. *Foreign Affairs* **85**(4): 17–32.

Raknerud, A and Hegre, H (1997) 'The hazard of war: reassessing the evidence for the democratic peace'. *Journal of Peace Research* **34**(4): 385–404.

Rankin, M B (2005) *Extending the Limits or Narrowing the Scope? Deconstructing the OAU Refugee Definition Thirty Years on. New Issues in Refugee Research.* Working Paper

No. 113. UNHCR Evaluation and Policy Analysis Unit, Geneva.

Rechkemmer, A (2005) *UNEO – Towards and International Environment Organization. Approaches to a Sustainable Reform of Global Environmental Governance.* Nomos, Baden-Baden.

Regan, P and Norton, D (2005) 'Greed, grievance, and mobilization in civil wars'. *Journal of Conflict Resolution* **49**(3): 319–36.

Reinhardt, D and Rolf, C (2006) *Humanitäre Hilfe und vergessene Katastrophen: UN-Weltgipfel und neue Finanzierungsmechanismen.* INEF Policy Brief 1/2006. Institut für Entwicklung und Frieden (INEF), Duisburg.

Rest, A (1998) 'Zur Notwendigkeit eines internationalen Umweltgerichtshofes'. In Hafner, G (ed) *Liber amicorum Ignaz Seidl-Hohenveldern.* Kluwer, The Hague, pp575–91.

Reuters Foundation (2000a) 'Venezuela's Chavez to Probe Rights Abuse Claims'. ReliefWeb website, http://www.reliefweb.int/rw/rwb.nsf/db900SID/OCHA-64C4BG? (viewed 27. June 2006).

Reuters Foundation (2000b) 'Eastern India Floods Claim More Victims'. ReliefWeb website, http://www.reliefweb.int/rw/rwb.nsf/db900SID/OCHA-64BVVF?OpenDocument&cc=ind&rc=3 (viewed 27. June 2006).

Reuveny, R (2005) *Environmental Change, Migration and Conflict: Theoretical Analysis and Empirical Explorations. International Workshop Paper: Human Security and Climate Change, Norway.* School of Public and Environmental Affairs Indiana University, Bloomington, Indiana.

Reynal-Querol, M (2002) 'Ethnicity, political systems, and civil wars'. *Journal of Conflict Resolution* **46**(1): 29–54.

Reynal-Querol, M (2005) 'Does democracy preempt civil wars?' *European Journal of Political Economy* **21**: 445–65.

Richerzhagen, C (2007) *Capacity for Global Environmental Governance. China and the Climate Change Problem.* Deutsches Institut für Entwicklungspolitik (DIE), Bonn.

Richter, R E (2000) 'Umweltflüchtlinge in Westafrika. Ursachen, Ausmaß und Perspektiven'. *Gegraphische Rundschau* **52**(11): 12–7.

Ridley, J K, Huybrechts, P, Gregory, J M and Lowe, J A (2005) 'Elimination of the Greenland ice sheet in a high CO_2 climate'. *Journal of Climate* **18**: 3409–27.

Rignot, E, Casassa, G, Gogineni, P, Krabill, W, Rivera, A and Thomas, R (2004) 'Accelerated ice discharge from the Antarctic Peninsula following the collapse of Larsen B ice shelf'. *Geophysical Research Letters* **31**(18): L18401, doi:10.1029/2004GL020697.

Riofrío, G (1996) 'Lima – Mega-city and mega-problem'. In Gilbert, A (ed) *The Mega-City in Latin America.* UNU Press, Tokyo, pp155–72.

Risse, T (2002) 'Transnational actors in world politics'. In Carlsnaes, W, Risse, T and Simmons, B A (eds) *Handbook of International Relations.* Sage, London, pp255–74.

Roberts, K (2007) 'Latin America's populist revival'. *SAIS Review* **27**(1): 3–15.

Roehder, K (2004) *Entwicklungspolitische Handlungsfelder im Kontext erodierender Staatlichkeit in Subsahara–Afrika.* Discussion Paper 5. Deutsches Institut für Entwicklungspolitik (DIE), Bonn.

Rohde, R A (2006) 'Saffir-Simpson Hurricane Intensity Scale'. University of California website, http://www.globalwarmingart.com/wiki/Image:Tropical_Storm_Map_png (viewed 14. May 2007).

Root, T L, Price, J T, Hall, K R, Schneider, S H, Rosenzweig, C and Pounds, J A (2003) 'Fingerprints of global warming on wild animals and plants'. *Nature* **421**: 57–60.

Rosenau, J N and Czempiel, E-O (Hrsg.) (1992) *Governance Without Government: Order and Change in World Politics. Cambridge Studies in International Relations.* Cambridge University Press, Cambridge, New York.

Ross, M L (2001) 'Does oil hinder democracy?' *World Politics* **53**(3): 325–6.

Ross, M L (2004a) 'How do natural resources influence civil war: evidence from thirteen cases'. *International Organization* **58**(Winter): 35–67.

Ross, M L (2004b) 'What do we know about natural resources and civil war?' *Journal of Peace Research* **41**(3): 337–56.

Rotberg, R I (2003) 'Failed states, collapsed states, weak states: causes and indicators'. In Rotberg, R I (ed) *State Failure and State Weakness in a Time of Terror.* Brookings Institution Press, Washington, DC, pp1–25.

Russett, B and O'Neal, J R (2001) *Triangulating Peace: Democracy, Interdependence, and International Organizations.* Norton, New York.

Russett, B, O'Neal, J R and Davis, D R (1998) 'The third leg of the Kantian tripod: international organizations and militarized disputes, 1950–1985'. *International Organizations* **52**(3): 441–67.

Rydgren, J (2004) 'Mechanims of exclusion: ethnic discrimination in the Swedish labour market'. *Journal of Ethnic and Migration Studies* **30**(4): 697–716.

Sadoff, C W and Grey, D (2002) 'Beyond the river: the benefits of cooperation on international rivers'. *Water Policy* **4**: 389–403.

Sahay, M (1987) 'Bullets for Bihar's hungry'. Calcutta, India: The Statesman Weekly 29.08.87.

Salati, E (1987) 'Amazônia: um ecossistema ameaçado'. In Kohlhepp, G and Schrader, A (eds) *Homem e natureza na Amazônia.* University Tübingen, Institute of Geography, Tübingen, pp33.

Salehyan, I (2005) *Refugees, Climate Change, and Instability. Human Security And Climate Change. An International Workshop at Holmen Fjord Hotel, Asker, Near Oslo, 21–23 June 2005.* University Of California, San Diego.

Sambanis, N (2001) 'Do ethnic and nonethnic civil wars have the same causes? A theoretical and empirical inquiry (part I)'. *Journal of Conflict Resolution* **45**(3): 259–82.

Sample, K and Zovatto, D (Hrsg.) (2006) *Democracia en la Región Andina – Los Telones de Fondo.* International Institute for Democracy and Electoral Assistance (IDEA), Lima.

Santos, F D, Forbes, K and Moita, R (2002) *Climate Change in Portugal: Scenarios, Impacts and Adaptation Measures. SIAM Project Report.* Gradiva, Lisbon.

SATP (South Asia Terrorism Portal) (2003) 'South Asia Intelligence Review: South Asia Assessment 2003'. SATP website, http://www.satp.org/satp/org/southasia (viewed 3. February 2007).

Scaife, A A, Knight, J R, Vallis, G K and Folland, C K (2005) 'A stratospheric influence on the winter NAO and North Atlantic surface climate'. *Geophysical Research Letters* **32** (L18715): doi:10.1029/2005GL023226.

Scambos, T A, Bohlander, J A, Shuman, C A and Skvarca, P (2004) 'Glacier acceleration and thinning after ice shelf collapse in the Larsen B embayment, Antarctica'. *Geophysical Research Letters* **31**(18): 1–4.

Schär, C, Vidale, P L, Lüthi, D, Frei, C, Häberli, C, Liniger, M A and Appenzeller, C (2004) 'The role of increasing temperature variability in European summer heatwaves'. *Nature* **427**: 332–6.

Scheffer, M, Brovkin, V and Cox, P M (2006) 'Positive feedback between global warming and atmospheric CO_2 concentration inferred from past climate change'. *Geophysical Research Letters* **33**: L10702, doi:10.1029/2005GL025044.

Schellnhuber, H J, Crutzen, P J, Clark, W C and Hunt, J (2005) 'Earth system analysis for sustainability'. *Environment* **47**(8): 11–25.

Schlesinger, W H, Reynolds, J F, Cunningham, G L, Huenneke, L F, Jarrell, W M, Virginia, R A and Whitford, W G (1990) 'Biological feedbacks in global desertification'. *Science* **247**: 1043–8.

Schlichte, K (2005) 'Gibt es überhaupt "Staatszerfall"? Anmerkungen zu einer ausufernden Debatte'. *Berliner Debatte Initial* **16**(4): 74–84.

Schmeidl, S (1997) 'Exploring the causes of forced migration'. *Social Science Quarterly* **78**(2): 284–308.

Schmittner, A (2005) 'Decline of marine ecosystem caused by a reduction in the Atlantic overturning circulation'. *Nature* **434**: 628–33.

Schmitz, A (2004) 'Turkmenistan: Der privatisierte Staat'. In Schneckener, U (ed) *States at Risk. Fragile Staaten als Sicherheits- und Entwicklungsproblem. SWP Studie*. Stiftung Wissenschaft und Politik (SWP), Berlin, pp147–69.

Schneckener, U (2004) *States at Risk. Fragile Staaten als Sicherheits- und Entwicklungsproblem. Diskussionspapier der Forschungsgruppe Globale Fragen der SWP*. Berlin: SWP.

Scholz, I (2006) 'Neue Trends in der globalen Umweltsituation'. In Stiftung Entwicklung und Frieden (SEI) (ed) *Globale Trends 2007. Frieden, Entwicklung, Umwelt*. Fischer, Frankfurt/M., pp401–21.

Scholz, I and Bauer, S (2006) 'Klimawandel und Desertifikation'. In Klingebiel, S (ed) *Afrika-Agenda 2007: Ansatzpunkte für den deutschen G8-Vorsitz und die EU-Präsidentschaft*. Deutsches Institut für Entwicklungspolitik (DIE), Bonn, pp63–70.

Scholz, U (1992) 'Transmigrasi – ein Desaster? Probleme und Chancen des indonesischen Umsiedlungsprogramms'. *Geographische Rundschau* **44**(1): 33–9.

Schradi, J (2006) 'Wo bleibt die zivile Krisenprävention. Grüne Kritik am Weißbuch zu den künftigen Aufgaben der Bundeswehr'. *Eins: Entwicklungspolitik. Information Nord-Süd* **13/14**: 46ff.

Schröter, D., Cramer, W., Leemans, R., Prentice, C., Araújo, M. B., Arnell, N. W., Bondeau, A., Bugmann, H., Carter, T. R., Gracia, C., de la Vega–Leinert, A. C., Erhard, M., Ewert, F., Glendining, M., House, J. I., Kankaanpää, S., Klein, R. J. T., Lavorel, S., Lindner, M., Metzger, M. J., Meyer, J., Mitchell, T. D., Reginster, I., Rounsevell, M., Sabaté, S., Sitch, S., Smith, B., Smith, J., Smith, P., Sykes, M. T., Thonicke, K., Thuiller, W., Tuck, G., Zaehle, S. und Zierl, B. (2005) 'Ecosystem Service Supply and Vulnerability to Global Change in Europe'. Science **25**: 1333–7.

Schubart, H (1983) 'Ecologia e utilização das florestas'. In Salati, E, Schubart, H, Junk, W and Oliveira, A (eds) *Amazônia. Desenvolvimento, Integração e Ecologia*. CNPq editora brasiliense, São Paulo.

Schumann, H (2007) 'Treibhaus made in China'. Hamburg, *Welt am Sonntag* (04.03.07).

Schwartz, P and Randall, D (2003) 'An Abrupt Climate Change Scenario and Its Implications for United States National Security'. Grist Environmental News & Commentary website, http://www.gristmagazine.com/pdf/AbruptClimateChange2003.pdf (viewed 25. April 2007).

Senat der Bundesforschungsanstalten im Geschäftsbereich des Bundesministeriums für Verbraucherschutz Ernährung und Landwirtschaft (2005) *Forschungsreport 1/2005. Schwerpunkt: Klimawandel und die Folgen*. Bundesministerium für Verbraucherschutz, Ernährung und Landwirtschaft (BMVEL), Berlin.

Seneviratne, S I, Lüthi, D, Litschi, M and Schär, C (2006) 'Land-atmosphere coupling and climate change in Europe'. *Nature* **443**: 205–9.

SEPA (State Environmental Protection Administration) (2006) 'China Green National Accounting Study Report 2004'. SEPA website, http://english.sepa.gov.cn/zwxx/xwfb/200609/t20060908_92580.htm (viewed 27. March 2007).

Shein, K A, Waple, A M, Diamond, J and Levi, J M (2006) 'State of the climate 2005'. *Bulletin of the American Meteorological Society* **87**: 6–102.

Shiklomanov, I (2000) 'Appraisal and assessment of world water resources'. *IWRA Water International* **25**(1): 11–32.

Sidikov, B (2006) 'Die Gefahr der politischen Instabilität wächst. Das Projekt Seidenstraße'. *Das Parlament* **56**(30/31): 11–2.

SIPRI (Stockholm International Peace Research Institute) (2006) *Yearbook 2006: Armaments, Disarmament and International Security*. SIPRI, Stockholm, Oxford.

Smakhtin, V, Revenga, C and Döll, P (2004) 'A pilot global assessment of environmental water requirements and scarcity'. *Water International* **29**(3): 307–17.

Smith, D (2004) 'Trends and causes of armed conflict'. In Austin, A, Fischer, M and Ropers, N (eds) *Transforming Ethnopolitical Conflict: The Berghof Handbook*. VS Verlag, Wiesbaden, pp2–14.

Sommer, G and Fuchs, A (2004) *Krieg und Frieden – Handbuch der Konflikt- und Friedenspsychologie*. Beltz, Weinheim.

Spanger, H J (2002) *Die Wiederkehr des Staates. Staatszerfall als wissenschaftliches und entwicklungspolitisches Problem*, HSFK–Report 1. Hessische Stiftung Friedens- und Konfliktforschung, Frankfurt/M.

Spencer, D (2001) 'The future of agriculture in Sub Sahara Africa'. *IFPRI*: 107–9.

Sprinz, D (forthcoming) 'Long-Term Policy Problems: Definition, Origin, and Responses'. In Wayman, F, Williamson, P and Bueno de Mesquita, B (eds) *Prediction: Breakthrough in Science, Markets, and Politics*. University of Michigan Press, Ann Arbor, MI,

Sprinz, D F (2003) 'Internationale Regime und Institutionen'. In Hellmann, G, Wolf, K D and Zürn, M (eds) *Die neuen Internationalen Beziehungen. Forschungsstand und Perspektiven in Deutschland*. Nomos, Baden-Baden, pp251–73.

Steiner, A (2007) *UNDP-UNEP Cooperation and the Launch of the Joint UNDP-UNEP Poverty-Environment Facility*. Statement by Achim Steiner, Executive Director, United Nations Environment Programme, UNEP GC-24 Special Event on UNDP-UNEP Cooperation. United Nations Environment Programme (UNEP), Nairobi.

Stern, N (2006) *The Economics of Climate Change. The Stern Review*. HM Treasury, London.

Sternfeld, E (2006) 'Umweltsituation und Umweltpolitik in China'. *Aus Politik und Zeitgeschichte* **49**: 27–34.

Sterzel, T (2004) *Correlation Analysis of Climate Variables and Wheat Yield Data on Various Aggregation Levels in Germany and the EU-15 Using GIS and Statistical Methods, With a Focus on Heat Wave Years*. Diploma Thesis. Potsdam Institute for Climate Impact Research (PIK), Humboldt University, Potsdam, Berlin.

Steward, F (2004) 'Horizontale Ungleichheit als Ursache von Bürgerkriegen'. In Kurtenbach, S and Lock, P (eds) *Kriege als (Über)Lebenswelten: Schattenglobalisierung, Kriegsökonomien und Inseln der Zivilität*. Dietz, Bonn, pp122–41.

Stockle, C O, Dyke, P T, Williams, J R, Jones, C A and Rosenberg, N J (1992) 'A method for estimating the direct and climatic effects of rising atmospheric carbon dioxide on growth and yield of crops: Part II – sensitivity analysis at three sites in the Midwestern USA.' *Agricultural Systems* **38**: 239–56.

Stroh, K (2005) 'Der Konflikt um das Wasser des Nils'. In Imbusch, P and Zoll, R (eds) *Friedens- und Konfliktforschung. Eine Einführung*. Verlag für Sozialwissenschaften, Wiesbaden, pp289–310.

Study Group on Europe's Security Capabilities (2004) *A Human Security Doctrine for Europe. The Barcelona Report of the Study Group on Europe's Security Capabilities*. Study Group on Europe's Security Capabilities, Barcelona.

Südasien Info (2006) 'Datenübersichten 2006 zu Ländern Südasiens'. Südasien-Informationsnetz e. V. website, http://www.suedasien.info/laenderinfos (viewed 3. February 2007).

Suliman, M (1994) *Ecology, Politics and Violent Conflict*. Zed Books, London.

Swatuk, L A (2005) 'Environmental security'. In Betsill, M M, Hochstetler, K and Stevis, D (eds) *International Environmental Politics*. Palgrave Macmillan, Basingstoke, UK,

Swatuk, L A (2007) 'Regionalexpertise: Southern Africa, Environmental Change and Regional Security: An Assessment. Expertise for the WBGU Report "World in Transition: Climate Change as a Security Risk"'. WBGU website, http://www.wbgu.de/wbgu_jg2007_ex07.pdf

SZ (Süddeutsche Zeitung) (07.06.2006) 'Landlosen-Proteste: 300 Bauern stürmen brasilianisches Parlament'. Süddeutsche Zeitung website, http://www.sueddeutsche.de/ausland/artikel/513/77436/article.html (viewed 7. June 2006).

SZ (Süddeutsche Zeitung) (17.04.2007) *Erderwärmung erstmals Thema im UN-Sicherheitsrat. EU: Klimawandel gefährdet den Weltfrieden*. Munich: SZ.

Tadross, M, Jack, C and Hewitson, B (2005) 'On RCM-based projections of change in southern Arican summer climate'. *Geophysical Research Letters* **32**(L23713): doi:10.1029/2005GL024460.

Tänzler, D, Schinke, B and Bals, C (2006) *Is there "Climate Security" for India? Tipping Points as Driver of Future Environmental Conflicts*. Background Paper Prepared for the Workshop "Klimawandel und Sicherheit – Fallbeispiel Indien", December 2006, Potsdam Institute für Climate Impact Research. Potsdam Institute for Climate Impact Research (PIK), Potsdam.

Tao, F, Yokozawa, M, Hayashi, Y and Lin, E (2003) 'Future climate change, the agricultural water cycle, and agricultural production in China'. *Agriculture, Ecosystems and Environment* **95**: 203–15.

Tetzlaff, R (1999) 'Der Wegfall effektiver Staatsgewalt in den Staaten Afrikas'. *Die Friedens-Warte* **74**(3): 307–30.

Thomas, C D, Cameron, A, Green, R E, Bakkenes, M, Beaumont, L J, Collingham, Y C, Erasmus, F N, Ferreira de Siqueira, M, Grainger, A, Hannah, L, Hughes, L, Huntley, B, van Jaarsveld, A S, Midgley, G F, Miles, L, Ortega-Huerta, M A, Townsend Peterson, A, Phillips, O L and Williams, S E (2004) 'Extinction risk from climate change'. *Nature* **427**: 145–8.

Thuiller, W, Lavorel, S, Araùjo, M B, Sykes, M T and Prentice, I C (2005) 'Climate change threats to plant diversity in Europe'. *Proceedings of the National Academy of Sciences of the United States of America* **102**: 8245–50.

Tränhard, D (2005) 'Entwicklung durch Migration: ein neuer Forschungsansatz'. *APUZ* **27**: 3–11.

Tuchman Mathews, J (1989) 'Redefining security'. *Foreign Affairs* **68**(2): 162–77.

Tull, D M (2004) 'Dimensionen und Ursachen von Flucht und Migration'. In Ferdowsi, M A (ed) *Afrika – ein verlorener Kontinent?* Bayerische Landeszentrale für politische Bildungsarbeit, Munich, pp126–50.

Tull, D M (2006) 'China's engagement in Africa: scope, significance and consequences'. *Journal of Modern African Studies* **44**(3): 459–79.

UK Foreign & Commonwealth Office (2007) *Margaret Beckett at UN Security Council Climate Change Debate (17.04.2007)*. UK Foreign & Commonwealth Office, London.

Ulph, A and Ulph, D (1997) 'Global warming, irreversibility and learning'. *The economic journal: The Journal of the Royal Economic Society* **442**: 636–50.

Umbach, F (2006) 'Die USA könnten in der Region geschwächt werden. Russland und China – strategische Partner oder dauerhafte Gegner'. *Das Parlament* **30/31**(56): 9.

UN (United Nations) (2004) *A More Secure World: Our Shared Responsibility. Report of the High–level Panel on Threats, Challenges and Change*. UN, New York.

UN (United Nations) (2005) *In Larger Freedom: Towards Development, Security and Human Rights For All. Report of the Secretary-General*. Addendum. A/59/2005/Add.2. UN, New York.

UN (United Nations) (2006a) *Global Survey of Early Warning Systems. An Assessment of Capacities, Gaps and Opportunities Towards Building a Comprehensive Global Early Warning System for all Natural Hazards. A Report Prepared at the Request of the Secretary-General of the United Nations*. UN, New York.

UN (United Nations) (2006b) *Delivering as One. Report of the Secretary-General's High-Level Panel on System-Wide Coherence*. UN, New York.

UN (United Nations) (2007) *Confronting Climate Change. Avoiding the Unmanageable and Managing the Unavoidable*. UN, New York.

UN DESA (United Nations Department of Economic and Social Affairs) (2005) *World Population Prospects. The 2004 Revision. Highlights*. UN Population Division, Department of Economic and Social Affairs, New York.

UN DESA (United Nations Department of Economic and Social Affairs) (2006) 'Population, Resources, Environment and Development. The 2005 Revision'. UN website, http://unstats.un.org (viewed 7. May 2007).

UN Millennium Project (2005) *Investing in Development: A Practical Plan to Achieve the Millennium Development Goals*. United Nations (UN), New York.

UNCCD (United Nations Convention to Combat Desertification) (1994) *United Nations Convention to Combat Desertification in Countries Experiencing Serious Drought and/or Desertification, Particularly in Africa*. UNCCD Secretariat, Bonn.

UNCTAD (United Nations Conference on Trade and Development) (2006) *World Investment Report 2005*. UNCTAD, Geneva.

UNDP (United Nations Development Programme) (Hrsg.) (1994) *Bericht über die menschliche Entwicklung 1994*. Deutsche Gesellschaft für die Vereinten Nationen (DGVN), Bonn.

UNDP (United Nations Development Programme) (2002) *China Human Development Report 2002*. UNDP, New York.

UNDP (United Nations Development Programme) (2005a) *China Human Development Report 2005*. UNDP, New York.

UNDP (United Nations Development Programme) (2005b) *Human Development Report 2005: International Cooperation at a Crossroads – Aid, Trade and Security in an Unequal World*. UNDP, New York.

UNDP (United Nations Development Programme) (2006) *Human Development Report 2006. Beyond Scarcity: Power, Poverty and the Global Water Crisis*. Macmillan, New York.

UNEP (United Nations Environment Programme) (Hrsg.) (2002a) *Global Environment Outlook 3. Past, Present and Future Perspectives*. UNEP, Nairobi.

UNEP (United Nations Environment Programme) (2002b) *African Environment Outlook. Past, Present and Future Perspectives*. UNEP, Nairobi.

UNEP (United Nations Environment Programme) (2006) *Adaptation and Vulnerability to Climate Change: The Role of the Finance Sector*. CEO Briefing. UNEP, Nairobi.

UNEP (United Nations Environment Programme) (2007) *President's Summary of the Discussions by Ministers and Heads of Delegation at the Twenty-Fourth Session of the Governing Council/Global Ministerial Environment Forum of the United Nations Environment Programme, Nairobi, 5.–9. February 2007*. UNEP, Nairobi.

UNESCO (United Nations Educational Scientific and Cultural Organization) (2003) *Water for People, Water for Life. The United Nations World Water Development Report 2003*. UNESCO, Paris.

UNESCO (United Nations Educational Scientific and Cultural Organization) (2006) *Water. A Shared Responsibility. The United Nations World Water Development Report 2*. UNESCO, Paris.

UNFPA (United Nations Population Fund) (2006) *State of World Population 2006. A Passage to Hope. Women and International Migration*. UNFPA, New York.

UNHCR (United Nations High Commissioner for Refugees) (2006a) 'Internally Displaced People. Questions and Answers'. UNHCR website, http://www.unhcr.org/cgi-bin/texis/vtx/basics/opendoc.pdf?tbl=BASICS&id=405ef8c64 (viewed 5. December 2006).

UNHCR (United Nations High Commissioner for Refugees) (2006b) *The State of the World's Refugees 2006: Human Displacement in the New Millennium*. Oxford University Press, Oxford, New York.

UNHCR (United Nations High Commissioner for Refugees) (2007) *Supplementary Appeal for Darfur. Protection and Assistance to Refugees and IDPs in Darfur*. UNHCR, New York.

UNISDR (United Nations International Strategy for Disaster Reduction) (2004) *Living with Risk: A Global Review of Disaster Reduction Initiatives*. UNISDR, New York.

UNISDR (United Nations International Strategy for Disaster Reduction) (2006) 'Terminology: Basic Terms of Disaster Risk Reduction'. UNISDR website, http://www.unisdr.org/eng/library/lib-terminology-eng%20home.htm (viewed 5. December 2006).

UNPD (United Nations Population Division) (2002) *International Migration 2002*. UNPD, New York.

UNPD (United Nations Population Division) (2005) 'Urban Agglomerations 2005'. UNPD website, http://www.un.org/esa/population/publications/WUP2005/2005urban_agglo.htm (viewed 05. December 2006).

UNSG (United Nations Secretary General) (2005) *In Larger Freedom: Towards Development, Security and Human Rights for all. Report of the Secretary-General*. UN, New York.

UNU–EHS (United Nations University Institute For Environment And Human Security) (2005) *As Ranks of 'Environmental Refugees' Swell Worldwide, Calls Grow For Better Definition, Recognition, Support*. Press Release. UNU-EHS, Tokyo.

Urdal, H (2005) 'People vs. Malthus: population pressure, environmental degradation, and armed conflict revisited'. *Journal of Peace Research* **42**(4): 417–34.

USDA (U.S. Department of Agriculture Natural Resources Conservation Service) (1998) *Soil Map and Soil Climate Map*. USDA Soil Survey Division, World Soil ResourcesWashington, DC.

USGS (U.S. Geological Survey) (2005) *Hurricane Hazards – A National Threat. Fact Sheet 2005–3121*. USGS, Washington, DC.

van Creveld, M (1998) *Die Zukunft des Krieges*. Mürmann, Munich.

van de Walle, N (2001) *African Economies and the Politics of Permanent Crisis, 1979–1999*. Cambridge University Press, Cambridge, New York.

van de Walle, N (2005) *Overcoming Stagnation in Aid-Dependent Countries*. Brookings Institution Press, Washington, DC.

Vasquez, J A (2000) 'What do we know about war'. In Vasquez, J A (ed) *What Do We Know About War*. University Press of America, Lanham, MD, pp335–70.

Verwimp, P (2002) 'Agricultural Policy, Crop Failure and the 'Ruriganiza' Famine (1989) in Southern Rwanda: A Prelude to Genocide?' University Leuven website, http://www.econ.kuleuven.be/ces/discussionpapers/Dps02/Dps0207.pdf (viewed 26. June 2006).

von Trotha, T (2000) 'Die Zukunft liegt in Afrika. Vom Zerfall des Staates, von der Vorherrschaft der konzentrischen Ordnung und vom Aufstieg der Parastaatlichkeit'. *Leviathan* **28**(2): 253–79.

von Winter, T (2004) *Multilateralismus und Unilateralismus in der Außenpolitik der USA. Info Brief WF II – 117/04*. Wissenschaftliche Dienste des Deutschen Bundestages, Berlin.

Waever, O (1995) 'Securitization and desecuritization'. In Lipschutz, R D (ed) *On Security*. Columbia University Press, New York, pp46–86.

Wagner, C (2006a) 'Eine funktionierende Anarchie: Die Indische Union und ihr politisches System'. *Das Parlament* (7./14. August): 2.

Wagner, C (2006b) *Indien als strategischer Partner der USA*. SWP-Aktuell 13. Stiftung Wissenschaft und Politik (SWP), Berlin.

Waichman, A V, Römbke, J, Ribeiro, M O A and Nailson, C S N (2002) 'Use and fate of pesticides in the Amazon State, Brazil'. *Environmental Sciences and Pollution Research* **9**: 423–8.

Walter, B F (2004) 'Does conflict beget conflict? Explaining recurring civil war'. *Journal of Peace Research* **41**(3): 371–88.

Walther, G-R, Post, E, Convey, P, Menzel, A, Parmesan, C, Beebee, T J C, Fromentin, J-M, Hoegh-Guldberg, O and Bairlein, F (2002) 'Ecological responses to recent climate change'. *Nature* **416**: 389–95.

Ward, M D and Gleditsch, K S (2002) 'Location, location, location: an MCMC approach to modeling the spatial context of war and peace'. *Political Analysis* **10**: 244–60.

WBGU (German Advisory Council on Global Change) (1995) *World in Transition: The Threat to Soils. 1994 Report*. Economica, Bonn.

WBGU (German Advisory Council on Global Change) (1997) *World in Transition: The Research Challenge. 1996 Report*. Springer, Berlin, Heidelberg, New York.

WBGU (German Advisory Council on Global Change) (1998) *World in Transition: Ways Towards Sustainable Management of Freshwater Resources. 1997 Report*. Springer, Berlin, Heidelberg, New York.

WBGU (German Advisory Council on Global Change) (2000) *World in Transition: Strategies for Managing Global Environmental Risks. 1998 Report*. Springer, Heidelberg, Berlin, New York.

WBGU (German Advisory Council on Global Change) (2001) *World in Transition: New Structures for Global Environmental Policy. 2000 Report*. Springer, Berlin, Heidelberg, New York.

WBGU (German Advisory Council on Global Change) (2002) *Charging the Use of Global Commons. Special Report 2002*. WBGU, Berlin.

WBGU (German Advisory Council on Global Change) (2003) *Climate Protection Strategies for the 21st Century: Kyoto and beyond. Special Report 2003*. WBGU, Berlin.

WBGU (German Advisory Council on Global Change) (2004) *World in Transition: Towards Sustainable Energy Systems. 2003 Report*. Earthscan, London.

WBGU (German Advisory Council on Global Change) (2005) *World in Transition: Fighting Poverty Through Environmental Policy. 2004 Report*. Earthscan, London.

WBGU (German Advisory Council on Global Change) (2006) *The Future Oceans – Warming Up, Rising High, Turning Sour. Special Report 2006*. WBGU, Berlin.

WBGU (German Advisory Council on Global Change) (2007) *New Impetus for Climate Policy: Making the Most of Germany's Dual Presidency. Policy Paper 5*. WBGU, Berlin.

WCD (World Commission on Dams) (2000) *Dams and Development. A New Framework for Decision-Making*. Earthscan, London.

Webster, P J, Holland, G J, Curry, J A and Chang, H-R (2005) 'Changes in tropical cyclone number, duration, and intensity in a warming environment'. *Science* **309**: 1844–6.

Webster, P J, Magana, V O, Palmer, T N, Shukla, J, Tomas, R A, Yanai, M and Yasunari, T (1998) 'Monsoons: processes, predictability, and the prospects for prediction'. *Journal of Geophysical Research* **103**(C7): 14451–510.

Weinlich, S (2006) 'Weder Feigenblatt noch Allheilmittel: Die neue Kommission für Friedenskonsolidierung der Vereinten Nationen'. *Vereinte Nationen* **54**: 2–11.

Wenk, K (2007) 'Chinas Kader wollen Klimaschützer werden'. *Welt am Sonntag* 25.2.07.

Wenzel, H J (2002) 'Umweltflüchtlinge oder Umweltmigranten? Umweldegradation, Verwundbarkeit und Migration/ Flucht im subsaharischen Afrika'. In Oltmer, J (ed) *Migrationsforschung und interkulturelle Studien. IMIS-Schriften Band 11*. Institut für Migrationsforschung und Interkulturelle Studien (IMIS), Osnabrück, pp287–311.

Wesner, F and Braun, A J (2006) *Chinas Energiediplomatie: Kooperation oder Konkurrenz in Asien? Aus chinesischen Fachzeitschriften*. SWP-Zeitschriftenschau 5. Stiftung Wissenschaft und Politik (SWP), Berlin.

White House (2002) *The National Security Strategy of the United States of America*. The White House, Washington, DC.

White House (2006) *The National Security Strategy of the United States of America*. The White House, Washington, DC.

WHO – World Health Organisation and UNICEF – United Nations Childrens Fund (2004) *Meeting the MDG Drinking Water and Sanitation Target. A Mid-Term Assessment of Progress. Joint Monitoring Programme for Water Supply and Sanitation*. WHO and UNICEF, New York.

WI (Wuppertal Institute for Climate Environment and Energy) (2004) *Wege von der nachholenden zur nachhaltigen Entwicklung*. WI, Wuppertal.

WI (Wuppertal Institute for Climate Environment and Energy) (2005) *Fair Future*. Wuppertal Institut, Wuppertal.

Wiggerthale, M (2004) 'Development instead of free trade. Time to turn around'. In Heinrich Boell Foundation (ed) *Liberalisation of Agricultural Trade – The Way Forward for Sustainable Development? Global Issue Papers*. vol 13. Heinrich Böll Foundation, Berlin, pp2–16.

Willmann, K (2006) 'Der Chemieunfall von Songhua und das Potenzial für soziale Proteste'. *China Aktuell* **1**: 57–65.

Winters, L A and Yusuf, S (Hrsg.) (2007a) *Dancing with Giants: China, India, and the Global Economy*. World Bank, Institute of Policy Studies, Washington, DC and Singapore.

Winters, L A and Yusuf, S (2007b) 'Introduction: dancing with giants'. In Winters, L A and Shahid, Y (eds) *Dancing with Giants: China, India, and the Global Economy*. World Bank, Institute of Policy Studies, Washington, DC und Singapur, pp1–34.

Wodinski, M (2006) *Karten zu Umweltparametern. GIS-I. Expertise for the WBGU Report "World in Transition: Climate Change as a Security Risk"*. Unpublished data. WBGU, Berlin.

Wolf, A T (2006) 'A Long Term View of Water and Security: International Waters, National Issues, and Regional Tensions. A Report to the German Advisory Council on Global Change (WBGU). Expertise for the WBGU Report "World in Transition: Climate Change as a Security Risk"'. WBGU website, http://www.wbgu.de/www.wbgu_jg2007_ex08.pdf

Wolf, A T, Kramer, A, Carius, A and Dabelko, G D (2005) *Managing Conflict and Cooperation. State of the World 2005. Redefining Global Security*. World Resources Institute (WRI), Washington, DC.

Wolf, A T, Yoffe, S B and Giordano, M (2003) 'International waters: identifying basins at risk'. *Water Policy* **5**(1): 29–60.

World Bank (2005a) *Pakistan Water Resources Assistance Strategy*. The World Bank, Washington, DC.

World Bank (2005b) 'Worldwide Governance Indicators Country Snapshot'. World Bank website, http://info.worldbank.org/governance/kkz2005/ (viewed 28. March 2007).

World Bank (2005c) 'Carbon Finance Annual Report 2005: Carbon Finance for Sustainable Development'. World Bank website, http://carbonfinance.org/docs/2005_CFU_Annual_Report.pdf (viewed 15. Februar 2007).

World Bank (2006a) *World Development Indicators 2006. Economy, States & Markets*. The World Bank, Washington, DC.

World Bank (2006b) 'Which Countries are LICIUS?' World Bank website, http://www.worldbank.org/ieg/licus/licus06_map.html (viewed 7. May 2007).

World Bank (2006c) *Can South Asia End Poverty in a Generation?* The World Bank, Washington, DC.

World Bank (2006d) 'Bangladesh Country Overview 2006'. The World Bank website, http://www.worldbank.org (viewed 27. October 2006).

World Bank (2006e) 'Europe and Central Asia – Country Briefs '. Updated September 2006. World Bank website, http://web.worldbank.org/WBSITE/EXTERNAL/COUNTRIES/ECAEXT/0,,contentMDK:20113477~menuPK:265246~pagePK:146736~piPK:226340~theSitePK:258599,00.html (viewed 13. June 2007).

Worldwatch Institute (2005) *State of the World 2005: Redefining Global Security*. Worldwatch Institute, Washington, DC.

Worldwatch Institute (2006) *State of the World 2006. Special Focus: China and India*. Earthscan, London, New York.

WTO (World Trade Organisation Committee on Agriculture) (2005a) *Proposal by the African Group in the Context of the Review of all Special and Differential Treatment Provisions by the Committee on Trade and Development in Special Session*. Document G/AG/20 of 15 July 2005. WTO, Geneva.

WTO (World Trade Organisation Committee on Agriculture) (2005b) *General Council Overview of WTO Activities 2005*. Document G/L/746 of 26 September 2005. WTO, Geneva.

WTO (World Trade Organisation Committee on Agriculture) (2006) *Summary Report of the Meeting Held on 27 January 2006*. Document G/AG/R/45 of 13 April 2006. WTO, Geneva.

WWF (World Wide Fund) (2006) *Drought in in the Mediterranean: WWF Policy Proposal*. WWF, Zurich.

Xoplaki, E, Gonzàlez-Rouco, J F, Luterbacher, J and H., W (2004) 'Wet season Mediterranean precipitation variability: influence of large-scale dynamics and trends'. *Climate Dynamics* **23**: 63–78.

Yepes, G and Ringskog, K (2002) *Estudio de Offerta y Demanda Servicios de Agua Potable y Alcantarillado*. Servicio de Agua Potable y Alcantarillado de Lima, Ministerio de Economia y Finanzas, Lima.

Young, O R (1999) *Governance in World Affairs*. Cornell University Press, Ithaca, NY.

Young, O R (2002) *The Institutional Dimensions of Environmental Change. Fit, Interplay, and Scale*. MIT Press, Cambridge, MA.

Zaman, M Q (1991) 'Rivers of life: living with floods in Bangladesh'. *Asian Survey* **33**(10): 985–96.

Zeng, N (2003) 'Drought in the Sahel'. *Science* **302**: 999–1000.

Zhang, R and Delworth, T L (2005) 'Simulated tropical response to a substantial weakening of the Atlantic thermohaline circulation'. *Journal of Climate* **18**: 1853–60.

Zickfeld, K, Knopf, B, Petoukhov, V and Schellnhuber, H J (2005) 'Is the Indian summer monsoon stable against global change?' *Geophysical Research Letters* **32**: L15707.

Zickfeld, K, Levermann, A, Morgan, M G, Kuhlbrodt, T, Rahmstorf, S and Keith, D (2007) 'Expert judgements on the response of the Atlantic meridional overturning circulation to climate change'. *Climatic Change* DOI 10.1007/s10584-007-9246-3.

Zürn, M (1998a) 'The rise of international environmental politics. A review of current research'. *World Politics* **50**: 617–49.

Zürn, M (1998b) *Regieren jenseits des Nationalstaates. Denationalisierung und Globalisierung als Chance*. Suhrkamp, Frankfurt/M.

Zürn, M (2002) 'From interdependence to globalization'. In Carlsnaes, W, Risse, T and Simmons, B A (eds) *Handbook of International Relations*. Sage, London, pp235–54.

Zwally, H J, Abdalati, W, Herring, T, Larson, K, Saba, J and Steffen, K (2002) 'Surface melt-induced acceleration of Greenland ice-sheet flow'. *Science* **297**(5579): 218–22.

Glossary

Action Plan 'Crisis Prevention'
The German federal government's action plan on 'Civilian Crisis Prevention, Conflict Resolution and Post-Conflict Peace Building', adopted in May 2004, sets out the principles and strategies of German policy on →crisis prevention. The Action Plan gives concrete form to the methodological approaches applied in crisis prevention and renders them operational. Ways of consolidating and restructuring existing institutions and of deploying available instruments in a coherent way are highlighted in order to strengthen the capacity of the government in this sphere. Crisis prevention in this context is to be understood in its widest sense, including conflict resolution and peace building.

Adaptation to climate change
Adaptations in natural or human systems to actual or expected climatic changes. Various types of adaptation can be distinguished, including anticipatory and reactive, private and public, autonomous and planned.

Annex I countries
Group of countries listed in Annex I to the →United Nations Framework Convention on Climate Change (UNFCCC). It includes all OECD countries apart from Mexico and South Korea, as well as the eastern European states and Russia. Annex I countries have committed under the UNFCCC to adopt a leading role in the reduction of greenhouse gases. Moreover, most of the Annex I countries have entered into binding reduction commitments under the →Kyoto Protocol. Such countries are listed in Annex B to the Kyoto Protocol.

Anomie
Situation in which there is minimal social order or none at all, and in which rules and norms exert little influence.

Clean Development Mechanism (CDM)
One of the flexible mechanisms introduced by the →Kyoto Protocol, which allows an investor to carry out emissions-reducing projects in a →developing or →newly industrializing country and to receive tradable certificates for this, which an industrialized country can offset against its reduction commitments.

Common Foreign and Security Policy (CFSP)
The CFSP was established as the second pillar of the European Union by the Maastricht Treaty of 1992. It sets out a framework for cooperation among the EU countries in matters of foreign and security policy. In contrast to common trade policy and development policy, which likewise regulate the external relations of the EU, but form part of the first pillar, the CFSP is an instrument of cooperation among the Member States at government level, but has no direct legal force.

Comprehensive security
Comprehensive security emphasizes that uncertainty, instability and violence are not only brought about by military aggression, but rather may have complex political, economic, socio-cultural and ecological origins. →Security policy viewed in this way is no longer merely a question of military capacity, but is based on the ability to defuse crises at the earliest possible stage using civil and, if necessary, military instruments. Alongside classical foreign and economic policy, civil means increasingly include development and environmental policy measures.

Conflict
Conflict describes disputes and disagreements of any kind over contrary interests between individual or collective actors. In the context of this report, conflict is conceived of as an intensification of →crisis. As such, it covers (mainly) violent disputes between two or more conflicting parties and can lead to social →destabilization.

Conflict constellation
Causal linkages at the interface of environment and society, whose dynamics can lead to social destabilization or violence. A conflict constellation identifies →key factors that determine the emergence of →cri-

sis and →conflict in the respective sphere. A causal chain identifies the sequence, interactions and stages of escalation of a possible conflictual development. The process of identifying and characterizing conflict constellations makes use of projections of anticipated climatic changes, past experience with environmental conflicts, the findings of research on the causes of war and conflict, and analyses of state fragility and multipolarity.

Convention
The term refers to a treaty under international law adopted multilaterally. These include the regimes developing under the United Nations umbrella, such as the three 'Rio Conventions': the →United Nations Framework Convention on Climate Change, the Convention to Combat Desertification and the Convention on Biological Diversity.

Crisis
A crisis is a difficult situation that signifies a potential turning point in a critical development. Crisis does not necessarily lead to →conflict, violence or even warfare. However, it does pose considerable challenges to the capacity of those affected to find solutions and manage crises.

Crisis and conflict prevention
Crisis and conflict prevention comprises policy measures and strategies that are aimed at avoiding →crises and →conflicts or escalation of the same. However, crisis and conflict prevention does not mean lack of conflict. What is important is that conflicts within and between states be dealt with non-violently. Crisis prevention has begun to play an ever greater role in →security policy – increasingly in the civilian context as well.

Dangerous climate change
This expression refers to the ultimate objective of the →United Nations Framework Convention on Climate Change, which is to 'prevent dangerous anthropogenic interference with the climate system'. No agreement has yet been reached within the Convention process on which level of climate change is to be considered dangerous. WBGU has proposed the following →guard rail: to limit global warming to 2 °C relative to the pre-industrial value and the rate of warming to 0.2 °C per decade in order to avoid dangerous climate change.

Desertification
Desertification is the degradation of land resources in arid, semi-arid and dry sub-humid zones, and may be caused by a variety of factors, including climate change and human activities.

Destabilization
Destabilization can be understood as a process that (gradually) causes an originally stable political and social situation to break down. Stability refers in general to the durability of political institutions and to robust social structures. Thus, destabilization can occur as a result of →crises and (violent) →conflicts and, in extreme cases, lead to the collapse of the societal order.

Developing countries
These are countries whose standard of living is far lower than in Europe (excluding Eastern Europe), North America and Oceania (Australia, New Zealand and Japan). The World Bank classifies national economies on the basis of per-capita income according to the following thresholds: low income group = below US$875 in 2005; lower middle income group = US$876–3,465; upper middle income group = US$3,466–10,725; high income group = above US$10,726. According to the →UNDP and →OECD, 137 countries from the first three categories are considered developing countries.

Development cooperation (DC)
This comprises all the inputs provided through technical, financial and human resources cooperation. DC is carried out by private and public agencies in industrialized and developing countries. DC may take the form of material assistance (grants or concessionary loans) or provision of non-material support (e.g. expertise/training).

Disaster
A disaster is a severe disruption in the functioning of a society, with far-reaching humanitarian, material, economic or environmental damage that exceeds the coping capacity of the society in question. Disasters may, for example, be a result of natural events such as earthquakes, but they may also occur as a result of industrial or transport accidents. The occurrence of such an event in itself does not necessarily lead to a disaster.

Disaster prevention
Disaster prevention is a conceptual approach that brings together a variety of elements aimed at preventive action to mitigate and avoid hazards. These include actions to improve risk awareness, enhance institutional capacities, implement appropriate methods of spatial planning and establish and maintain appropriate early warning systems.

Doha Round
The Doha Round refers to a set of tasks drawn up by the economics and trade ministers of the WTO mem-

ber countries in 2001 at the 4th Ministerial Conference in Doha, work on which was to have been concluded by 2005. The primary aim of the Doha Round was to work on the problems of developing countries, such as access to industrialized countries' markets via dismantling import quotas and tariffs and reducing agricultural subsidies in the industrialized countries. Following suspension of the negotiations for a time, they have once again resumed, but differences of opinion among the WTO members have so far prevented their completion.

Drought
During a drought, the level of precipitation is significantly lower than levels normally recorded, and this causes a severe hydrological imbalance that has a detrimental effect on terrestrial production systems.

ECOSOC
The Economic and Social Council of the United Nations (ECOSOC) is one of the six principal organs of the UN. Its task is to coordinate UN activities in the sphere of economic and social policy and to report on the social state of the world.

Environment and conflict research
Environment and conflict research is dedicated to examining the interconnections between environmental degradation and the genesis of →conflict. The beginnings of scientific research into environmental conflicts can be traced back to the early 1970s. Systematic empirical studies and research programmes were undertaken during the 1990s, for instance in Toronto and Zurich.

Environmental migration
→Environmentally induced migration

Environmental refugees
Can be understood as people who have been forced to leave their traditional habitat, temporarily or permanently, because of an environmental change. The term is disputed both politically and legally. Under the Geneva Refugee Convention, 'refugees' are people who had to leave their country of residence because of well-founded fear of being persecuted for reasons of race, religion, nationality, membership of a particular social group or political opinion. As these criteria do not apply to environmentally induced migration, those affected are not protected by the guarantees accorded to refugees in international law.

Environmentally induced migration
The term migration covers in general all forms of displacement and flight that take place voluntarily or involuntarily and across or within national borders.

The term 'environmental migration' places the focus on the precipitating factor, i.e. an environmental change as the trigger of migration.

Food security
A situation in which all people at all times have physical, social and economic access to adequate, secure and nutritious food that meets their physiological requirements, accords with their food habits and preferences and guarantees an active and healthy life.

Forecast
Forecasts (as opposed to →scenarios) are assertions derived from scientific research regarding the probable course of future events in a given timeframe.

Gini coefficient
The Gini coefficient is a statistical measure used to represent various kinds of unequal distribution. The coefficient can be used, for example, to specify numerically the unequal distribution of income and wealth. The value of the coefficient can lie anywhere between 0 and 1. The closer the Gini coefficient is to 1, the larger the inequality. The Gini index is found by multiplying the Gini coefficient by 100.

Gini index
→Gini coefficient

Global governance
There is no universally agreed definition of the concept of global governance; it is used normatively by some, and analytically by others. It generally denotes the deepening of international cooperation and the creation and reinforcement of multilateral regimes in order to master global challenges. Global governance involves, in particular, cooperation between states and non-government actors from the local to the global level. The notion of an architecture of global governance refers to corresponding multilateral international structures.

Globalization
This refers to the process of increasing transnational interconnectedness in every sector (economy, politics, culture, environment, communications, etc.) between individual and collective actors as well as between public and private actors. The main drivers behind globalization are seen as being technological change, especially in communications and transportation, and political decision-making aimed at liberalizing world trade. The sovereignty of nation states becomes subject to constraint as a result of globalization.

Good governance
A concept developed in the late 1980s by the World Bank. Good governance aims to promote and safeguard 'good' state systems of governance and regulation. The normative concept is underpinned by methodologies which make it possible to assess the quality of political leadership and governance. In →developing countries good governance is often a precondition of →development cooperation inputs (conditionality).

Governance capacities
→Governance structures

Governance structures
Governance, or governance structure, generally describes the system through which a political entity exerts control and establishes rules. The concept arose in distinction to the term government and is intended to convey the fact that political control is enacted not only hierarchically by the state, but also by private actors such as interest-based organizations. Governance capacity refers to the facility for governance provided by functioning institutions and regulatory systems.

Gross Domestic Product (GDP)
Currently the most frequently used indicator of the total output of an economy over a specific period of time. GDP is the sum of the market values, or prices, of all final goods and services produced within a country, minus the value of intermediate inputs.

Gross National Income (GNI)
An important indicator of the total output of an economy over a specific period of time. GNI is distinct from →Gross Domestic Product in that it comprises the total value of goods and services produced within a country together with net income (including interest and dividends) received from those institutions and individuals in other countries whose headquarters or place of residence are located in the economy in question. GNI is also known as Gross National Product (GNP).

Group of Eight (G8)
The G8 is an informal forum of heads of state and government. Initially a purely economic summit of the world's leading economies in 1975, it has since become one of the most important forums of global policy-making. The G8 comprises Germany, France, Great Britain, Italy, Japan, the USA, Canada and (since 1998) Russia. The European Union is also represented. Non-member countries are invited to participate in expanded dialogue sessions. These 'outreach countries' comprise the five most important →newly industrializing countries, China, India, South Africa, Brazil and Mexico. The G8 together with the outreach countries are sometimes referred to as the G8+5.

Guard rail
A concept introduced by WBGU. Guard rails are quantifiable limits to damage whose transgression would entail intolerable consequences today or in the future, so that even major utility gains could not compensate for such damage. They demarcate the non-sustainable domains of development within the people-environment system, so that sustainable pathways run within the domain defined by these guard rails. One example is WBGU's climate protection guard rail, which demarcates a mean global warming of more than 2 °C from pre-industrial levels and a rate of temperature change of more than 0.2 °C per decade as the boundary of →dangerous climate change.

Human Development Index (HDI)
Since 1990 the HDI has attempted to illustrate the status of human development in the countries of the world by means of a specially devised measure. Unlike the country comparison of the World Bank, it takes account not only of the gross domestic product (GDP) per inhabitant of a country, but also of life expectancy and level of education, measured using the rate of literacy within the population and the rate of school enrolment. The HDI is published annually in the Human Development Report (HDR), which is issued by the →United Nations Development Programme (UNDP).

Human rights
Describe a conceptual scheme in which all people are accorded universal rights. The existence of human rights is now recognized in principle by almost all countries, although the substance and scope of these rights are disputed. The key international source for their existence and content are the treaties of the United Nations. Alongside the Universal Declaration of Human Rights of 1948, the main human rights instruments are the International Covenant on Civil and Political Rights and the International Covenant on Economic, Social and Cultural Rights. In addition there are a large number of conventions that regulate the protection of specific human rights in detail, e.g. the Geneva Refugee Convention and regional human rights agreements.

Human security
The concept of →security is traditionally used in relation to collective actors such as societies and states. In contrast to this, the concept of human security

focuses on the security needs of humans. Security is no longer seen simply as 'freedom from fear', but also as 'freedom from want'. In this perspective, economic and environmental crises or pandemics such as AIDS are to be considered just as much as security risks as violent conflicts and wars. Introduced by →UNDP in 1994, the concept has since been crucial in shaping the international security debate.

Industrialized countries
WBGU understands industrialized countries as meaning those developed countries which are not classed as →developing countries or →newly industrializing countries. This classification is not made primarily on the basis of the industrialization of a national economy, but mainly according to per-capita income. A related term is →OECD states.

Integrated Coastal Zone Management (ICZM)
ICZM is a dynamic, multidisciplinary and iterative process for the sustainable management of coasts. The goal is to reconcile the various economic, ecological, social and cultural claims and entitlements within the limits set by natural dynamics. The process embraces all steps from information and data procurement through planning and decision-making to implementation and monitoring. Great attention is paid to stakeholder participation in ICZM.

Integrated water resources management (IWRM)
IWRM seeks to develop and to manage water, land or soil and the associated resources in a cross-sectoral, participatory process in a sustainable manner. The aim is to move beyond isolated, sectoral and inefficient systems of use. IWRM is a relatively new concept that has been developed and advanced mainly in international research and water policy forums.

Interdependence
In international relations, this describes the phenomenon of increasing global interconnections between states, economies and societies and their growing complexity. Complex interdependence raises the threshold at which conflicts become violent and encourages international cooperation. The rapid growth of international institutions confirms this institutionalist perspective, which is widespread in theories of International Relations.

Internally Displaced Persons (IDPs)
If →migration takes place within states, this may be termed internal migration. In cases where internal migrants are forced to leave their place of origin for reasons that essentially also apply to refugees within the meaning of the Refugee Convention, such migrants are termed Internally Displaced Persons (IDPs).

International regimes
International regimes are intergovernmental institutions established to address transboundary issues. They involve principles, standards, rules and decision-making procedures, and are based upon formal or informal intergovernmental agreements.

Key factors
Factors of importance to the emergence of violence within the context of climate-induced environmental changes. They serve to analyse causal interrelations and cause-effect mechanisms within a →conflict constellation. The recommendations for action presented in this report address those factors which can be influenced by human action.

Kyoto Protocol
An agreement under international law supplementing the →United Nations Framework Convention on Climate Change, setting greenhouse gas emissions reduction targets for industrialized countries. The countries listed in Annex B to the Protocol are obliged to reduce their greenhouse gas emissions by around 5 per cent from the base year 1990 in the commitment period 2008–2012. The Protocol has been ratified by 174 states to date. Among the Annex B countries, the USA and Australia have not yet ratified.

Migration
→Environmental induced migration

Millennium Development Goals (MDGs)
The MDGs are set out in the United Nations Millennium Declaration and encapsulate the main outcomes of the World Conferences held in the previous decade. The MDGs comprise a set of eight international development goals together with a framework of 18 targets and 48 indicators to specify and operationalize the goals and measure progress. For most of the goals and targets, quantitative criteria and a timeframe were also adopted, generally to 2015 (the baseline year is 1990).

Mitigation of climate change
Mitigation embraces all measures that either limit greenhouse gas emissions or amplify their absorption in 'sinks'.

Multilateralism
Institutionalized cooperation between states in international organizations (such as the UN) or other regulatory mechanisms (e.g. →conventions). Uni-

lateralism, by contrast, refers to a one-sided policy approach oriented towards the national interests of a state. In this case, multilateral solutions to problems are sought only if they appear necessary to protect a state's own interests.

Multipolarity, multipolar world order
Multipolarity refers to a world order shaped by the competition between major powers in the international system. It exists in contrast to unipolarity, in which a single extremely powerful actor (hegemon) determines global politics. The international conflict of systems between East and West during the Cold War represented a bipolar world order dominated by the USA and the Soviet Union.

Natural disasters
The term natural disasters refers to →disasters that can occur without human influence (e.g. earthquakes, tsunamis or volcanic eruptions), but it also includes disasters where interaction between human activities and the environment plays a role (e.g. floods, droughts or cyclones). The extent of the humanitarian, material, economic and environmental damage incurred as a result of such disasters also depends on the types of human activities and form of social organization in place.

New wars
This concept represents an attempt to describe the transformation of armed violence in a changed (global) political context. In contrast to 'classical' interstate wars, the 'new wars' are characterized by the fact that they contain elements of organized crime and violations of international law; in addition, the distinction between public and private combat units, political and economic actors and interests becomes increasingly blurred.

Newly industrializing countries
This term is applied to developing countries undergoing a successful process of catch-up industrial development and which are thus poised to become →industrialized countries. Their social development indicators, such as literacy, infant mortality, life expectancy or civil society development, may lag far behind economic indicators. In recent years interest has centred particularly on the Asian newly industrializing countries, also known as Asian drivers.

Official Development Assistance (ODA)
Official Development Assistance is defined as the resources provided, for the purposes of development, by the members of the OECD's Development Assistance Committee (DAC) to developing countries on a bilateral basis or through multilateral institutions. Under this definition, the financial resources provided by, for instance, the new and upcoming →newly industrializing countries such as China or India do not count as ODA. Efforts are in progress to integrate these countries within the DAC guidelines.

Organization for Economic Cooperation and Development (OECD)
OECD is an international organization headquartered in Paris. It is also described as the organization of 'first world' countries, as almost all of its 30 member countries are industrialized countries. The aims of the OECD are to help its member countries to optimize their economic development and continually raise living standards, to promote economic growth both in its member countries and in →developing countries, and to facilitate expansion of world trade.

Poverty
State in which a person's physical existence is in jeopardy. WBGU applies a broad concept of poverty, which it defines as a lack of access rights and entitlements. In addition to low per-capita income or unequal income distribution, poverty also means inadequate food, poor health and healthcare, a lack of access to education, and an absence of social capital and opportunities for participation. There are other definitions of poverty. According to the World Bank, a person commanding over less than US$1 a day (measured in purchasing power parities) lives in absolute poverty. Other sources set the level at less than US$2 a day.

'Responsibility to protect' principle of the UN
Against the backdrop of the failure of the international community to respond adequately to intrastate conflicts such as those in Rwanda and the former Yugoslavia, a new concept related to security and international law has evolved over the last few years, known as 'responsibility to protect'. While it still holds that the state is responsible for protecting its own population, in cases of a serious breakdown of national responsibility, the international community – especially the United Nations – should have a duty to act, in order to protect the civilian population from serious attacks (genocide, ethnic cleansing and other crimes against humanity).

Scenario
Scenarios – in contrast to →forecasts – provide a plausible description of how the future might look based on analysis of a coherent and consistent set of assumptions, trends, relationships and driving forces. In the present WBGU report, narrative scenarios have been developed that describe a variety of paths

towards the future as plausible 'storylines', with the aim of giving a broader view of the whole bandwidth of possible developments and identify pointers for setting the direction of policy.

Security
In the realm of international politics security refers classically to the inviolability of territorially organized sovereign nation states within the system of international law. Security policy includes all those measures enacted by a state or a group of states aimed at averting or mitigating dangers from outside, generally military aggression. This classical security concept has since been complemented and in some ways replaced by the new, →comprehensive security concept.

Security policy
→Security

SRES scenarios
SRES scenarios are emissions scenarios developed for the Second Report on Emission Scenarios of the Intergovernmental Panel on Climate Change (IPCC) and, among other things, were used as a basis for climate projections. They take account of the full range of potential developments in the 21st century in the spheres of population growth, economic and social development, technological change, resource consumption and environmental management. The scenarios are organized in the dimensions A and B (strong economic growth versus sustainability) and 1 and 2 (globalization versus regionalization), so that, in combination, four scenario families result: A1 (rapid growth), B1 (global sustainability), A2 (regionalized economic development), B2 (regional sustainability).

State fragility
→Weak and fragile states

Syndrome concept
A scientific concept for the transdisciplinary characterization and analysis of global change developed by WBGU. Key elements of the syndrome analysis approach, in addition to the syndromes themselves, are the global network of interrelations, comprising trends and their interactions, and the →guard rails.

Tipping points of the climate system
Strongly non-linear responses by system components are often referred to as 'tipping points' in the climate system. This term is used to denote the behaviour of the system when a critical threshold has been crossed, triggering runaway changes that are then very difficult to control.

Tropical cyclones
Tropical cyclones are extreme low-pressure systems that only develop over large bodies of water with a surface temperature in excess of 26 °C. They are associated with high wind speeds and severe storms and can give rise to storm surges in the vicinity of coasts. Depending on the region in which they originate, severe tropical cyclones are also referred to as hurricanes (Atlantic, north Pacific), cyclones (Indian Ocean, Gulf of Bengal, south-eastern Pacific) or typhoons (north-western Pacific). Tropical cyclones are classified into categories based on their intensity using the Saffir-Simpson Hurricane Scale. It starts with Category 1, which refers to wind speeds of more than around 120km/h, and goes up to Category 5, which denotes wind speeds of more than 250km/h.

UN Security Council
The UN Security Council is the most powerful of the six principal organs of the United Nations. It comprises five permanent and ten non-permanent member countries. The Council's prime responsibility is to maintain international peace and security; to fulfil this task, the member states have assigned the Security Council exclusive authority to impose sanctions. In April 2007, for the first time in its history, the UN Security Council discussed an environmentally induced security risk, namely the issue of climate change.

UNDP
→ United Nations Development Programme

UNEP
→ United Nations Environment Programme

Unilateralism
→Multilateralism

Unipolarity
→Multipolarity

United Nations Development Programme (UNDP)
The United Nations Development Programme was established in 1965 and has its headquarters in New York. It is the principal organization of the UN system for →development cooperation and, in addition, is the UN's coordinating body for the →Millennium Development Goals. UNDP supports its partner countries by providing policy advice and assistance with developing and enhancing capacity. The organization also coordinates UN development activities on the ground. UNDP publications include the annual Human Development Report, which also provides information on the →Human Development Index (HDI).

United Nations Environment Programme (UNEP)
UNEP was established in 1972 in the wake of the UN Conference on the Human Environment in Stockholm as a subsidiary organ of the UN General Assembly. It has its headquarters in Nairobi, Kenya. Its aim is to identify and analyse regional and global environmental problems, and to develop and coordinate the environmental activities of the United Nations. UNEP's funding comes from an Environment Fund made up of voluntary contributions from member countries.

United Nations Framework Convention on Climate Change (UNFCCC)
The UNFCCC was adopted in 1992 and entered into force in 1994. With 191 states, its membership is almost universal. The ultimate objective is stabilization of greenhouse gas concentrations in the atmosphere at a level that would prevent dangerous anthropogenic interference with the climate system (→dangerous climate change). Such a level should be achieved within a timeframe sufficient to allow ecosystems to adapt naturally to climate change, to ensure that food production is not threatened and to enable economic development to proceed in a sustainable manner. The →Kyoto Protocol, adopted in 1997, sets out binding commitments to reduce greenhouse gas emissions.

User charges
This fiscal instrument entails raising a charge for the use of global common goods, such as international airspace or the oceans. Through the payment that has to be made, the scarcity of a resource and the costs of its provision are signalled to the user. It has both allocation effect and a financing function, as the revenue from the charge must be deployed to mitigate the damage associated with the use of the resource.

Vulnerability
This denotes the susceptibility of a social group or (environmental) system to →crises and pressures. A distinction can be made between social and biophysical vulnerability.

War, research on the causes of
Research on the causes of war is a strand of political science research that was originally focused on the causes of organized interstate violence. It now deals in general with the factors that affect the genesis and intensification of wars and armed conflicts, whereby a distinction is drawn between conflicts within states (intrastate) and conflicts between states (interstate).

Water scarcity
Various concepts of water scarcity relate to physical water scarcity, where water withdrawal for human uses exceeds 40% of the available renewable water resources, or the available water resources are less than 1,000m^3 per capita per year. Economic water scarcity affects regions in which there is no physical water scarcity, but lack of investment in water infrastructure means that people's water requirements cannot be adequately met. Environmental water scarcity is present when withdrawal of water resources for human use is so great that it threatens the integrity of ecosystems.

Weak and fragile states
States are considered weak or fragile when their institutions are not (adequately) capable of carrying out key state functions effectively. This applies to maintaining the state's monopoly on violence within and outside its borders, the provision of basic public goods such as infrastructure, health care and education along with the adoption and enforcement of public law (rule of law).

World Trade Organization (WTO)
WTO was established in 1994 as the umbrella organization of the multilateral trade agreements GATT, GATS and TRIPS. The objective of WTO is to reduce barriers to trade and thereby liberalize international trade. The organization currently has 151 members. The current round of negotiations (→Doha Round) began in 2001 and has not yet been concluded.

Index

A
Aceh 107
Africa 70–71, 72, 94–95, 100, 160, 174, 182, 186
 – North Africa 80, 94, 101, 124–125, 133–135
 – southern Africa 59, 64, 100, 101, 138–140, 169
agricultural policy 80, 100, 191
agricultural production 70, 75, 94–98, 100, 102, 133, 135, 159, 211
 – world agricultural markets 95, 98, 102, 201
agriculture 59, 69, 74, 79, 89, 97, 98, 101, 126, 134, 142, 143–144, 150, 153, 178, 182, 185, 201; *see also* irrigation: irrigated farming
 – productivity 66, 70, 93–94, 125, 133
air pollution 45, 50, 110
Alaska 56, 133
Algeria 29, 124, 134
Amazon rainforest 67, 75–76
Amazon region 153–154
Andes region 82, 87, 151–154, 190
Angola 45, 138, 140
Annex I countries 50, 173, 209
Aral Sea 89, 142
Arctic 132–133
Arctic Ocean 132
Argentina 151, 152

B
Bahamas 150
balance of power 39, 52, 164
Bangladesh 33, 106, 122–124, 143–145
Barbados 150
Barcelona Process 126, 135
Belize 151
biodiversity; *cf* biological diversity
Biodiversity Convention; *cf* Convention on Biological Diversity (CBD)
bioenergy
 – bioethanol 155
 – energy crops 96, 102, 201
biological diversity 31, 66, 156, 162
 – hotspots 139
biosphere 57, 75, 154, 178
Bolivia 82, 86, 151–153
Botswana 101, 138, 140

Brazil 48, 154–156, 160, 169, 170

C
'coalition of the willing' 195
Cairo 135
Canada 48, 70, 133
carbon cycle 57, 75, 134
Caribbean 113–115, 149–151
Central Asia 86, 88, 141–142
cereals 26, 66, 71, 93–95, 99, 139
Chad 136–137
Chile 151–153
China 41, 46–48, 53, 59, 79, 88, 95–96, 110–113, 146, 166, 192
civil wars 27, 34, 88, 137, 145, 169, 196; *see also* wars
Clean Development Mechanism (CDM); *cf* Kyoto Protocol
climate change 55, 64, 77
 – hotspots 131, 163
climate policy 94, 157, 165, 167–168, 175, 181, 189, 191, 194, 198, 208, 213
 – pioneering role 168, 191, 194
climate protection 56, 75, 162, 166–168, 182, 191, 198, 208
 – mitigation costs 170, 179, 208
climate risks 55
climatic parameters 55
CO_2 fertilization effect 66
coastal erosion 122, 133, 150
coastal regions 63, 72, 87, 104, 110, 114, 122, 147, 185
Colombia 150–151
Commission on Sustainable Development (CSD); *cf* United Nations Commission on Sustainable Development (CSD)
Common Foreign and Security Policy (CFSP) 191, 194, 203; *see also* European Union (EU)
Conference on Security and Co-operation in Europe (CSCE) 194
conflict constellations 16, 77, 86, 116, 157–161, 184, 189
 – Climate-induced decline in food production 93
 – Climate-induced degradation of freshwater resources 79
 – Climate-induced increase in storm and flood disasters 103
 – Environmentally induced migration 116

conflicts 27, 66, 79, 89, 96–97, 106, 108, 120, 123, 127, 134, 142, 148, 182, 187, 190, 192; *see also* environmental conflicts
- Cyprus conflict 107
- defusing tensions 84, 98, 101, 103, 128
- North-South conflict 48, 171
- violent conflicts 25–29, 77, 79, 85–86, 89, 114, 121

consumption patterns 94, 98, 115
Convention on Biological Diversity (CBD) 197
corruption 43, 87, 99, 122, 142, 145
Costa Rica 42, 150
cotton 89, 138, 141
criminality 43, 88, 107, 108, 142, 145
- drugs trade 43, 87, 142, 145
- looting 35, 99, 106–107

crises 19, 31, 69, 98–99, 105, 109, 114, 121–122, 135, 137, 156, 157–159, 169, 179, 192, 200
- intervention 20, 44, 123, 175, 182, 195, 203, 205
- prevention 22, 179, 184, 203, 208, 211–212

Cuba 150–151
cyclones 15, 55, 59–60, 104, 122–124, 143–144, 149, 162, 177; *see also* hurricanes

D

Dar es Salaam 139
Darfur 137
deforestation; *cf* forests: deforestation
demographics 37, 121, 159
desertification 100, 103, 124–126; *see also* soils: degradation
Desertification Convention (UNCCD); *cf* United Nations Convention to Combat Desertification (UNCCD)
destabilization of societies and states 15–17, 27, 43, 77, 79, 87, 98, 106, 131, 157, 159, 165, 168–169, 179, 190, 213; *see also* stability
development assistance 21, 212; *see also* Official Development Assistance (ODA)
Development Assistance Committee (DAC); *cf* Organization for Economic Cooperation and Development (OECD)
development cooperation 49, 90, 169, 182, 199–202, 204, 210
development policy 103, 174, 190, 194, 200, 213
disasters; *see also* natural disasters
- disaster prevention 102, 115, 123, 175, 186, 202
- disaster risk management 70, 121–122, 158, 160, 162, 180, 186
- disaster risks 110, 115, 161, 187, 202

disease 20, 72, 80, 170, 184
- malaria 20, 72

distributional equity 83, 127, 152, 158
Doha Round 48, 102, 201; *see also* World Trade Organization (WTO)
Dominican Republic 150
drought 65, 75, 81, 86, 90, 136–138, 150, 154, 163, 170; *see also* natural disasters

E

early warning systems 120, 122, 192, 202, 207; *see also* disaster prevention
earthquakes 103, 107, 110; *see also* natural disasters
economic performance 36, 47, 97–98, 158
economic policy 19, 179, 213
economy 36, 70–72, 135, 158, 201
- economic development 36, 47, 78, 110, 135, 147, 148, 170, 181

ecosystems 20, 65, 67, 71, 79, 100, 122, 177, 190, 198
ecosystem services 64, 80, 139
Ecuador 82, 151–152
Egypt 125, 133–135
electricity generation 135, 142
El Niño/Southern Oscillation (ENSO) 75, 146, 154
emissions; *cf* greenhouse gases
energy 49, 71, 87, 89–90
energy policy 102, 166, 191, 194, 209
- Energy Policy for Europe 199

environmental conflicts 20, 25, 27–28, 31, 37, 184, 189; *see also* conflicts
- world map of environmental conflicts 28, 31

environmentally induced migration; *cf* migration
Ethiopia 45, 126–127, 135
ethnicity 120, 123, 125, 142, 159, 187
European policy 191
European Security Strategy 21, 169; *see also* security strategies
European Union (EU) 22, 102, 168, 174, 191, 199, 203, 210
extreme events 45, 55, 65, 71, 108, 127, 151, 163, 178, 207

F

failed states 42–43; *see also* state collapse
Falkenmark indicator 65; *see also* water shortage
financing 98, 121, 126, 174, 203, 208–210, 211–212
- funds 106, 123, 135, 196, 208, 210
- Least Developed Countries Fund 210
- microfinancing instruments 211

fisheries 70, 161
floods 26, 34, 69, 103–104, 108, 110, 114, 122, 144, 146, 150; *see also* storm and flood disasters
Food and Agriculture Organization of the United Nations (FAO); *cf* United Nations Food and Agriculture Organization
food crises 93, 98, 100, 102; *see also* crises
food production 69, 77, 93–96, 98, 135, 139, 141, 169, 186; *see also* food security
food security 94, 102, 125, 159, 173, 201
forecasts 55, 68, 77, 92, 97, 102, 118, 136, 201, 207
forestry 70, 134; *see also* forests
forests 68, 94, 100, 104, 115, 198, 209
- deforestation 75, 93, 100, 104, 115, 134, 155, 156, 209
- forest fires 134
- tropical forests 67, 154, 198

fragile states 43–47, 138, 162, 169, 183, 190, 202

Framework Convention on Climate Change; cf United Nations Framework Convention on Climate Change (UNFCCC)
France   126, 134, 136
freshwater; cf water

G

G20   48
Ganges   122, 143–144, 170, 190
General Agreement on Tariffs and Trade (GATT); see also World Trade Organisation (WTO)
Geneva Refugee Convention   118, 129, 205–206
geographical factors   38, 122, 159
German Action Plan 'Civilian Crisis Prevention'   193, 203
German Ministry for Economic Cooperation and Development (BMZ)   199, 203
glaciers   57, 82, 87–88, 147, 177
 – melting   65, 132, 141, 144, 151, 190
 – meltwater   63, 74, 81–82, 87, 143, 152
Global Commission on International Migration (GCIM)   206
Global Environment Facility (GEF)   199, 208
global governance   46, 51–54, 131, 156, 191, 213
good governance   78, 102, 115, 140, 202
governance structures   35–36, 72, 85, 98, 115, 134, 142, 151
Greece   107, 134
greenhouse gases   56, 73–74, 92, 112, 124, 165, 166–167, 168, 171–173, 177, 181, 191
 – emissions reductions   56
 – greenhouse gas concentration   56, 66, 72, 167, 171, 198
Greenland   74, 133
Greenland ice sheet   57, 63, 72, 74, 132; see also ice sheets
Green Revolution   93, 100–101, 186
groundwater   64, 79, 80, 87, 95, 100, 104, 110, 134, 159, 162, 185
Guatemala   31, 150–151
Gulf of Mexico   104, 149–151
Guyana   154, 162

H

Haiti   31–33, 104, 150–151
health   72, 86, 110, 120, 142, 148
heatwaves   56, 72, 136, 146, 163; see also natural disasters
heavy rainfall events   104, 110, 142, 144, 146, 161; see also natural disasters and extreme events
Himalayas   82, 144, 163, 190
HIV/AIDS pandemic   139–140
Honduras   31, 34, 104, 114, 150
Human Development Index (HDI)   133, 135, 140, 147, 150
humanitarian assistance   41, 116, 195
human rights   49, 54, 126, 169, 173–174, 184, 195
human security; cf security
Human Security Doctrine for Europe   21, 212
hurricanes   104, 112–115, 144, 149–150; see also natural disasters and storm and flood disasters
 – Hurricane Catarina   60
 – Hurricane Katrina   22, 35, 41, 45, 71, 104, 113, 116; see also New Orleans
hydropower   71, 79, 87, 89, 90, 135, 155, 156, 160, 199

I

Iceland   133
ice sheets   57, 74, 132, 177; see also Greenland ice sheet
ICLEI; cf Local Governments for Sustainability network (ICLEI)
imports, dependency on   103, 202
India   35–36, 41, 47–50, 53, 93, 123–124, 143–145, 192
indigenous peoples   31, 133, 153, 156
Indonesia   45, 46, 105, 107, 128
Indus   145
insurance sector   71, 113, 211
interdependence of states   39, 52, 159
Intergovernmental Panel on Climate Change (IPCC)   56, 64, 73, 180, 204, 210
International Environmental Court   184
International Monetary Fund (IMF)   53, 103
International Organization for Migration (IOM)   118, 128, 211
International Strategy for Disaster Reduction (ISDR)   105, 207
irrigation   63, 69, 89, 91, 94, 141, 147, 152
 – irrigated farming   79, 82, 89–90, 94, 100, 139, 145
islands   63, 67, 70, 113, 114, 148, 185
Italy   134

K

Kashmir   107, 145
Kazakhstan   45, 88, 90, 141–142
key factors   77, 82, 99, 108, 127, 157; see also conflict constellations
Kyoto Protocol   50, 166, 191, 198, 210
 – Clean Development Mechanism (CDM)   50, 208, 210
Kyrgyzstan   88, 141, 142

L

land distribution   31, 99, 100, 103
landless   98, 123; see also land distribution
land reforms   103
land use   31, 66–68, 75, 96, 134, 144, 150, 166, 172, 178, 202
Lesotho   138–139
Libya   124, 134–135
Lima   86–88, 153
Local Governments for Sustainability network (ICLEI)   115

M

Madagascar   138
Malawi   45, 138–139
meat consumption   93, 96
Mediterranean region   59, 64, 122, 124, 133–135
Middle East   45, 69, 80, 94, 132, 175

migration   27, 31, 72, 77, 86, 98, 101, 116–123, 135, 150, 159–160, 163, 174, 188, 204–205
 – environmentally induced migration   77, 116–118, 120, 123, 124–128, 206
 – internal migration   31, 117, 137, 140, 146, 149, 213
military measures   196
Millennium Development Goals (MDGs)   72, 80, 91, 189, 200, 210
Millennium Ecosystem Assessment (MA)   94, 180
millet   66, 137
mineral resources   19, 37, 49, 101, 126
Mongolia   29, 147–148
monsoon   55, 59, 74, 78, 122, 143–144, 190
Morocco   124, 134–135, 204
Mozambique   138, 140
multilateralism   52–53, 167, 173, 191–192
multipolarity   41, 45–46, 52–54, 192

N
Namibia   59, 138, 140
National Adaptation Programmes of Action   210
National Security Strategy of the USA; *cf* USA
natural disasters   26, 69, 103–108, 110, 116, 150, 202
 – emergency aid   106, 114, 123, 127, 186, 202
nature conservation   115, 140, 201
newly industrializing countries   41, 48, 93, 103, 115, 159, 166, 170, 187
New Orleans   45, 72, 104–106, 113; *see also* natural disasters *and* hurricanes: Hurricane Katrina
New Partnership for Africa's Development (NEPAD)   101
Nicaragua   34–35, 104, 150–151
Nile Delta   125, 134–135, 190
non-refoulement, principle of   206
North Atlantic Current   26, 63, 73, 78
North Atlantic Treaty Organisation (NATO)   19, 23, 212
Norway   43, 133, 212

O
Official Development Assistance (ODA)   210
Okavango   140
Organization for Economic Cooperation and Development (OECD)   91, 95, 203, 209
 – Development Assistance Committee (DAC)   44, 203, 210, 212
 – OECD countries   42, 93
Organization of African Unity (OAU)   205
 – OAU Refugee Convention   206

P
Pakistan   33, 69, 107, 143–145
Panama   150
Paris Declaration on Aid Effectiveness   210
peace   107, 195
Peru   82, 86–87, 151–153, 196
population density   37, 69, 96, 108, 125, 144, 159

population growth   27, 37, 82, 87, 94, 100, 121, 125, 138, 163
poverty   28, 72, 83, 108, 110, 125, 185, 189, 195
 – absolute poverty   110, 137, 204
 – extreme poverty   109, 110, 147
poverty reduction   22, 90, 159, 195, 199
 – Poverty Reduction Strategy Papers (PRSPs)   199, 202
Poverty Reduction Strategy Papers (PRSPs); *cf* poverty reduction
power distribution   39, 159
precipitation   57–59, 75, 81–83, 132, 134, 136, 141, 144, 146, 154, 201; *see also* heavy rain events
public goods   89, 99, 102

R
Rabat Action Plan   205
refugees   117, 120, 128, 190, 205–206
 – climate refugees   116, 174
 – internally displaced persons   117, 137, 190
regime change   38, 122
Republic of South Africa   139–140
research
 – climate research   45, 177
 – conflict research   25–30, 35, 103, 121, 187, 188
 – environmental conflict research   25–29, 103
 – environmental research   29, 40, 162, 179, 183
 – long-term perspective   180
resource conflicts   31, 87, 133, 141; *see also* conflicts
responsibility to protect   174, 196
rice   66, 141, 144
risk assessments   73
Russia   45, 110, 112, 126, 132–133, 148

S
Sahara   59, 126
Sahel region   31, 59, 125, 135
Saint Kitts and Nevis   150
Scandinavia   70, 132
scenarios   45, 94, 136, 138, 141, 166–168, 172, 175, 185
 – fictitious confrontation scenario   77, 87, 89, 100, 111, 123, 125
 – fictitious cooperation scenario   77, 88, 90, 101, 112, 114, 123, 126
 – model scenario   55–58
 – narrative scenarios   16, 77–78, 100, 110, 122
 – SRES scenarios   60, 64, 94, 134, 165
sea-level rise   60, 63, 69, 74, 104, 122, 150, 177
sea ice   132–133
 – retreat of the Arctic sea ice   132
sea temperatures   59, 81, 163
security   19, 44, 106, 168
 – human security   20, 22, 28, 131, 140, 173
 – wider security concept   20
Security Council; *cf* United Nations Security Council
security policy   19–22, 132, 169, 174–175, 191, 198, 200, 211
security risks   22, 77–78, 128, 175, 179, 189, 196, 200, 204,

208, 211
security strategies   19, 21, 175, 190, 192; *see also* USA: National Security Strategy of the USA
Senegal   29, 124, 136
Siberia   56
social tensions   54, 106, 108
– state of emergency   86
soils   96, 124, 178
– degradation   31, 55, 68, 86, 95–96, 100, 134, 147, 152
– overgrazing   69, 134, 136, 139, 144
– permafrost   56, 132–133, 169
– salinization   69, 89, 95, 134, 141, 144, 162
Somalia   42–43, 137
South Africa; *cf* Africa: southern Africa *and* Republic of South Africa
South America   31, 43, 69, 151, 162, 190
South Asia   64, 72, 84, 94, 101, 162
southern Europe   84, 125, 134–136
soya   66, 133, 155–156
Spain   116, 121, 134, 204
Spitzbergen   133
Sri Lanka   45, 107, 145
stability   136, 202
– political stability   84–86, 98, 102, 106, 121, 125, 145
– social stability   78, 99, 108, 148, 200
– state stability   42, 142, 157, 169
state collapse   169; *see also* failed states
state constitution   36, 157
state responsibility   188, 205
storm and flood disasters   26, 31, 34, 69, 103, 110, 114–115, 186, 202; *see also* natural disasters *and* extreme events *and* hurricanes
Subarctic   132–133
subsidence   104, 110, 115, 161
Sudan   124, 136–137
Suriname   154
sustainability   29, 85, 199, 204
Swaziland   138, 140
syndromes   27–28

T
Taiwan   46, 148
Tajikistan   88, 141–142
Tanzania   138–139
temperature rise   15, 55–56, 66–67, 70–71, 81, 94, 132, 136, 143, 149, 151, 154, 168–169, 180, 182, 198
terrorism   15, 21, 22, 145, 169, 175, 191, 195
tipping points   78, 178
tourism   70, 82, 135, 150, 161
trade policy   48, 213
transboundary water cooperation   186
transport policy   191
Trinidad and Tobago   150
tsunamis   103, 107; *see also* natural disasters
Tunisia   124, 134
Turkey   107

Turkmenistan   88–89, 90

U
United Nations (UN)   22, 53, 167, 173, 184, 195–196
United Nations Commission on Sustainable Development (CSD)   197
United Nations Convention to Combat Desertification (UNCCD)   197
United Nations Development Programme (UNDP)   196–197, 207
United Nations Economic and Social Council (ECOSOC)   197, 198
United Nations Environment Programme (UNEP)   178, 184, 196
United Nations Food and Agriculture Organization (FAO)   94–95, 100, 102
United Nations Framework Convention on Climate Change (UNFCCC)   50, 197–198; *see also* Kyoto Protocol
– compliance mechanisms   198
United Nations General Assembly (UNGA)   195, 197, 212
United Nations High Commissioner for Refugees (UNHCR)   196, 206
United Nations Security Council   50, 195
USA   41, 49–51, 53, 71, 107, 113, 145, 150–151, 166, 191–192
– National Security Strategy of the USA; *see also* security strategies
Uzbekistan   45, 89, 141–142

V
vegetation   66–68, 75, 134, 136, 141, 155, 178
Venezuela   35, 107, 115, 150–152
violence   77, 84–86, 90, 96, 98, 100–103, 106–107, 125, 137, 145

W
warming; *cf* temperature rise
wars   20, 31, 33, 38, 51; *see also* conflicts
– 'new wars'   23, 43, 169
– 'war on terror'   51, 142, 195
– 'water wars'   29, 79
water
– degradation of freshwater resources; *cf* conflict constellations: degradation of freshwater resources
– desalination   65, 80, 88, 185
– freshwater   64–67, 73, 79, 95, 159, 185
– hydrological cycle   59, 64, 95
– integrated water resources management (IWRM)   80, 88, 185
– privatization in the water sector   39, 86, 185
– transboundary water cooperation   92, 185, 186
– virtual water   91, 185
– water balance   57–58, 79, 152, 180, 185
– water pollution   79

– water resources   64, 139, 143, 152, 200
water availability   64, 79, 81–83, 90, 95, 138, 147, 152, 160, 185, 201
water conflicts   84–85, 135–136, 145, 153, 201; *see also* conflicts
water crises   80, 82, 84, 87, 88, 90–91, 200; *see also* crises
water scarcity   31, 65, 80, 84, 87, 90, 134, 200
water shortage   64, 84, 88, 90, 124, 160, 185
water stress   64–65, 132, 138, 149
weak states   42, 90, 134, 169, 174, 193, 202; *see also* fragile states
wheat   56, 66, 139, 144, 155
wider concept of security; *cf* security
World Bank   45, 53, 85, 103, 122, 182
world economy   17, 46–47, 71, 166, 170–171, 190, 192, 213
World Food Programme (WFP)   103, 197
world order   41, 46–47, 52–54, 168, 189, 191–192, 194
World Trade Organization (WTO)   103, 184

Z
Zambia   138, 140
Zimbabwe   45, 138, 140